W9-DAC-288

Molecular Breeding of Forage and Turf

Toshihiko Yamada • German Spangenberg
Editors

Molecular Breeding of Forage and Turf

The Proceedings of the 5th International Symposium
on the Molecular Breeding of Forage and Turf

 Springer

Editors
Prof. Toshihiko Yamada
Hokkaido University
Field Science Center for
Northern Biosphere
Kita 11, Nishi 10
Kita-ku
Sapporo 060-0811
Japan

Prof. German Spangenberg
Victorian AgriBiosciences
Centre
Dept. Primary Industries
La Trobe R&D Park
1 Park Drive
Bundoora, VIC 3083
Australia

ISBN: 978-0-387-79143-2   e-ISBN: 978-0-387-79144-9
DOI: 10.1007/978-0-387-79144-9

Library of Congress Control Number: 2008932631

© 2009 Springer Science + Business Media, LLC
All rights reserved. This work may not be translated or copied in whole or in part without the written
permission of the publisher (Springer Science+Business Media, LLC, 233 Spring Street, New York,
NY 10013, USA), except for brief excerpts in connection with reviews or scholarly analysis. Use in
connection with any form of information storage and retrieval, electronic adaptation, computer
software, or by similar or dissimilar methodology now known or hereafter developed is forbidden.
The use in this publication of trade names, trademarks, service marks, and similar terms, even if they
are not identified as such, is not to be taken as an expression of opinion as to whether or not they are
subject to proprietary rights.

Printed on acid-free paper

springer.com

Preface

Grassland produces feed for livestock, maintains soil fertility, protects and conserves soil and water resources, creates a habitat for wildlife, and provides recreational spaces for sports and leisure while simultaneously maintaining sustainable economic outputs. Turf species similarly contribute considerably to our environment by adding beauty to surroundings, providing a safe playing surface for sports and recreation, and preventing erosion. In addition to food and environment, bio-energy is a global concern related to these species. Renewable biomass energy is increasingly being accepted as a possible alternative to fossil fuels and some forages are promising for energy crops.

Breeding programs in forages have produced improvements in both forage yield and quality. Forage and turf in the future must utilize resources (nutrients and water) more efficiently and must also confer measurable benefits in terms of environmental quality and renewable energy. With a widening range of traits, techniques for more accurate, rapid and non-invasive phenotyping and genotyping become increasingly important. The large amounts of data involved require good bioinfomatics support. Data of various kinds must be integrated from an increasingly wide range of sources such as genetic resources and mapping information for plant populations through to the transcriptome and metabolome of individual tissues. The merging of data from disparate sources and multivariate data-mining across datasets can reveal novel information concerning the biology of complex.

Previous International Symposium on the Molecular Breeding of Forage and Turf (MBFT) Symposia were held in Japan in 1998, Australia in 2000, the USA in 2003 and the UK in 2005. On this occasion the 5th MBFT was held in Sapporo, Japan in 2007. The 5th MBFT was hosted by the Hokkaido University in cooperation with the National Agricultural Research Center for Hokkaido Region and the National Institute of Livestock and Grassland Science in the National Agriculture and Food Research Organization. Attendees included breeders, geneticists, molecular biologists, agronomists and biochemists from 19 countries. The program featured plenary addresses from leading international speakers, selected oral presentations, volunteered poster presentations, as well as tours of the National Agricultural Research Center for Hokkaido Region, Rakuno Gakuen University and Sapporo Dome.

This book includes papers from the plenary lectures and selected oral presentations of the Conference. A wide variety of themes are included and a collection of authoritative reports provided on the recent progress and understanding of molecular technologies and their application in plant breeding. Almost all relevant areas in molecular breeding of forage and turf, from gene discovery to the development of improved cultivars, are discussed in the proceedings.

The 5th MBFT and the publication of this book, *Molecular Breeding of Forage and Turf*, have been supported by National Agricultural Research Center for Hokkaido Region; National Institute of Livestock and Grassland Science; Sustainability Governance Project, Hokkaido University; Alumni Association, Faculty of Agriculture, Hokkaido University; Japan Grassland Agriculture and Forage Seed Association; Japan Livestock Technology Association; Green Techno Bank; The Akiyama Foundation; The Kajima Foundation; Japan Plant Science Foundation; Novartis Foundation Japan for the Promotion of Science; Life Science Foundation of Japan; Supporting Organization for Research of Agricultural and Life Science (SORALS); The Kao Foundation for Arts and Sciences; The Suginome Memorial Foundation; Sapporo International Communication Plaza Foundation; Hokuren Federation of Agricultural Cooperatives; Snow Brand Seed Co., Ltd.; Toyota Motor Corporation; Monsanto Company; Syngenta Seeds K.K.; Japan Turfgrass II; Nippon Medical & Chemical Instruments Co., Ltd.; Applied Biosystems Japan Ltd.; Nihon SiberHegner Co., Ltd.; Nikon Instech Co., Ltd.; HUB Co., Ltd.; Mutoh Co., Ltd.; Imuno Science Co., Ltd.

We thank Mervyn Humpherys, German Spangenberg, Reed Barker, Andy Hopkins, Odd Arne Rognli, Hitoshi Nakagawa of the International Organizing Committee, as well as Toshinori Komatsu, Yoshio Sano, Yoh Horikawa, Hajime Araki, Akira Kanazawa, Toshiyuki Hirata, Yoshihiro Okamoto, Kazuhiro Tase, Kenji Okumura, Sachiko Isobe, Hiroyuki Tamaki, Ryo Akashi, Masayuki Yamashita, Yoshiaki Nagamura, Tadashi Takamizo, Makoto Kobayashi, Masumi Ebina, Makoto Yaneshita, of the Local Organizing Committee for their contributions to the success of the Conference. We also thanks following scientists for their critical reviewing of the manuscripts of this book:

Michael Abberton, Toshio Aoki, Ian Armstead, Reed Barker, Philippe Barre, Susanne Barth, Faith Belanger, Yves Castonguay, Hiroyuki Enoki, Sachiko Isobe, Bryan Kindiger, Takako Kiyoshi, Sohei Kobayashi, Steven Larson, Dariusz Malinowski, Maria Monteros, Kenji Okumura, Juan Pablo Ortiz, Mark Robbins, Odd Arne Rognli, Isabel Roldan-Ruiz, Malay Saha, Christopher Schardl, Leif Skøt, Satoshi Tabata, Tadashi Takamizo, Hiroyuki Tamaki, Ken-ichi Tamura, Scott Warnke, Yan Zhang.

We thank Jinnie Kim and Jillian Slaight of Springer for their assistance and cooperation in the publication of this book. Finally, we express our gratitude to the authors whose dedication and work made this book possible.

Toshihiko Yamada
German Spangenberg

February 2008

Contents

Molecular Breeding to Improve Forages for Use in Animal and Biofuel Production Systems

Joseph H. Bouton

The Samuel Roberts Noble Foundation, 2510 Sam Noble Parkway, Ardmore, OK 73401, USA, jhbouton@noble.org

Abstract. Forage cultivars with positive impacts on animal production are currently being released using traditional plant breeding approaches. Molecular breeding is a relatively new term that describes the use of genomic and transgenic biotechnologies in conjunction with traditional breeding. Traits currently under investigation via these biotechnologies include herbicide tolerance, drought tolerance, resistance to disease and insect pests, tolerance to acid, aluminum toxic and/or saline soils, tolerance to cold or freezing injury, expression of plant genes controlling nodulation and nitrogen fixation, increasing nutritional quality via down regulation of lignin genes, flowering control, and reducing pasture bloat via incorporation of genes to express condensed tannins. Molecular breeding approaches are expensive, and in the case of transgenics, controversial, requiring much planning and even partnerships or consortia with others to defray cost, and overcome a "valley of death" for commercialization due to patent and regulatory issues. Trait incorporation via molecular breeding being conducted by the Consortium for Alfalfa Improvement is discussed as an example of this type of research approach. The future of molecular breeding in forage crops is bright, but is tied to funding, and in the case of transgenics, also lies in the hands of regulatory agencies and their ability to establish a fair process to evaluate real versus perceived risks. Finally, the use of forages as cellulosic biofuel crops offers new molecular breeding opportunities, especially for value added traits such as enhanced biomass and fermentation efficiency. The main criteria for any biofuel crops are high yields achieved with low input costs in an environmentally friendly manner. By this definition, many high yielding, currently grown, perennial forages are good candidates as biofuel crops especially if they can be delivered to a biorefinery as cheaply as possible.

T. Yamada and G. Spangenberg (eds.), *Molecular Breeding of Forage and Turf,*
doi: 10.1007/978-0-387-79144-9_1, © Springer Science + Business Media, LLC 2009

Introduction

New forage cultivars, developed through plant breeding, have a long history of positively impacting forage and livestock systems. Traditional breeding methods of hybridization and selection have always been, and still continue to be, used. However, forage improvement programs have entered the biotechnology era by the use of molecular biology tools (Brummer et al. 2007). Molecular breeding is therefore a relatively new term that describes the use of genomic and transgenic biotechnologies in conjunction with traditional breeding.

Genomics research received great publicity with the successful completion of the human genome sequencing project. Plant species were next with rice (*Oryza sativa* L.) and *Medicago truncatula* Gaertn., an annual relative of cultivated alfalfa (*Medicago sativa* L.), now being sequenced and used as a reference species for grasses and legumes, respectively. The sequencing data for these reference species, combined with high throughput machinery and data analysis (e.g. bioinformatics), allows more accurate determinations of species relationships and gene expression. From this understanding, new and innovative methods for improving forage crops are evolving.

Transgenics involve the movement of specific and useful genes into the crop of choice and this approach is sometimes referred to as genetic engineering. Scientists using this approach have already shown success in introducing genes which make many important row crops resistant to insects, viruses, and herbicides. The transgenic approach has also been very useful in creating unique plants that allow basic research to be conducted on physiological and biochemical pathways.

Why Molecular Breeding?

An ability to easily manipulate and control genes is fundamental to plant breeding. This is shown historically by the formula P=G+E+GE or Phenotype = Genotype + Environment + Genotype x Environment. Therefore, the genotype or G provides the best estimate for the genes involved in the phenotypic trait or traits being investigated and expressed. However, it is only a general estimate. Molecular tools available through genomics and transgenics offer a powerful ability to move from simply estimating to more accurately measuring G and even ways to manipulate the actual genes. Combining traditional plant breeding with these molecular tools should assist with making progress in cultivar development.

The Samuel Roberts Noble Foundation's Forage Improvement Division, as probably with most organizations, uses a model of combining traditional breeding approaches with molecular tools to incorporate useful genes (Fig. 1). In this approach, the basic five steps of the cultivar development process, (1) clearly defining objectives, (2) collecting and developing parental germplasm, (3) conducting the actual breeding and selection to produce an experimental cultivar, (4) extensive testing program to prove the worth of this cultivar, and (5) final release and commercialization, proceed as they always have. However, sometimes the traits are very complex to locate and manipulate, or possibly not even contained in a species' primary germplasm. When this happens, biotechnology approaches are an option for trait incorporation and/or validation through more efficient gene discovery, tagging, and even genetic engineering.

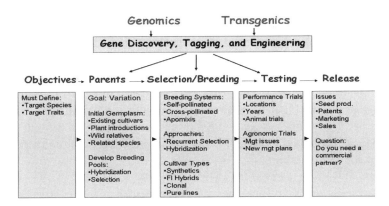

Fig. 1 A cultivar development model demonstrating the traditional steps in the process, and how and where the new transgenic and genomic biotechnologies will likely impact that process

Current Molecular Breeding

Biotechnology research in all forage crops, especially to study and/or incorporate complex traits, is in a time of increased emphasis and success throughout the world. For example, at the International Symposia on Molecular Breeding of Forage and Turf (MBFT) held at Victoria, Australia in 2000, at Dallas, Texas, in 2003, and at Aberystwyth, Wales in 2005, there were hundreds of scientists in attendance from many countries. Research talks at these meetings are found in the proceedings on many aspects of basic biotechnology in forage grasses and legumes as well as excellent keynote presentations on molecular breeding by Professors Spangenberg, Dixon, and Lübberstedt (Spangenberg 2001; Hopkins et al. 2003; Humphreys 2005), respectively. This current MBFT conference in Sapporo, Japan provides a similar venue.

Some of the research areas traditionally receiving emphasis at MBFT are accurate genomics techniques to more rapidly identify and manipulate important genes (molecular markers and marker assisted selection breeding); tolerance to biotic and abiotic stresses; flowering control; plant-symbiont relations; breeding for animal, human and environmental welfare; transgenics; bioinformatics; population genetics; genomics of the model legume *M. truncatula*; field testing and risk assessment as well as intellectual property rights. These symposia, and many others like them such as The North American Alfalfa Improvement Conference (NAAIC), are proof that research in this area is intense and growing for all forage crops.

Specific biotech traits currently under investigation and reported at the current and past MBFT symposia include herbicide tolerance, drought tolerance, resistance to disease and insect pests, tolerance to acid, aluminum toxic and/or saline soils, tolerance to cold or freezing injury, expression of plant genes controlling nodulation and nitrogen fixation, increasing nutritional quality via down regulation of lignin genes, flowering control, and reducing pasture bloat via incorporation of genes to express condensed tannins.

These traits are therefore ones that breeders have made little progress for improvement through conventional breeding. Another aspect is the high potential impact for farmers if these traits can be incorporated into cultivars. This type of impact would justify the use of biotechnologies even when considered against the issues surrounding their use.

Considerations and Issues

There are several issues to consider when deciding to use biotechnologies especially in cultivar development programs. The first is cost. Compared to the traditional model for cultivar development, everything is more costly with molecular breeding. That one has to recover these costs through the sale of seed, a notoriously low margin product, provides less incentive for using molecular breeding by many commercial seed companies. Second, one must have freedom to operate for all enabling technologies and patents involved in the process; especially for using transgenics. This again is a cost issue, but can become a legal issue if all the proper patents and licenses are not put in place. Third, the regulatory costs for transgenic traits are problematic and rising. For example, applications for de-regulation of the Roundup Ready (RR) gene were submitted in the USA for a turf and a forage crop, creeping bentgrass (*Agrostis palustris* Hud.) and alfalfa. Although the RR gene is a 1990s technology that is currently found in millions of acres of corn (*Zea mays* L.) and soybean (*Glycine max* L.), crops also fed to livestock, only alfalfa was de-regulated. However, RR alfalfa has been re-regulated and still not being sold (Tietz 2007), and RR creeping bentgrass has yet to be de-regulated (it has also been in the application process longer than any crop to date). This slow progress is not encouraging for production of transgenic cross-pollinated forage or turf crops. These delays are due, in part, to initial estimations that pollen can flow for extreme distances and into related weedy species potentially causing herbicide resistant weeds to develop. Whatever the reasons, these delays further adds to the regulatory costs, and if these two applications are not finally successful, it could set a negative precedent for the future of transgenic, cross-pollinated, perennial forages. Additionally, these problems with transgenic development have created further negative public perception for use of other biotech developed traits and methods.

All molecular breeding approaches are therefore expensive, and in the case of transgenics, controversial, creating for many a "valley of death" for the commercialization. Although there is no problem for conducting basic molecular biology research by creating unique plants to study, there may need to be new models created and used to overcome these inherent problems.

Models for Using Transgenics

Due to the cost and controversy of using transgenics, it is usually the option of last resort. However, although transgenic biotechnologies provide very powerful and useful alternatives to not having the trait altogether, the main question is this: Is the trait of such value and impact that it will justify a transformation approach? If the answer to this question is yes, then a good model of how to do this is the Consortium for Alfalfa Improvement (CAI).

The CAI is composed of researchers from Noble Foundation, the U.S. Dairy Forage Research Center (USDFRC) in Madison, WI, and Forage Genetics International (FGI), a commercial alfalfa research and seed company. Therefore, these three organizations have complementary strengths coming together to improve important characteristics in alfalfa. The main steps for using transgenics that are all covered by at least one of the CAI partners include (1) investigating and obtaining the requisite biotech pieces (including freedom to operate for patents on genes and enabling technologies), (2) trait development including introgression into commercially viable cultivars, proof of concept studies, and animal testing, and (3) commercialization including regulatory trials.

The first initiative by the CAI focused on improving protein utilization and cell wall digestibility via lignin reduction, and the second was expression of condensed tannins to reduce pasture bloat and improve protein utilization in ruminant animals. Therefore, the consortium's overall goal is to re-design alfalfa as the major forage source. This would be of such impact as to justify use of any biotechnologies. It also brings to bear additional resources to leverage with those existing for each organization as a separate entity. The CAI is therefore a good model of what may need to be done to justify the costs and reduce risks when using transgenics.

Other Options

On its face, transgenics simply create unique variation not found in the primary germplasm. However, it is the cost and controversy of using transgenics that cause most of its problems. Therefore, are there other, less controversial and costly approaches to creating unique variation?

Stebbins (1950) wrote that three main driving forces in the evolution of higher plants were inter-specific hybridization, mutation with Mendelian segregation, and polyploidy. For example, crop plants such as wheat (*Triticum aesativum* L.) evolved with inter-specific hybridization and polyploidy; while in alfalfa, polyploidy underpinned its development as an autotetraploid. Gene mutations that control traits such as yield, maturity, seed size, flower color, disease resistance, etc. have always been recognized in the primary germplasm of all crop plants. These same driving forces therefore underpin the basic approaches used by most plant breeders in the modern era with hybridization and selection for the natural mutations being the most used.

In the forage crop bermudagrass (*Cynodon dactylon* L.), however, hybridization, including inter-specific hybridization, was successful in producing vegetatively propagated, clonal F_1 cultivars that are currently planted on millions of hectares in the southeastern USA, and many other areas in the sub-tropics (Burton and Hanna 1995). In the case of the bermudagrass hybrid, "Coastal", it was unique enough to be used as a parent to produce other hybrids. So, as this example indicates, inter-specific hybridization provides unique plants, that if their propagation methods are worked out, can become cultivars themselves or used as parents to produce other unique plants. There are now several forage species, such as the clovers (*Trifolium* spp.), where the phylogenetic relationships among species are being examined through molecular markers (Ellison et al. 2006). These species are therefore good candidates for an inter-specific hybridization approach due to an improved ability to predict the success of each potential cross. It is also a good example of how a genomics based approach can be successfully employed in inter-specific hybridization.

Other avenues to create unique variation in plants that are possibly less controversial than transgenics include somatic hybridization and selection via somaclonal variation. One problem with producing inter-specific hybrids is that reproductive barriers prevent embryo or endosperm development. Therefore, somatic hybridization, or fusion of protoplasts under tissue culture conditions, offers a method to overcome these barriers and create unique inter-specific, and possibly inter-generic, hybrids (Arcioni et al. 1997). Likewise, when growing any cells in tissue culture, stable genetic changes are common, leading to unique cell to cell variation called somaclonal variation. Therefore, somaclonal variation offers another, and safer, form of mutation breeding (Evans 1989). If the tissue culture media also contains a specific stress or toxin, then the cells are simultaneously selected for ability to grow in these conditions. Further selection is then practiced among the regenerated plants for the desired changes. Its best application to conventional breeding occurs when the best available germplasm is used to begin the process.

So, does using inter-specific hybridization, or even somaclonal variation and somatic hybridization, create variation as useful to breeders as transgenics? This is a legitimate question because these approaches are generally less expensive, and surely less controversial, than transgenics.

Future of Molecular Breeding

As stated above, there is no problem for using biotechnologies to create unique plants in order to conduct basic plant molecular biology research. This is important work that will not only continue, but increase in scope. However, the irony is not whether basic biotechnology research is increasing in forages, because it is, but whether useful biotechnology traits can be delivered directly to the farmer in an improved cultivar.

The future of molecular breeding for cultivar development in forage crops is also tied to funding, and in the case of transgenics, lies in the hands of regulatory agencies and their ability to establish a fair process to evaluate real versus perceived risks. Consortia of various partners like those described for the CAI will also be important to bring the fruits of these new technologies to researchers and farmers alike. However, it is hoped that more funding will be available to help the regulatory agencies in assessing the question of real versus perceived risks. At the end of the day, these agencies will need to make decisions on what are the real risks, establish a rigorous regulatory process to assess these risks, oversee the regulatory process in a fair manner, and make a decision! We can all then move forward based strictly on the value of the traits to the well-being of the environment, the farmer, agriculture, and all citizens.

The use of forages as cellulosic biofuel crops also now offers new molecular breeding opportunities, especially for value added traits such as enhanced biomass and fermentation efficiency. The main criteria for any biofuel crops are high yields achieved with low input costs in an environmentally friendly manner. By this definition, several high yielding, perennial forages are good candidates as biofuel crops. However, the initial requirement of low cost of the delivered feedstock may be the greatest hurdle for breeders and growers to overcome for most forages.

In addition to their direct use as cellulosic feedstock, the evolving biofuel industry has created other opportunities for forage and pasture crops, such as an increased need for high value forage finishing systems created by expensive feed grain prices due ironically to the current use of corn grain as a main ethanol producing feedstock.

Biofuels

Biofuels include ethanol, biodiesel, and other hydrocarbons achieved either through a fermentation or gasification process using plant biomass as a "feedstock". However, this current discussion will concentrate mainly on forages for use as cellulosic feedstock to produce ethanol.

Cellulosic ethanol is ethanol produced from cellulosic material (e.g. all plant parts especially stems, leaves, seedheads, etc.). Cellulosic feedstocks are generally comprised of three components: cellulose (~44%), hemicellulose (~30%) and lignin (~26%). The cellulose and hemicellulose provide a rich supply of carbohydrates that are ultimately used to produce ethanol. Sources of cellulosic material include grasses, wood and wood residue, and crop residues such as corn stover and wheat straw. However, ethanol produced from any feedstock, corn grain, perennial grasses, wheat straw, etc. is all chemically identical.

The technology to create cellulosic ethanol is becoming closer to reality. Many companies world-wide are in the later stages of development and entering the early stages of commercial scale-up into ethanol plants (also called biorefineries). Though most of the pieces are in place, the key is to continue to make ethanol production more cost-effective and economically competitive.

A biorefinery produces fuel-grade ethanol, and that ethanol is then blended in a percentage with gasoline to make a finished motor fuel. Commonly, we hear about E10 (10% ethanol/90% gasoline) and E85 (85% ethanol/15% gasoline). It is unlikely most vehicles will run on pure ethanol anytime soon.

At this time, there are not many service stations selling fuel grade ethanol at the pump. This is one of the national issues concerning its use and adoption especially in the USA where there are over 150,000 outlets – gas stations and convenience stores – and fewer than 1,000 sell ethanol.

Based on current estimates, cellulosic feedstocks are far better than grain in producing ethanol. Cellulosic feedstocks are estimated to produce approximately five times more energy that corn grain. Further, cellulosic feedstocks are intended to have a broader range of adaptability to poorer soils, which would allow them to be grown in regions that cannot support large-scale grain production. The cellulosic feedstocks being considered are crop residues, perennial crops such as grasses and trees, animal manures, and even municipal waste. For the perennial grasses, the main ones being investigated are switchgrass (*Panicum virgatum* L.), giant miscanthus (*Miscanthus* × *giganteus* Greef & Deuter ex Hodkinson & Renvoize), and giant reed (*Arundo donax* L.).

Switchgrass as a Biofuel Crop

Although any high yielding perennial forage will suffice, switchgrass is being investigated as one of the main perennial biomass species for cellulosic ethanol production in the USA. This is because it is a perennial grass native to the prairies of North America that was also identified by the United States Department of Energy (DOE) as a primary target for development as a dedicated energy crop because of its potential for high fuel yields, drought tolerance, and ability to grow well on marginal cropland without heavy fertilizing or intensive management (McLauglin and Kszos 2005).

The initial DOE program to evaluate and develop switchgrass as a bio-energy crop was recently reviewed and demonstrated that switchgrass has potential as an alternative to corn for ethanol production and as a supple-ment for coal in electricity generation (McLaughlin and Kszos 2005). The program identified the best varieties and management practices to optimize productivity, while concurrently developing a research base for long-term improvement through breeding and sustainable production in conventional agro-ecosystems. Gains through plant breeding were found for switchgrass yield to exceed that of corn. Significant carbon sequestration was projected for soils under switchgrass that should improve both soil productivity and nutrient cycling. Co-firing switchgrass with coal will also reduce green-house gas production. Finally, collaborative research with industry included fuel production and handling in power production, herbicide testing and licensing, release of new cultivars, and genetic modifications for chemical co-product enhancement.

More research will need to be conducted on crops like switchgrass that incorporate biotechnologies. In the USA, the DOE-USDA biomass genomics research program announced projects with switchgrass as one of the main species (USDA, DOE News Release; URL: http://genomicsgtl.energy.gov/ research/DOEUSDA/), that along with a new bioenergy center concentrating on improving switchgrass recalcitrance (DOE Bioenergy Research Center Fact Sheets; URL: http://www.science.doe.gov/News_Information/News_Room/ 2007/Bioenergy_Research_Centers/DOE%20BRC%20fact%20sheet%20 final%206-26-07.pdf), are important developments. The future is therefore bright for switchgrass as a dedicated energy crop with millions of hectares projected to be planted in order to meet DOE goals.

Other Forages as Biofuel Feedstock

Again, the main criteria for any biofuel crops are high biomass yields achieved with low input costs in an environmentally friendly manner. It is also important that these crops have alternate uses besides feedstock for biorefineres such as forage for livestock. This is why switchgrass is a very good choice. By this definition, the traditional, high yielding forages like bermudagrass, tall fescue (*Festuca arundinacea* Schreb.), ryegrasses (*Lolium* spp.), red clover (*Trifolium pratense* L.), white clover (*Trifolium repens* L.), and alfalfa are also good candidates. However, the requirement of low cost of the delivered feedstock, possibly as low as $50USD per US ton, is the greatest hurdle for growers of these crops to overcome.

For high value crops like alfalfa to be used, the harvested product needs to be divided into components, such as leaves and stems, and using the leaves to produce high value meal and the stems for sale to a biorefinery. If co-products such as pharmaceuticals are simultaneously extracted from the leaf material, this allows the economics of using alfalfa as a biofuel crop to work even better.

It is possible that each specific geographic region will have its own cropping system(s) based on several adaptive crops to supply a local biorefinerery. Co-cropping alfalfa or tall fescue with switchgrass to achieve an off-season supply of biomass, or inter-cropping switchgrass with alfalfa or clovers to supply nitrogen into the production system are good examples of how this could work.

Summary and Conclusions

Molecular breeding is important and will be used extensively in future forage research efforts. The overall participation and depth of research in this area as presented at this and past MBFT meetings supports this fact. Transgenics will have a big role to play in future forage cultivar deve-lopment efforts, but other approaches such as inter-specific hybridization and somatic hybridization need to be re-examined for potential use.

Molecular breeding also needs to develop from a platform of good conventional breeding and include supporting agronomic research and partnering with commercial industry where appropriate.

Future problems for molecular breeding in forages include high development costs, poor breeding histories and the polyploid nature of the main species, accurate phenotyping for most of the genomics based approaches, and freedom to operate, regulatory, and public perception issues for transgenics. To overcome these problems, development and regulatory costs will need to be funded by government grants and organizational consortia. The regulatory agencies will likewise need to establish a fair system that separates real from perceived risk.

All biofuel industries will be local with their own cropping systems, but high yielding forage crops will play a large role as feedstocks for this emerging industry. Initial context for the biofuels industry is for cheaply produced feedstock and this is why switchgrass is being touted as an initial dedicated crop. The future context is unclear, but should involve value-added feedstocks. The main traits to be improved are increased biomass yield, reduced input costs, and reduced chemical recalcitrance. Molecular breeding is therefore poised to make positive impacts in the biofuel feedstock development area.

References

Arcioni S, Damiani F, Mariani A, Pupilli F (1997) Salinity and aluminum. In: McKersie BD, Brown DCW (eds) Biotechnology and the Improvement of Forage Legumes. CAB International, Wallingford, UK, pp 61–89

Brummer EC, Bouton JH, Sledge M (2007) Biotechnology and molecular approaches to forage improvement. In: Barnes RF, Nelson CJ, Moore KJ, Collins M (eds) Forages – The Science of Grassland Agriculture, 6th edition, Volume II, Blackwell Publishing, Ames, IA, USA, pp 439–451

Burton GW, Hanna WW (1995) Bermudagrass. In: Barnes RF, Miller DA, Nelson CJ (eds) Forages – An Introduction to Grassland Agriculture, 5th edition, Volume 1, Iowa State University Press, Ames, Iowa, USA, pp 421–429

Ellison NA, Liston A, Steiner JJ, Williams WM, Taylor NL (2006) Molecular phylogenetics of the clover genus (*Trifolium*-Leguminosae). Mol Phylogenet Evol 39:688–705

Evans DA (1989) Somaclonal variation – genetic basis and breeding applications. Trends in Genet 5:46–50

Hopkins A, Wang Z, Mian R, Sledge M, Barker R (2003) Molecular breeding of forage and turf. Kluwer, Dordrecht, the Netherlands

Humphreys MO (2005) Molecular breeding for genetic improvement of forage crops and turf. Wageningen Academic Publishers, Wageningen, the Netherlands

McLaughlin SB, Kszos LA (2005) Development of switchgrass (*Panicum virgatum*) as a bioenergy feedstock in the United States. Biomass Bioenergy 28:515–535

Spangenberg G (2001) Molecular breeding of forage crops. Kluwer, Dordrecht, the Netherlands

Stebbins GL (1950) Variation and evolution in plants. Columbia University Press. New York

Tietz N (2007) No guarantees. Hay & Forage Grower 22(6):4–12 (Also available online at http://hayandforage.com/mag/farming_no_guarantees/)

DREB Regulons in Abiotic-Stress-Responsive Gene Expression in Plants

Kazuko Yamaguchi-Shinozaki[1,2,4] and Kazuo Shinozaki[3,4]

[1]Laboratory of Plant Molecular Physiology, Graduate School of Agricultural and Life Sciences, The University of Tokyo, 1-1-1, Yayoi, Bunkyo-ku, Tokyo 113-8657, Japan, kazukoys@jircas.affrc.go.jp
[2]Biological Resources Division, Japan International Research Center for Agricultural Sciences (JIRCAS), Tsukuba, Ibaraki 305-8686, Japan
[3]RIKEN Plant Science Center, Tsurumi, Yokohama, Kanagawa 230-0045, Japan, sinozaki@rtc.riken.jp
[4]CREST, Japan Science and Technology Corporation (JST), Kawaguchi, Saitama 332-0012, Japan

Abstract. Plant growth and productivity is affected by various abiotic stresses such as drought, high salinity, and low temperature. Expression of a variety of genes is induced by these stresses in various plants. In the signal transduction network from perception of stress signals to stress-responsive gene expression, various transcription factors and *cis*-acting elements in the stress-responsive gene expression function for plant adaptation to environmental stresses. The dehydration-responsive element (DRE)/C-repeat (CRT) *cis*-acting element is involved in osmotic- and cold-stress-inducible gene expression. Transcription factors that bind to the DRE/CRT were isolated and named DREB1/CBF and DREB2. DREB1/CBF regulon is involved in cold-stress-responsive gene expression, whereas, DREB2 is involved in osmotic-stress-responsive gene expression. Recently, we highlight transcriptional regulation of gene expression in response to drought and cold stresses, with particular emphasis on the role of DREB regulon in stress-responsive gene expression.

Function of Drought Stress-Inducible Genes

Drought, high salinity, and freezing are environmental conditions that cause adverse effects on the growth of plants and the productivity of crops. Plants

T. Yamada and G. Spangenberg (eds.), *Molecular Breeding of Forage and Turf,*
doi: 10.1007/978-0-387-79144-9_2, © Springer Science + Business Media, LLC 2009

respond and adapt to these stresses to survive under stress conditions at the molecular and cellular levels as well as at the physiological and biochemical levels. Expression of a variety of genes is induced by these abiotic stresses (Thomashow 1999; Shinozaki and Yamaguchi-Shinozaki 2000; Bray et al. 2000; Zhu 2002; Yamaguchi-Shinozaki and Shinozaki 2006). Transcriptome analysis using microarray technology has proven to be very useful for the discovery of many stress-inducible genes involved in stress response and tolerance (Shinozaki et al. 2003). Numerous genes that are induced by various abiotic stresses have been identified using various microarray systems (Seki et al. 2002; Fowler and Thomashow 2002; Kreps et al. 2002; Maruyama et al. 2004; Vogel et al. 2005).

Genes induced during stress conditions are thought to function not only in protecting cells from stress by the production of important metabolic proteins but also in the regulation of genes for signal transduction in the stress response. Thus, these gene products are classified into two groups (Seki et al. 2002; Fowler and Thomashow 2002; Kreps et al. 2002). The first group includes proteins that probably function in stress tolerance, such as chaperones, LEA (late embryogenesis abundant) proteins, osmotin, antifreeze proteins, mRNA binding proteins, key enzymes for osmolyte biosynthesis such as proline, water channel proteins, sugar and proline transporters, detoxification enzymes, enzymes for fatty acid metabolism, proteinase inhibitors, ferritin and lipid-transfer proteins. Some of these stress-inducible genes that encode proteins such as key enzymes for osmolyte biosynthesis, LEA proteins and detoxification enzymes have been overexpressed in transgenic plants and have been found to produce stress-tolerant phenotypes in the transgenic plants (Holmberg and Bulow 1998; Cushman and Bohner 2000). These results indicate that the gene products of the stress-inducible genes really function in stress tolerance.

The second group contained protein factors involved in further regulation of signal transduction and gene expression that probably function in stress response. They included various transcription factors suggesting that various transcriptional regulatory mechanisms function in the drought-, cold- or high-salinity-stress signal transduction pathways (Seki et al. 2003). The others were protein kinases, protein phosphatases, enzymes involved in phospholipids metabolism, and other signaling molecules such as calmodulin-binding protein and 14-3-3 proteins. At present the function of most of these genes are not fully understood. It is important to elucidate the role of these regulatory proteins for further understanding of plant responses to abiotic stress.

DREB Regulons in *Arabidopsis*

The promoter of a drought-, high-salinity- and cold-inducible gene, RD29A/COR78/LTI78, in *Arabidopsis* contains a major cis-acting element, the dehydration-responsive element (DRE)/C-repeat (CRT), that is involved in stress-inducible gene expression and its consensus was G/ACCGAC. DRE functions in one of the ABA-independent pathways in response to drought, high-salinity and cold stresses (Shinozaki and Yamaguchi-Shinozaki 2000). cDNAs encoding DRE binding proteins, *DREB1/CBF*, and *DREB2*, have been isolated by using yeast one-hybrid screening (Stockinger et al. 1997; Liu et al. 1998). These proteins contained the conserved DNA-binding domain found in the ERF and AP2 proteins. These proteins specifically bind to the DRE sequence and activate the expression of genes driven by the DRE sequence.

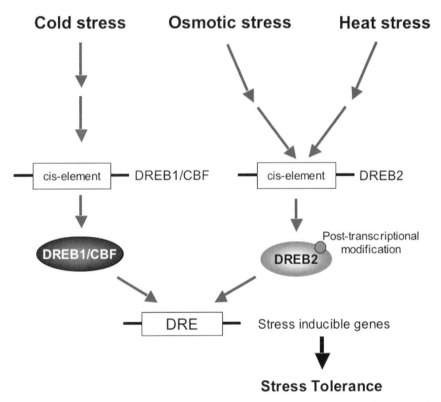

Fig. 1 A model of the induction of abiotic-stress-inducible genes that have the DRE cis-element in their promoters. Two different type DRE-binding proteins, DREB1/CBF and DREB2, distinguish different signal transduction pathways. DREB1/CBF-type transcription factors function in response to cold, and DREB2-type transcription factors function in drought and heat stresses

In *Arabidopsis*, three DREB1/CBF proteins are encoded by genes that lie in tandem on chromosome 4 in the order of *DREB1B/CBF1*, *DREB1A/CBF3* and *DREB1C/CBF2* (Gilmour et al. 1998; Liu et al. 1998). *Arabidopsis* also contains two DREB2 proteins, DREB2A and DREB2B (Liu et al. 1998). Expression of the *DREB1/CBF* genes is induced by cold, but not by dehydration and high-salinity stresses (Liu et al. 1998; Shinwari et al. 1998). By contrast, expression of the *DREB2* genes is induced by dehydration and high-salinity stresses but not by cold stress (Fig. 1; Liu et al. 1998; Nakashima et al. 2000). Later, Sakuma et al. (2002) reported three novel *DREB1/CBF*-related genes and six novel *DREB2*-related genes that were not expressed at high levels under various stress conditions. However, one of the *CBF/DREB1* genes, *CBF4/DREB1D* is induced by osmotic stress, suggesting the existence of crosstalk between the CBF/DREB1 and the DREB2 pathways (Haake et al. 2002).

DREB1/CBFs, Major Transcription Factors in Cold-Responsive Gene Expression

Transgenic *Arabidopsis* plants overexpressing *CBF1/DREB1B* under control of the cauliflower mosaic virus (CaMV) 35S promoter showed strong tolerance to freezing stress (Jaglo-Ottosen et al. 1998). Overexpression of the *DREB1A/CBF3* under the control of the CaMV 35S promoter also increased the tolerance to drought, high-salinity, and freezing stresses (Liu et al. 1998; Kasuga et al. 1999; Gilmour et al. 2000). Six genes have been identified as the target stress-inducible genes of DREB1A using RNA gel blot analysis (Kasuga et al. 1999). By using microarray analyses, more than 40 target genes of DREB1/CBF have been identified (Seki et al. 2001; Fowler and Thomashow 2002; Maruyama et al. 2004; Vogel et al. 2005). Most of these target genes contained the DRE or DRE-related core motifs in their promoter regions (Maruyama et al. 2004). These gene products are transcription factors, phospholipase C, RNA-binding protein, sugar transport protein, desaturase, carbohydrate metabolism-related proteins, LEA proteins, KIN (cold-inducible) proteins, osmoprotectant biosynthesis-protein, protease inhibitors, and so on. Many of them were proteins known to function against stress and were probably responsible for the stress tolerance of the transgenic plants. However, overexpression of the DREB1A protein also severely retarded growth under normal growth conditions. Use of the stress-inducible *rd29A* promoter instead of the constitutive 35S CaMV promoter for the overexpression of DREB1A minimizes negative effects on plant growth (Kasuga et al. 1999).

DRE has been shown to function in gene expression in response to stress in tobacco plants, which suggests the existence of similar regulatory systems in tobacco and other crop plants (Yamaguchi-Shinozaki and Shinozaki 1994). The DRE-related motifs have been reported in the promoter region of cold-inducible *Brassica napus* and wheat genes (Jiang et al. 1996; Ouellet et al. 1998). Additionally, the changes that occur in the *Arabidopsis* metabolome in response to cold were examined and the role of the CBF/DREB1 cold response pathway were assessed (Cook et al. 2004). On the other hand overexpression of the *Arabidopsis DREB1/CBF* genes in transgenic *B. napus* or tobacco plants induced expression of orthologs of *Arabidopsis* DREB1/CBF-targeted genes and increased the freezing tolerance of transgenic plants (Jaglo et al. 2001; Kasuga et al. 2004). These observations suggest that the DREB1/CBF regulon can be used to improve the tolerance of various kinds of agriculturally important crop plants to drought, high-salinity and freezing stresses by gene transfer.

Interestingly, Zhang et al. (2004) reported that tomato, a chilling sensitive plant, encodes three *DREB1/CBF* homologs, *LeCBF1-3*, that are present in a tandem array in the genome. Only the tomato *LeCBF1* gene was found to be cold-inducible. Constitutive overexpression of *LeCBF1* in transgenic *Arabidopsis* plants induced expression of DREB1/CBF-targeted genes and increased freezing tolerance. These results clearly indicated that *LeCBF1* encodes a functional homolog of the *Arabidopsis* DREB1/CBF proteins. Overexpression of *Arabidopsis CBF1/DREB1B* in tomato has been shown to increase the chilling and drought tolerance of transgenic tomato plants (Hsieh et al. 2002a,b). However, constitutive overexpression of either *LeCBF1* or *Arabidopsis DREB1A* in transgenic tomato plants did not increase freezing tolerance (Zhang et al. 2004). microarray analysis only identified four genes that were induced 2.5-fold or more in the *LeCBF1* or *DREB1A* overexpressing plants. Three out of the four identified genes were putative members of the tomato DREB1/CBF regulon as they were also upregulated in response to low temperature and they concluded that an intact CBF/DREB1 cold response pathway is present in tomato but the tomato CBF/DREB1 regulon differs from that of *Arabidopsis* and appears to be considerably smaller and less diverse in function.

In rice, four CBF/DREB1 homologues and one DREB2 homologous genes, *OsDREB1A*, *OsDREB1B*, *OsDREB1C* and *OsDREB1D*, and *OsDREB2A*, respectively have been isolated (Dubouzet et al. 2003). Overexpression of OsDREB1A in transgenic *Arabidopsis* resulted in improved high-salinity and freezing stress tolerance. A DREB1/CBF-type transcription factor, ZmDREB1A was also identified in maize (Qin et al. 2004). The ZmDREB1A was shown to be involved in cold-responsive gene expression, and the overexpression of this gene in *Arabidopsis* resulted in improved stress tolerance to drought and freezing. These results indicate that similar regulatory systems are conserved in monocots as well as dicots. Pellegrineschi et al. (2004) showed that overexpression of DREB1A/CBF3 driven by the stress-inducible rd29A promoter in transgenic wheat improved drought stress tolerance. Oh et al. (2005) reported that constitutive overexpression of DREB1A using the 35S promoter in transgenic rice resulted in increased stress tolerance to drought and high salinity. Similarly, Ito et al. (2006) also developed transgenic rice plants that constitutively expressed *DREB1A* or *OsDREB1A* genes. In this work, these factors in transgenic rice elevated tolerance to drought, high salinity, and low-temperature. These observations suggest that the DREB regulon can be used to improve the tolerance of various kinds of agriculturally important crop plants to drought, high-salinity and freezing stresses by gene transfer.

DREB2, Major Transcription Factors
in Osmotic-Responsive Gene Expression

The DREB2A protein has a conserved ERF/AP2 DNA-binding domain and recognizes the DRE sequence like DREB1A (Liu et al. 1998). Among the eight DREB2-type proteins, DREB2A and DREB2B are thought to be major transcription factors that function under drought and high-salinity stress conditions (Nakashima et al. 2000; Sakuma et al. 2002). However, overexpression of DREB2A in transgenic plants neither caused growth retardation nor improved stress tolerance, suggesting that the DREB2A protein requires post-translational modification such as phosphorylation for its activation (Liu et al. 1998), Nevertheless, the activation mechanism of the DREB2A protein has not yet been elucidated. Domain analysis of DREB2A using *Arabidopsis* protoplasts revealed that a negative regulatory domain exists in the central region of DREB2A and deletion of this region transforms DREB2A to a constitutive active form. Overexpression of the constitutive active form of DREB2A (DREB2A-CA) resulted in growth retardation in transgenic *Arabidopsis* plants. These transgenic plants revealed significant tolerance to drought stress but only slight tolerance to freezing. Microarray analyses of the transgenic plants revealed that DREB2A regulates expression of many drought-inducible genes. However, some genes downstream of DREB2A are not downstream of DREB1A, which also recognizes DRE but functions in cold-stress-responsive gene expression (Sakuma et al. 2006a). The genes downstream of DREB2A play an important role in drought stress tolerance, but alone are not sufficient to withstand freezing stress.

Figure 2 shows a model of the induction of genes regulated by DREB1A and DREB2A under drought, high salinity, and cold stress conditions. The *DREB1A* gene is induced by cold stress and *DREB2A* is induced by drought and high-salinity stresses. Modification of the DREB2A protein such as phosphorylation may be necessary for its activation under drought and high-salinity stress conditions. The genes downstream of the DREB proteins are categorized into three groups. The first group consists of downstream genes shared by DREB1A and DREB2A; most of these have ACCGACNT in their promoter regions. The second group consists of DREB1A-specific downstream genes; these genes have A/GCCGACNT in their promoters. The third group consists of DREB2A-specific downstream genes; we found ACCGACNA/G/C frequently in their promoter regions. These different downstream genes between the DREB1A and DREB2A proteins result in different stress tolerance to cold and drought in plants (Sakuma et al. 2006a).

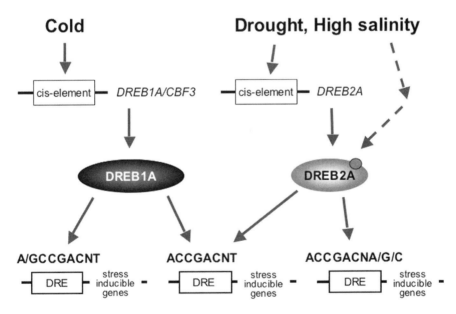

Fig. 2 Model of the induction of genes regulated by DREB1A and DREB2A under drought, high-salinity and cold stress conditions.
The genes downstream of the DREB proteins are categorized into three groups. The middle group contains downstream genes shared by DREB1A and DREB2A. The other groups consist of DREB1A- and DREB2A-specific downstream genes.

Synthetic green fluorescent protein (sGFP) gave a strong signal in the nucleus under unstressed control conditions when fused to the DREB2A-CA but only a weak signal when fused to the full-length DREB2A, suggesting that the constitutive active form of DREB2A is more stable than the full-length DREB2A protein in the nucleus of the transgenic plants. The full-length DREB2A protein containing the PEST sequence in the negative regulatory domain may be degraded rapidly by the ubiquitin-proteasome system, whereas the DREB2A-CA protein may have a long lifetime in the nucleus. Recently, *Arabidopsis* cDNAs for DREB2A interacting proteins that function as RING E3 ligases were identified and shown to negatively regulate plant drought stress responsive gene expression (Qin et al. 2008).

Dual Function of DREB2A in Osmotic-Stress- and Heat-Stress-Responsive Gene Expression

Microarray analysis of transgenic *Arabidopsis* overexpressing *DREB2A-CA* indicated that the overexpression of *DREB2A-CA* induces not only drought and salt responsive genes but also heat-shock related genes (Sakuma et al. 2006b). Moreover, transient induction of the *DREB2A* and *DREB2B* was shown to occur rapidly by heat-shock stress and that the sGFP-DREB2A protein accumulates in nuclei of heat shock stressed cells. Thirty-six DREB2A upregulated genes that exhibited remarkable expression ratios greater than eight times were classified into three groups based on their expression patterns: genes induced by heat shock, genes induced by drought stress and genes induced by both heat shock and drought stress. An *Arabidopsis* heat shock factor, *HsfA3* is one of the most up-regulated heat-inducible genes in transgenic plants overexpressing *DREB2A-CA*. *HsfA3* is directly regulated by DREB2A under heat stress indicating that further transcriptional cascades in response to heat stress are expected downstream of DREB2A. DREB2A upregulated genes were downregulated in DREB2A knockout mutants under stress conditions. Thermotolerance was significantly increased in the plants overexpressing *DREB2A-CA* and decreased in the *DREB2A* knockout plants. These results indicate that DREB2A functions in both osmotic and heat-shock stress responses.

Recently, DREB2C was reported to be induced by mild heat stress. Constitutive expression of DREB2C led to enhanced thermotolerance in transgenic lines of Arabidopsis. Microarray analyses of transgenic plants revealed that DREB2C also regulates expression of several heat stress-inducible genes that contain elements in their promoters indicating that DREB2C is a regulator of heat stress tolerance in *Arabidopsis* (Lim et al. 2007).

A *DREB2* homolog from maize, *ZmDREB2A*, whose transcripts were accumulated by cold, dehydration, salt and heat stresses in maize seedlings. Unlike *Arabidopsis DREB2A*, *ZmDREB2A* produced two forms of transcripts and quantitative RT-PCR analyses demonstrated that only the functional transcription form of *ZmDREB2A* was significantly induced by stresses. Moreover, the ZmDREB2A protein exhibited considerably high transactivation activity compared with DREB2A in *Arabidopsis* protoplasts, suggesting that protein modification is not necessary for ZmDREB2A to be active. Constitutive or stress inducible expression of *ZmDREB2A* resulted in an improved drought stress tolerance in plants. Microarray analyses of transgenic plants overexpressing *ZmDREB2A* revealed that in addition to genes encoding LEA proteins, some genes related to heat shock and detoxification were also upregulated. Furthermore, overexpression of *ZmDREB2A* enhanced thermotolerance in transgenic plants, implying that ZmDREB2A also play a dual functional role in mediating both water-stress- and heat-stress-responsive gene expressions (Qin et al. 2007).

Conclusions and Future Perspectives

Many plant genes are regulated in response to abiotic stresses, such as dehydration and cold, and their gene products function in stress response and tolerance. In the stress-responsive gene expression, various transcription factors and *cis*-acting elements in the stress-responsive promoters function not only as molecular switches for gene expression but also as terminal points of stress signals in the signaling processes. Timing of stress-responsive gene expression is regulated by a combination of transcription factors and *cis*-acting elements in stress-inducible promoters. DRE is one of major *cis*-acting elements in abiotic stress-inducible gene expression. DRE functions in crosstalk between drought/salinity stress response and cold stress response. DRE also function as a heat-shock stress responsive *cis*-acting elements. Combinations of the DRE sequence and transcription factors are important to determine crosstalk in stress signaling pathways.

Abiotic stresses affect plant growth and development, such as flowering timing and cell growth. This indicates crosstalk between environmental stress signal and plant growth. Plant hormones are thought to be involved in these crosstalk events. Transcription is important in regulation of plant development and environmental interactions, which may be affected by cross talk in transcription regulatory networks. Crosstalk between signal transduction pathways will be one of important subjects in the near future.

Negative regulation as well as positive regulation is important for gene expression. The degradation of transcription factor proteins plays an important role in the negative regulation of gene expression. Specific E3 ligase proteins are involved in stabilization of some transcription factors in stress response. Recently, RNA interference or mRNA degradation has been suggested to function in stress responsive gene expression. Complex regulation of gene expression may cause complex and flexible responses of plant to abiotic stresses.

References

Bray E, Bailey-Serres J, Weretilnyk E (2000) Responses to abiotic stresses. In: Buchanan BB, Gruissem W, Jones RL (eds) Biochemistry and Molecular Biology of Plants, American Society of Plant Physiologists, Rockville, pp 1158–1203

Cook D, Fowler S, Fiehn O, Thomashow MF (2004) A prominent role for the CBF cold response pathway in configuring the low-temperature metabolome of Arabidopsis. Proc Natl Acad Sci USA 101: 15243–15248

Cushman JC, Bohner HJ (2000) Genome approaches to plant stress tolerance. Curr Opin Plant Biol 3: 117–124

Dubouzet JG, Sakuma Y, Ito Y, Kasuga M, Dubouzet EG, Miura S, Seki M, Shinozaki K, Yamaguchi-Shinozaki K (2003) OsDREB genes in rice, *Oryza sativa* L., encode transcription activators that function in drought-, high-salt- and cold-responsive gene expression. Plant J 33: 751–763

Fowler S, Thomashow MF (2002) Arabidopsis transcriptome profiling indicates that multiple regulatory pathways are activated during cold acclimation in addition to the CBF cold response pathway. Plant Cell 14: 1675–1690

Gilmour SJ, Zarka DG, Stockinger EJ, Salazar MP, Houghton JM, Thomashow MF (1998) Low temperature regulation of *Arabidopsis* CBF family of AP2 transcriptional activators as an early step in cold-induced COR gene expression. Plant J 16: 433–442

Gilmour SJ, Sebolt AM, Salazar MP, Everard JD, Thomashow MF (2000) Over-expression of the Arabidopsis CBF3 transcriptional activator mimics multiple biochemical changes associated with cold acclimation. Plant Physiol 124: 1854–1865

Haake V, Cook D, Riechmann JL, Pineda O, Thomashow MF, Zhang JZ (2002) Transcription factor CBF4 is a regulator of drought adaptation in Arabidopsis. Plant Physiol 130: 639–648

Holmberg N, Bulow L (1998) Improving stress tolerance in plants by gene transfer. Trends Plant Sci 3: 61–66

Hsieh TH, Lee JT, Yang PT, Chiu LH, Charng YY, Wang YC, Chan MT (2002a) Heterology expression of the Arabidopsis C-repeat/dehydration response element binding factor 1 gene confers elevated tolerance to chilling and oxidative stresses in transgenic tomato. Plant Physiol 129: 1086–1094

Hsieh TH, Lee JT, Charng YY, Chan MT (2002b) Tomato plants ectopically expressing Arabidopsis CBF1 show enhanced resistance to water deficit stress. Plant Physiol 130: 618–626

Ito Y, Katsura K, Maruyama K, Taji T, Kobayashi M, Shinozaki K, Yamaguchi-Shinozaki K (2006) Functional analysis of rice DREB1/CBF-type transcription factors involved in cold-responsive gene expression in transgenic rice. Plant Cell Physiol 47: 141–153

Jaglo KR, Kleff S, Amundsen KL, Zhang X, Haake V, Zhang JZ, Deits T, Thomashow MF (2001) Components of the Arabidopsis C-repeat/dehydration-responsive element binding factor cold-response pathway are conserved in *Brassica napus* and other plant species. Plant Physiol 127: 910–917

Jaglo-Ottosen KR, Gilmour SJ, Zarka DG, Schabenberger O, Thomashow MF (1998) *Arabidopsis* CBF1 overexpression induces *cor* genes and enhances freezing tolerance. Science 280: 104–106

Jiang C, Iu B, Singh J (1996) Requirement of a CCGAC *cis*-acting element for cold induction of the *BN115* gene from winter *Brassica napus*. Plant Mol Biol 30: 679–684

Kasuga M, Liu Q, Miura S, Yamaguchi-Shinozaki K, Shinozaki K (1999) Improving plant drought, salt, and freezing tolerance by gene transfer of a single stress-inducible transcription factor. Nat Biotechnol 17: 287–291

Kasuga M, Miura S, Shinozaki K, Yamaguchi-Shinozaki K (2004) A combination of the Arabidopsis *DREB1A* gene and stress-inducible rd29A promoter improved drought- and low-temperature stress tolerance in tobacco by gene transfer. Plant Cell Physiol 45: 346–350

Kreps JA, Wu Y, Chang HS, Zhu T, Wang X, Harper JF (2002) Transcriptome changes for Arabidopsis in response to salt, osmotic, and cold stress. Plant Physiol 130: 2129–2141

Lim CJ, Hwang JE, Chen H, Hong JK, Yang KA, Choi MS, Lee KO, Chun WS, Lee SY, Lim CO (2007) Over-expression of the Arabidopsis DRE/CRT-binding transcription factor DREB2C enhances thermotolerance. Biochem Biophys Res Commun 362: 431–436

Liu Q, Kasuga M, Sakuma Y, Abe H, Miura S, Goda H, Shimada Y, Yoshida S, Shinozaki K, Yamaguchi-Shinozaki K (1998) Two transcription factors, DREB1 and DREB2, with an EREBP/AP2 DNA binding domain separate two cellular signal transduction pathways in drought- and low-temperature-responsive gene expression, respectively, in *Arabidopsis*. Plant Cell 10: 391–406

Maruyama K, Sakuma Y, Kasuga M, Ito Y, Seki M, Goda H, Shimada Y, Yoshida S, Shinozaki K, Yamaguchi-Shinozaki K (2004) Identification of cold-inducible downstream genes of the *Arabidopsis* DREB1A/CBF3 transcriptional factor using two microarray systems. Plant J 38: 982–993

Nakashima K, Shinwar ZK, Sakuma Y, Seki M, Miura S, Shinozaki K, Yamaguchi-Shinozaki K (2000) Organization and expression of two *Arabidopsis DREB2* genes encoding DRE-binding proteins involved in dehydration- and high-salinity-responsive gene expression. Plant Mol Biol 42: 657–665

Oh SJ, Song SI, Kim YS, Jang H, Kim SY, Kim M, Kim YK, Nahm BH, Kim, JK (2005) Arabidopsis CBF3/DREB1A and ABF3 in transgenic rice increased tolerance to abiotic stress without stunting growth. Plant Physiol 138: 341–351

Ouellet F, Vazquez-Tello A, Sarhan F (1998) The wheat wcs120 promoter is cold-inducible in both monocotyledonous and dicotyledonous species. FEBS Lett 423: 324–328

Pellegrineschi A, Reynolds M, Pacheco M, Brito RM, Almeraya R, Yamaguchi-Shinozaki K, Hoisington D (2004) Stress-induced expression in wheat of the *Arabidopsis thaliana DREB1A* gene delays water stress symptoms under greenhouse conditions. Genome 47: 493–500

Qin F, Sakuma Y, Li J, Liu Q, Li YQ, Shinozaki K, Yamaguchi-Shinozaki K (2004) Cloning and functional analysis of a novel DREB1/CBF transcription factor involved in cold-responsive gene expression in *Zea mays* L. Plant Cell Physiol 45: 1042–1052

Qin F, Kakimoto M, Sakuma Y, Maruyama K, Osakabe Y, Tran LS, Shinozaki K, Yamaguchi-Shinozaki K (2007) Regulation and functional analysis of *ZmDREB2A* in response to drought and heat stresses in *Zea mays* L. Plant J 50: 54–69

Qin F, Sakuma Y, Tran L-SP, Maruyama K, Kidokoro S, Fujita Y, Fujita M, Umezawa T, Sawano Y, Miyazono K, Tanokura M, Shinozaki K, Yamaguchi-Shinozaki K (2008) *Arabidopsis* DREB2A-interacting proteins function as RING E3 ligases and negatively regulate plant drought stress-responsive gene expression. Plant Cell 20: 1693–1707

Sakuma Y, Liu Q, Dubouzet JG, Abe H, Shinozaki K, Yamaguchi-Shinozaki K (2002) DNA-binding specificity of the ERF/AP2 domain of *Arabidopsis* DREBs, transcription factors involved in dehydration- and cold inducible gene expression. Biochem Biophys Res Commun 290: 998–1009

Sakuma Y, Maruyama K, Osakabe K, Qin F, Seki M, Shinozaki K, Yamaguchi-Shinozak K (2006a) Functional analysis of an Arabidopsis transcription factor, DREB2A, involved in drought-responsive gene expression. Plant Cell 18: 1292–1309

Sakuma Y, Maruyama K, Qin F, Osakabe Y, Shinozaki K, Yamaguchi-Shinozaki K (2006b) Dual function of an *Arabidopsis* transcription factor DREB2A in water-stress- and heat-stress-responsive gene expression. Proc Natl Acad Sc USA 103: 18822–18827

Seki M, Narusaka M, Abe H, Kasuga M, Yamaguchi-Shinozaki K, Carninci P, Hayashizaki Y, Shinozaki K (2001) Monitoring the expression pattern of 1300 Arabidopsis genes under drought and cold stresses by using a full-length cDNA microarray. Plant Cell 13: 61–72

Seki M, Narusaka M, Ishida J, Nanjo T, Fujita M, Oono Y, Kamiya A, Nakajima M, Enju A, Sakurai T, Satou M, Akiyama K, Taji T, Yamaguchi-Shinozaki K, Carninci P, Kawai J, Hayashizaki Y, Shinozaki K (2002) Monitoring the expression profiles of 7000 *Arabidopsis* genes under drought, cold and high-salinity stresses using a full-length cDNA microarray. Plant J 31: 279–292

Seki M, Kamei A, Yamaguchi-Shinozaki K, Shinozaki K (2003) Molecular responses to drought, salinity and frost: common and different paths for plant protection. Curr Opin Biotechnol 14: 194–199

Shinozaki K, Yamaguchi-Shinozaki K (2000) Molecular responses to dehydration and low temperature: differences and cross-talk between two stress signaling pathways. Curr Opin Plant Biol 3: 217–223

Shinozaki K, Yamaguchi-Shinozaki K, Seki M (2003) Regulatory network of gene expression in the drought and cold stress responses. Curr Opin Plant Biol 6: 410–417

Shinwari ZK, Nakashima K, Miura S, Kasuga M, Seki M, Yamaguchi Shinozaki K, Shinozaki K (1998) An Arabidopsis gene family encoding DRE/CRT binding s involved in low-temperature-responsive gene expression. Biochem Biophys Res Commun 250: 161–170

Stockinger EJ, Gilmour SJ, Thomashow MF (1997) *Arabidopsis thaliana* CBF1 encodes an AP2 domain-containing transcriptional activator that binds to the C-repeat/DRE, a cis-acting DNA regulatory element that stimulates transcription in response to low temperature and water deficit. Proc Natl Acad Sci USA 94: 1035–1040

Thomashow MF (1999) Plant cold acclimation: freezing tolerance genes and regulatory mechanisms Annu Rev Plant Physiol Plant Mol Biol 50: 571–599

Vogel JT, Zarka DG, Van Buskirk HA, Fowler SG, Thomashow MF (2005) Roles of the CBF2 and ZAT12 transcription factors in configuring the low temperature transcriptome of *Arabidopsis*. Plant J 41: 195–211

Yamaguchi-Shinozaki K, Shinozaki K (1994) A novel *cis*-acting element in an *Arabidopsis* gene is involved in responsiveness to drought, low-temperature, or high-salt stress. Plant Cell 6: 251–264

Yamaguchi-Shinozaki K, Shinozaki K (2006) Transcriptional regulatory networks in cellular responses and tolerance to dehydration and cold stressed. Annu Rev Plant Biol 57: 781–803

Zhang JZ, Creelman RA, Zhu JK (2004) From laboratory to field. Using information from Arabidopsis to engineer salt, cold, and drought tolerance in crops. Plant Physiol 135: 615–621

Zhu J (2002) Salt and drought stress signal transduction in plants. Annu Rev Plant Biol 53: 247–273

Comparative Genomics in Legumes

Steven Cannon

USDA-ARS and Department of Agronomy, Iowa State University, Ames, IA, USA, steven.cannon@ars.usda.gov

Abstract. The legume family will soon include three sequenced genomes. The majority of the euchromatic portions of the model legumes *Medicago truncatula* and *Lotus japonicus* have been sequenced in clone-by-clone projects, and the sequencing of the soybean genome is underway in a whole-genome shotgun project. Genome-wide sequence-based comparisons between three genomes with common ancestry at less than ~50 million years will enable us to infer many features of the ancestral genome, to trace evolutionary differences such as rates of particular gene- or transposon-family expansions or losses, to better understand processes of genome remodeling that follow polyploidy, and to transfer knowledge to other crop and forage legumes. Comparisons among these genomes show a lack of large-scale genome duplications within the *Lotus* or *Medicago* genomes following separation of those lineages approximately 40 mya, evidence of an older shared polyploidy event, and clear evidence of a more recent duplication in soybean following the separation from the *Medicago* and *Lotus* common ancestor at approximately 50 mya. In contrast to the extensive rearrangements observed in the Arabidopsis genome, the *Lotus* and *Medicago* genomes have retained substantial gene collinearity, at the scale of whole chromosomes or chromosome arms – good news for translational genomics across a broad spectrum of forage and crop legumes.

Introduction

The ability to sequence full plant genomes dramatically broadens opportunities in plant molecular research and breeding – and not just for sequenced species, but for many related agronomic crops. Much of the power in genome sequence is in the ability to combine it with a wide range other genomic resources such as dense genetic maps, quantitative trait loci (QTL) maps, transposon-tagged (or other) mutant libraries, bacterial artificial chromosome (BAC) libraries, microarrays, transcript sequences, chromatin immunoprecipitation assays, and haplotype (association) maps.

Such genomic resources don't come cheaply, however, so tend to be developed for the best-funded model species. Many species important for forage and turf have received less intensive funding, and some have large, complex, recently duplicated genomes that make them unlikely targets for full genome sequencing. For example, the small-genome legume *Medicago truncatula* is being sequenced, while the recent tetraploid *Medicago sativa* is not; the small-genome model temperate grass *Brachypodium distachyon* is being sequenced, while the large-genome grasses such as perennial ryegrass are not.

Fortunately, evolutionary conservation of genomic sequence allows us to translate knowledge between species. The ability to make comparisons between genome sequences will be crucial for leveraging and exchanging knowledge learned in these model systems, and applying that knowledge to a wide range of agronomically important species. Sequence comparisons are also a key tool for understanding the evolutionary trajectories that give rise to new plant functions, structures, chemistries, and physiologies.

Legume Genome Sequencing Strategies and Status

The international *Medicago truncatuala* (*Mt*) genome sequencing consortium, initiated by early funding from the Samuel Roberts Noble Foundation, and now funded by the National Science Foundation (NSF) and the European Union, is scheduled to complete the euchromatic genome regions (16 arms of 8 chromosomes) by the end of 2008. This project is using a clone-by-clone approach, in which BACs, with average insert size of approximately 120 kb, are sequenced and used to extend BAC-contig tiling paths to produce increasingly large sequence contigs. Contigs are anchored and oriented using genetic markers developed from a large proportion of the BAC sequences. As of early 2007, BAC contigs and sequences cover approximately 60% of the major euchromatic regions of the *Mt* genome (Cannon et al. 2006 and unpublished data).

The *Lotus japonicus* (*Lj*) genome sequencing project is being carried out by the Kazusa DNA Research Institute in Japan. This project is also primarily using a clone-by-clone approach, sequencing transformation-competent artificial bacterial chromosomes (TACs), with average insert size of approximately 100 kb. The clone-by-clone sequence is also being augmented by a combination of whole genome shotgun (WGS) and low-coverage TAC sequencing. The *Lj* sequence coverage has spanned approximately 177 Mbp, or roughly 60% of the euchromatic regions of the *Lj* genome (Young et al. 2005; Cannon et al. 2006).

The *Glycine max* (*Gm*) genome is being sequenced primarily with a WGS approach, with sequence coming from a combination of random reads, paired fosmid ends, and paired BAC end sequences. Additionally, approximately 500 BACs will be sequenced to high coverage; these will be a mix of BACs selected for biological interest by members of the soybean research community, and to span gaps where necessary. This project is a combined effort of the U.S. Department of Energy's Joint Genome Institute (carrying out the sequencing), and the NSF and USDA-ARS (managing mapping and physical mapping components of the project). This project is scheduled for completion near the end of 2008.

Thus, three legume genomes will be at or near completion by 2009. Together, these three models span a great deal of taxonomic space in the legumes, representing more than 10,000 species, including most forage legumes. This "taxonomic space" is described next.

Legume Systematics and Consequences for Information Transfer to Forage Legumes

The legume family is extremely diverse, with around 20,000 species and 700 genera, found in every terrestrial environment (Doyle and Luckow 2003). The majority of species are in the papilionoid subfamily, with 476 genera and about 14,000 species (Lewis et al. 2003). The Mimosoideae subfamily contains 77 genera and around 3,000 species. The remainder fall in the caesalpinoideae subfamily – something of a polyphyletic grab-bag, with 162 genera and around 3,000 species, including diverse early-diverging legume taxa (Fig. 1).

The papilionoid subfamily includes the crop legumes and the major model legume species, and thus is the taxonomic space across which much of legume comparative genomics and "translational genomics" will take place. Most papilionoid species of agronomic interest fall within one of two large clades: first, the Hologalegina clade, containing most of the temperate herbaceous legumes (thus, the shorthand "temperate herbaceous legumes"), including clovers, vetches, pea, lentil, *Medicago*, and *Lotus*; and second, the Millettioid clade, mostly consisting of tropical and subtropical species, and including common beans, soybean, and cow-pea (Maddison and Schulz 1996–2006; Doyle and Luckow 2003; Doyle et al. 1997; Hu et al. 2000). Some commonly encountered genera in Hologalegina are *Vicia*, *Medicago*, *Pisum*, *Trifolium*, *Cicer*, *Lens*, *Astragalus*, *Wisteria*, *Lotus*, *Robinia*, and *Sesbania*. Some commonly encountered genera in the Millettioid clade are *Glycine*, *Phaseolus*, *Vigna*, *Erythrina* (coral bean), and *Apios americana* (groundnut), as well as some earlier-diverging clades, one with the eponymous genus *Milletia*, and the other with Indigofera (containing the shrub that was used to produce indigo dye). Beyond of these large clades, basal genera in the papilionoid subfamily include the "dalbergioid" clade, including numerous tropical trees (e.g. rosewood) and *Arachis* (peanut), and the "genistoid" clade, including *Lupinus* (lupine).

Fossil and molecular dating methods indicate that most morphological and species diversity in the legumes originated during a burst of speciation early in the Tertiary, ~60–50 mya (Lavin et al. 2005; Cronk et al. 2006). This is shortly after the Cretaceous, around the time of the major extinction event that ended the "age of the dinosaurs." This early radiation means that, perhaps surprisingly, many early-diverging genera – including those in the caesalpnoideae and mimosoideae – did not originate a great deal earlier than early-diverging lineages in the papilionoidae. Lavin et al. (2005) date the genistoid clade (*Lupinus*) at ~56 mya; the dalbergioid clade (*Arachis*) at ~55 mya; the millettoid clade (*Glycine*) at ~45 mya; and the Hologalegina clade (*Medicago*, *Lotus*) at ~51 mya. The *Glycine-Medicago* split occurred ~54 mya. And *Medicago* and *Lotus* separated early in Hologalegina, so they diverged at ~51 mya.

These dates and the likely early burst of legume speciation have important implications for comparisons between model legumes (*Glycine*, *Medicago*, *Lotus*, *Phaseolus* and *Pisum*) and other agronomic species. (*Phaseolus* and *Pisum* may be considered models even though they do not currently have genome sequencing projects: *Pisum*, as a long-time genetic and developmental model; and *Phaseolus*, as a model of seed biology and domestication patterns, among others (Gepts et al. 2005)). Comparisons between phaseolid species and *Medicago*, or between *Medicago* and *Lotus*, actually require traversing substantial evolutionary time (~50–55 million years to common ancestors). Additionally, evolutionary events that may have occurred "early" in the legumes (most prominently, nodulation or polyploidy) may actually have occurred within a relatively short evolutionary timeframe – of, say, ~10 million years in the early Cenozoic.

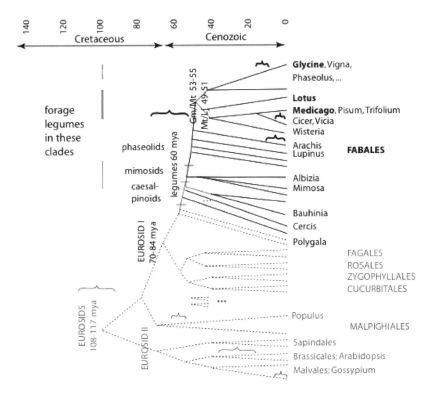

Fig. 1 Legume phylogeny, outgroups, and selected genomic features. Legume species now being sequenced are in *bold*. Outgroup plant orders are included to provide a frame of reference for Fabales (*in bold*). Difficulty of translational genomics is affected by taxonomic distance and by presence of whole genome duplications (WGD). Ranges for some known or suspected WGD are shown with braces. References for WGD and clade timings are given in Table 1. Note early radiation in the legumes, indicated by long branch terminal lengths for many lineages (data from Lavin et al. 2005). A small number of other lineages are included for reference, including Malpighiales (with poplar), Brassicales (with Arabidopsis)

Table 1 References for estimated dates of clades (A) and WGD (B)

A. Speciations	Example	Date (mya)	Ref.
Rosid I/Rosid II	Soybean/*Arabidopsis*	100–120	1, 2
Fabaceae/Salicaceae	Soybean/poplar	70–84	3, 4
Hologalegina/Millettoid	*Medicago*/soybean	54.3±0.6	5
Medicago/*Lotus*	*Medicago*/*Lotus*	50.6±0.8	5
Eudicot crown age	Ranunculus (buttercup)	125–147	3
Eurosid I crown age	Cucumber	70–84	3
Rosid crown age	Cercis (redbud)	108–117	3

B. Duplications (WGD)	Date (mya)	Ref.
Brassicaceae	24–40	5, 6
Salicaceae	60–65	1
Fabaceae	55–80	9
Glycine max	14.5	7
Glycine tomentella	<50 kya	8

References and notes in right-hand column are: (1) Tuskan et al. (2006); (2) Sanderson et al. (2004); (3) Wikstrom et al. (2001); (4) Lavin et al. (2005); (5) Blanc and Wolfe (2004); (6) Bowers et al. (2003); (7) Schlueter et al. (2004); (8) Rauscher et al. (2004); (9) extrapolated from Lavin et al. (2005 between papilionid crown and poplar/legume split; also 44–58 mya in Schlueter et al. (2004)

The practical implications of the nature of legume systematics for development of forage legumes are that, first, any legume species of interest will have sequenced legume models available for comparison; and second, at least some of those model species will be at a significant evolutionary distance from the species of interest. Most forage legumes (particularly alfalfa and the clovers) fall in the Hologalegina, and are most similar to the models *Mt* or *Lj*. Numerous phaseolid or Millettioid species also have potential as forage crops and green manures, particularly in tropical regions. Some examples include *Macroptilium bracteatum* (burgundy bean); *Mucuna pruriens* (velvet bean); *Clitoria ternatea* (butterfly pea) (http://www.tropicalforages.info). Of the sequenced models, these are most similar soybean. Other diverse legumes are also important components in many pasture systems, and of course in most native prairie and grassland ecosystems. In the Cesalpinoid subfamily, several *Chamaecrista* species are used in pastures, including *Chamaecrista fasciculata* (partridge pea) in North America, and *Chamaecrista rotundifolia* (round-leafed cassia) from

Mexico through South America (Partridge and Wright 1992; Ruthven 2006). *Chamaecrista fasciculata* has been studied as a model of population and ecotype response to climate change (Etterson and Shaw 2001), and is the subject of ongoing transcript profiling (Susan Singer, pers. comm.). In the mimosid subfamily, various species are important in tropical and subtropical areas, though more frequently as browse than as component of pasture systems.

Polyploidy and Consequences for Genome Comparisons

Definitions and History

The terms "polyploidy" "paleopolyploidy," and "whole genome duplication" (WGD) all point to the same process of doubling of chromosomal number, and are essentially interchangeable. Over time, with rearrangements and loss of genes and chromosomal segments, the genome "diploidizes," losing most evidence of the original duplication.

Plant genome comparisons have established that polyploidy occurred early in angiosperm evolution, and has occurred numerous times independently in subsequent lineages. Thus, most if not all angiosperms retain remnants of several rounds of WGD. (Masterson 1994; Bowers et al. 2003; De Bodt et al. 2005).

Polyploidy has far-reaching effects on a genome. It expands allelic variation and phenotypic diversity, it opens the door to functional divergence and innovation in metabolic and developmental pathways, and it creates a reproductive barrier and evolutionary bottleneck. Polyploidy also tends to have effects similar to heterosis, with transient increases in measures such as stature, total dry weight, and seed size (e.g. Guo et al. 1996; Bretagnolle and Thompson 2001; Birchler et al. 2003). This effect may be due in part to gene dosage effects: every gene is immediately present in at least two copies. Polyploidy is also of interest because it complicates gene positional comparisons between related species, whereas species with a shared polyploidy history are more likely to have simple chromosomal relationships. All of these characteristics make it important to determination the history of polyploidy events in the legumes.

Genome Duplications in the Legumes

In the legumes, most evidence points to one round of WGD very early in or shortly preceding the origin of the family. Studies in the 1990s of chromosomal correspondences within the soybean genome, using genetic marker comparisons, suggested that the soybean genome contained at least some regions present in more than two copies (Shoemaker et al. 1996; Yan et al. 2003). Self-comparisons of large ESTs data sets from either soybean or *Medicago* show a clear recent duplication in soybean (Schlueter et al. 2004; Blanc and Wolfe 2004). The basis for these studies is that silent-site mutations in homologous gene pairs give a distribution of changes per silent site (often called a "Ks" measurement). The older Ks peaks in soybean and *Mt* have been dated to ~44–64 mya. Schlueter et al. (2004) places a duplication event in *Mt* at ~58 mya, consistent with ~54 mya estimated by Lavin et al. (2005).

Using similar dating of gene pairs, but taking gene pairs from internal synteny blocks, Mudge et al. (2005) estimate the duplication in *Medicago* occurred at ~64 mya (0.79 synonymous substitutions per site), compared with and *Glycine/Medicago* separation at 48–50 mya. Similarly, Cannon et al. (2006) found a peak at 0.80 synonymous substitutions per site, corresponding to ~65 mya, using a whole-genome comparison of *Medicago* to itself. This is significantly before the split with *Lotus*, estimated in the same study at ~51 mya (0.64 synonymous substitutions per site) – consistent with the range 50.6±0.9 mya in Lavin et al. (2005). A duplication date of ~65 mya would place the duplication before the ~60 mya origin of the legumes proposed by Lavin et al. (2005) – though all of these molecular rate conversions need to be treated cautiously, as silent-site variation may not always be entirely silent, and rates should not be assumed to be the same in different lineages.

In addition to the early legume genomic duplication, polyploidy has occurred in various legume lineages. Alfalfa (*Medicago sativa*) and white clover (*Trifolium repens*) are recent allopolyploids (Bingham and McCoy 1988; Ellison et al. 2006), and polyploidy can be induced artificially in red clover (Taylor et al. 1963). Polyploidy has occurred several times in *Arachis* (peanut, possibly in the course of domestication by early agriculturists (Kochert et al. 1996). Polyploidy has also occurred early in the *Glycine* genus (Schlueter et al. 2004; Shoemaker et al. 1996; Lee et al. 1999; Yan et al. 2003).

Synteny

Macrosynteny in the Legumes

The term "synteny" was coined to describe genes on the same chromosome (regard-less of whether they show linkage in classical tests for recombination). More frequently now, the term is used to indicate conserved genes order between chromosomal regions (either between species or within a duplicated region of one genome).

With the summary by Choi et al. (2004) of more than a decade worth of synteny and comparative marker studies in the legumes, it became clear that synteny extended across broad swaths of diverse species in at least the Papilionoid subfamily. This finding of synteny on the basis of conserved marker order (Choi et al. 2004) is also supported by comparisons between draft sequence from the *Mt* and *Lj* genome projects, which show conservation of gene order spanning nearly the length of entire chromosomes in some cases – for example, between chromosomes *Mt* 1 and *Lj* 5 or *Mt* 2 and *Lj* 6 (Cannon et al. 2006). The conservation is disrupted by gene losses, duplications, and small inversions, but is unmistakable (Fig. 2). However, interestingly, this study finds very little synteny between *Mt* 6 and any *Lj* chromosome. The *Mt* 6 also has an unusually high transposon density, with several density peaks across the chromosome, perhaps indicating knobs such as seen on Arabidopsis chromosome 4. The *Mt* 6 also includes an unusually large cluster of genes from the TIR-NBS-LRR disease resistance family. The cause of these unusual patterns is not yet known – whether, for example, the chromosome is new, or is not removing transposons because of suppressed recombination, or is hypermethylated, as is the knob on Arabidopsis chromosome 4 (Gendrel et al. 2002).

Effect of Polyploidy on Synteny Comparisons

Our capacity to make use of knowledge gained from model legumes and apply it to other forage legumes will depend substantially on the complexities of the comparison genomes, and on the extent of rearrangements between the genomes. Perhaps the largest single source of difference between related genomes is polyploidy, which rapidly spurs other genomic rearrangements, segmental losses, and losses of interspersed genes.

Synteny studies can provide a means of testing hypothesized genome duplication histories. A comparison of ~1 Mbp of genomic sequence from two regions in soybean to *Medicago* and Arabidopsis showed the soybean region(s) matching one to two *Medicago* regions and two to four Arabidopsis regions (Mudge et al. 2005). Another study shows correspondences between *Lotus*, *Medicago*, poplar, and Arabidopsis in regions containing the *Lotus* SYMRK and *Medicago* NORK receptor kinase genes (Kevei et al. 2005). In this region, Kevei et al. (2005) find synteny with four regions in Arabidopsis and three in poplar. As with the regions described by Mudge et al. (2005), synteny is interrupted between any two regions by interspersed gene losses or local duplications.

Not all classes of genes respond to polyploidy in the same way. In some gene families such as transcription factors, most genes are retained, whereas other gene families (such as those involved in defense recognition) undergo rapid turnover (Cannon et al. 2004). In a more comprehensive analysis of Arabidopsis genes, Maere et al. (2005) argue that three whole-genome duplications in that genome have been directly responsible for >90% of the increase in transcription factors, signal transducers, and developmental genes in the last 350 million years. Chapman et al. (2006) propose that genes retained after polyploidy may buffer critical functions, and further, that gradual loss of this buffering capacity of duplicated genes may contribute to the cyclicality of genome duplication over time.

Microsynteny in the Legumes

Substantial microsynteny is seen in comparisons between *Medicago* and *Lotus* (Choi et al. 2004, 2006; Zhu et al. 2006; Kevei et al. 2005; Cannon et al. 2006), between *Medicago* and *Glycine* (Mudge et al. 2005), and between *Lotus* and *Glycine* (Hwang et al. 2006).

Quantifying synteny is complicated by tandem duplications and by gene-calling parameters and accuracy. For example, inclusion of coding sequences from transposons would decrease apparent synteny, as would counting of differential tandem expansions in one region vs. the other.

Defining "synteny quality" as twice the number of gene matches divided by the total number of genes in both segments (after excluding transposable elements and collapsing tandem duplications), Cannon et al. (2006) report that "synteny quality" for $Mt \times Lj$ is 62% for an extended syntenic block

(58/94 of genes exhibit corresponding homologs within these regions). This region is shown in Fig. 2. This figure also shows homoeologous segments from *Mt* and *Lj* self-comparisons. Synteny quality in the *Mt* × *Mt* comparison is just 36%, and is 30% in the *Lj* × *Lj* region. The synteny in the self-comparisons of either the *Mt* or *Lj* is highly degraded, consistent with a history of very early polyploidy. The synteny seen within *G. max* was variable in the regions examined by Schlueter et al. (2008), but at the high ends, was far less degraded than between any duplications within *Mt* or *Lj*. Again, this would be consistent with polyploidy in *G. max* much more recent than in the ancestral legume duplication.

Mudge et al. (2005) make a comparison of several corresponding regions in *Medicago*, soybean, and Arabidopsis. The comparison illustrates several important points. In one synteny comparison spanning ~400 kb from each of two homoeologous regions in *Medicago* and a corresponding soybean region, phylogenetic analysis of each gene in the region shows one of the *Medicago* homoeologs is more closely related to the soybean region (in other words, is separated by speciation and so is orthologous); and the other *Medicago* region separated much earlier and is paralogous to both *Medicago* and soybean regions. This clearly fits a model of an early legume polyploidy, significantly predating the soybean-*Medicago* split. Extent of microsynteny is consistent with this model. Synteny quality between the soybean and *Medicago* orthologous regions is 60% but between the soybean and *Medicago* paralogous regions is 27%. And between the *Medicago* homoeologs, the synteny quality is only 18% (only nine genes shared of approximately 50 in either *Medicago* homoeolog).

It may be significant in the Mudge et al. (2005) study that the synteny quality is lower in the *Medicago* paralogous regions than in the *Medicago–Glycine* homoeolog: within a single genome, selection pressure should be lowered for duplicated genes, so rate of loss of either gene should be higher than in the stochastic, independent losses between two different genomes. An important conclusion is that internal synteny, remaining after polyploidy, may be more difficult to detect than synteny between two species at separated by a similar amount of time. For example, we might expect much clearer synteny between *Glycine* and Phaseolus than between two *Glycine* homoeologs. Relatedly, synteny may turn out to be cleaner between *Medicago* and *Lotus* than between *Medicago* and *Glycine* (which has undergone polyploidy), even though all three species diverged in similar time frames (~40–50 mya).

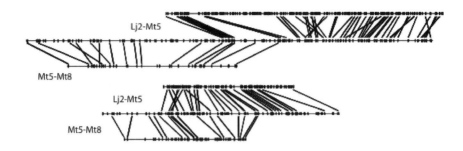

Fig. 2 Synteny in selected chromosomal regions between *Medicago* (*Mt*) and *Lotus* (*Lj*), and within a duplication in *Mt* compared with itself. *Top pair*: *Lj* 2 × *Mt* 5; *second pair*: *Mt* 5 × *Mt* 8; *third pair*: *Lj* 2 × *Mt* 5 (another region, within 1 Mbp of first regions); *last pair*: *Mt* 5 × *Mt* 8 (also within 1 Mbp of first regions). Note higher densities of collinear genes in the *Mt* × *Lj* comparison than in the internal genome duplication. Figure is adapted from Cannon et al. (2006)

Conclusions

Comparative genomics will be crucial for translating knowledge between models and crop species, enabling us to make best use of the enormously inventive germplasm and phenotypic variation across the legumes. There is a great deal of benefit in considering the legumes as a coherent, broad genetic system. This concept was stated succinctly in a report on the 2004 meeting on the "Legume Crops Genome Initiative" (LCGI): "Cross-legume genomics seeks to advance: (1) knowledge about the legume family as a whole; (2) understanding about the evolutionary origin of legume-characteristic features such as rhizobial symbiosis, flower and fruit development, and its nitrogen economy; and (3) pooling of genomic resources across legume species to address issues of scientific, agronomic, environmental, and societal importance." (Gepts et al. 2005). Comparative genomic techniques will be useful not solely as a means of positional cloning or gene-finding in related species, but also for elucidating how traits have evolved and continue to evolve. It is only by comparing nodulation in diverse species that we will learn how this important trait originated: whether once or several times; using what existing molecular machinery; etc. Similarly, comparisons will show how the diversity – and capacity for change – in defense response mechanisms evolved; and similarly for a large number of traits.

References

Bingham, ET, McCoy, TJ (1988) Cytology and cytogenetics of alfalfa. In: Alfalfa and Alfalfa improvement (eds. Hanson AA et al.), pp. 737–776. ASA, CSSA and SSSA, Madison, WI

Birchler, JA, Auger, DL, Riddle, NC (2003) In search of a molecular basis of heterosis. Plant Cell 15:2236–2239

Blanc, G, Wolfe, KH (2004) Widespread paleopolyploidy in model plant species inferred from age distributions of duplicate genes. Plant Cell 16:1667–1678

Bowers, JE, Chapman, BA, Rong, J, Paterson, AH (2003) Unravelling angiosperm genome evolution by phylogenetic analysis of chromosomal duplication events. Nature 422:433–436

Bretagnolle, F, Thompson, JD (2001) Phenotypic plasticity in sympatric diploid and autotetraploid *Dactylis glomerata*. Int J Plant Sci 162:309–316

Cannon, SB, Mitra, A, Baumgarten, A, Young, ND, May, G (2004) The roles of segmental and tandem gene duplication in the evolution of large gene families in *Arabidopsis thaliana*. BMC Plant Biol 4:10

Cannon, SB, Sterck, L, Rombauts, S, Sato, S, Cheung, F, Gouzy, JP, Wang, X, Mudge, J, Vasdewani, J, Scheix, T, Spannagl, M, Nicholson, C, Humphray, SJ, Schoof, H, Mayer, KFX, Rogers, J, Quetier, F, Oldroyd, GE, Debelle, F, Cook, DR, Retzel, EF, Roe, BA, Town, CD, Tabata, S, Van de Peer, Y, Young, ND (2006) Legume genome evolution viewed through the *Medicago truncatula* and *Lotus japonicus* genomes. Proc Natl Acad Sci USA 103: 14959–14964

Chapman, BA, Bowers, JE, Feltus, FA, Paterson, AH (2006) Buffering crucial functions by paleologous duplicated genes may contribute to cyclicality to angiosperm genome duplication. Proc Natl Acad Sci USA 103(8):2730–2735

Choi, H-K, Mun, J-H, Kim, D-J, Zhu, H, Baek, J-M, Mudge, J, Roe, BA, Ellis, N, Doyle, J, Kiss, GB, Young, ND, Cook, DR (2004) Estimating genome conservation between crop and model legume species. Proc Natl Acad Sci U S A 101:15289–15294

Choi, H-K, Luckow, MA, Doyle, JJ, Cook, DR (2006) Development of nuclear genederived markers linked to legume genetic maps. Mol Genet Genom 276:56–70

Cronk, Q, Ojeda, I, Pennington, RT (2006) Legume comparative genomics: progress in phylogenetics and phylogenomics. Curr Opin Plant Biol 9:99–103

De Bodt, S, Maere, S, Van de Peer, Y (2005) Genome duplication and the origin of angiosperms. Trends Ecol Evol 20:592–597

Doyle, JJ, Luckow, MA (2003) The rest of the iceberg. Legume diversity and evolution in a phylogenetic context. Plant Physiol 131:900–910

Doyle, JJ, Doyle, JL, Ballenger, JA, Dickson, EE, Kajita, T, Ohashi, H (1997) A phylogeny of the chloroplast gene rbcL in the Leguminosae: taxonomic correlations and insights into the evolution of nodulation. Am J Bot 84:541–554

Ellison, NW, Liston, A, Steiner, JJ, Williams, WM, Taylor, NL (2006) Molecular phylogenetics of the clover genus (Trifolium – Leguminosae). Mol Phylogenet Evol 39:688–705

Etterson, JR, Shaw, RG (2001) Constraint to adaptive evolution in response to global warming. Science 294:151–154

Gendrel, AV, Lippman, ZZ, Yordan, CC, Colot, V, Martienssen, R (2002) Heterochromatic histone H3 methylation patterns depend on the Arabidopsis gene DDM1. Science 297:1871–1873

Gepts, P, Beavis, WD, Brummer, EC, Shoemaker, RC, Stalker, HT, Weeden, NF, Young, ND (2005) Legumes as a model plant family. Genomics for food and feed report of the cross-legume advances through genomics conference. Plant Physiol 137:1228–1235

Guo, M, Davis, D, Birchler, JA (1996) Dosage effects on gene expression in a maize ploidy series. Genetics 142:1349–1355

Hu, J-M, Lavin, M, Wojciechowski, M, Sanderson, MJ (2000) Phylogenetic systematics of the tribe Millettieae (Leguminosae) based on trnK/matK sequences, and its implications for the evolutionary patterns in Papilionoideae. Am J Bot 87:418–430

Hwang, T-Y, Moon, J-K, Yu, S, Yang, K, Mohankumar, S, Yu, YH, Lee, YH, Kim, HS, Kim, HM, Maroof, MAS, Jeong, S-C (2006) Application of comparative genomics in developing molecular markers tightly linked to the virus resistance gene Rsv4 in soybean. Genome 49:380–388

Kevei, Z, Seres, A, Kereszt, A, Kalo, P, Kiss, P, Toth, G, Endre, G, Kiss, GB (2005) Significant microsynteny with new evolutionary highlights is detected between Arabidopsis and legume model plants despite the lack of macrosynteny. Mol Gen Genom 274:644–657

Kochert, G, Stalker, HT, Gimenes, M, Galgaro, L, Lopes, CR, Moore, K (1996) RFLP and cytogenetic evidence on the origin and evolution of allotetraploid domesticated peanut, *Arachis hypogaea* (Leguminosae). Am J Bot 83:1282–1291

Lavin, M, Herendeen, PS, Wojciechowski, MF (2005) Evolutionary rates analysis of Leguminosae implicates a rapid diversification of lineages during the Tertiary. Syst Biol 54:530–549

Lee, JM, Bush, A, Specht, JE, Shoemaker, R (1999) Mapping duplicate genes in soybean. Genome 42:829–836

Lewis, GP, Schrire, BD, Mackinder, BA, Lock, JM (2003) Legumes of the world. Royal Botanic Gardens, Kew, UK

Maddison, DR, Schulz, K-S (eds.) 1996–2006. The Tree of Life Web Project. Internet address: http://tolweb.org

Maere, S, De Bodt, S, Raes, J, Casneuf, T, Montagu, MV, Kuiper, M, Van de Peer, Y (2005) Modeling gene and genome duplications in eukaryotes. Proc Natl Acad Sci USA 102:5454–5459

Masterson, J (1994) Stomatal size in fossil plants: evidence for polyploidy in majority of angiosperms. Science 264:421–424

Mudge, J, Cannon, SB, Kalo, P, Oldroyd, GED, Roe, BA, Town, CD, Young, ND (2005) Highly syntenic regions in the genomes of soybean, *Medicago truncatula*, and *Arabidopsis thaliana*. BMC Plant Biol 5:15

Partridge, IJ, Wright, J (1992) The value of round-leafed cassia (*Cassia rotundifolia* cv. Wynn) in a native pasture grazed with steers in southeast Queensland. Trop Grassl 26:263–269

Rauscher, JT, Doyle, JJ, Brown, AH (2004) Multiple origins and nrDNA internal transcribed spacer homeologue evolution in the *Glycine tomentella* (Leguminosae) allopolyploid complex. Genetics 166:987–998

Ruthven, DC (2006) Grazing effects on forb diversity and abundance in a honey mesquite parkland. J Arid Environ 68:668–677

Sanderson, MJ, Thorne, JL, Wikstrom, N, Bremer, K (2004) Molecular evidence on plant divergence times. Am J Bot 91(10):1656–1665

Schlueter, JA, Dixon, P, Granger, C, Grant, D, Clark, L, Doyle, JJ, Shoemaker, RC (2004) Mining EST databases to resolve evolutionary events in major crop species. Genome 47:868–876

Schlueter, JA, Scheffler, BE, Jackson, S, Shoemaker, RC (2008) Fractionation of synteny in a genomic region containing tandemly duplicated gene across *Glycine max, Medicago truncatula,* and *Arabidopsis thaliana*. J Hered 99: 390–395

Shoemaker, RC, Polzin, K, Labate, J, Specht, J, Brummer, EC, Olson, T, Young, ND, Concibido, V, Wilcox, J, Tamulonis, JP, Kochert, G, Boerma, HR (1996) Genome duplication in soybean (*Glycine* subgenus *soja*). Genetics 144:329–338

Taylor, NL, Stroube, WH, Collins, GB, Kendall, WA (1963) Interspecific hybridisation of red clover (*Trifolium pratense* L.). Crop Sci 3:549–552

Tuskan, GA, DiFazio, S, Jansson, S, Bohlmann, J et al. (2006) The genome of black cottonwood (*Populus trichocarpa*). Science 313:1596–1604

Wikstrom, N, Savolainen, V, Chase, MW (2001) Evolution of the angiosperms: calibrating the family tree. Proc R Soc Lond Ser B 268:2211–2220

Yan, HH, Mudge, J, Kim, DJ, Larsen, D, Shoemaker, RC, Cook, DR, Young, ND (2003) Estimates of conserved microsynteny among the genomes of *Glycine max*, *Medicago truncatula* and *Arabidopsis thaliana*. Theor Appl Genet 106:1256–1265

Young, ND, Cannon, SB, Sato, S, Kim, DJ, Cook, DR, Town, CD, Roe, BA, Tabata, S (2005) Sequencing the genespaces of *Medicago truncatula* and *Lotus japonicus*. Plant Physiol 137:1174–1181

Zhu, H, Riely, BK, Burns, NJ, Ane, J-M (2006) Tracing nonlegume orthologs of legume genes required for nodulation and arbuscular mycorrhizal symbioses. Genetics 172:2491–2499

Development of *Trifolium occidentale* as a Plant Model System for Perennial Clonal Species

W.M. Williams[1], A.G. Griffiths[2], M.J.M. Hay[1], K.A. Richardson[1],
N.W. Ellison[1], S. Rasmussen[2], I.M. Verry[1], V. Collette[2], S.W. Hussain[1],
R.G. Thomas[1], C.S. Jones[2], C. Anderson[2], D. Maher[2], A.G. Scott[2],
K. Hancock[2], M.L. Williamson[1], J.C. Tilbrook[1], M. Greig[2] and A. Allan[2]

[1]AgResearch Grasslands, Private Bag 11008, Palmerston North, New Zealand,
warren.williams@agresearch.co.nz
[2]Pastoral Genomics, c/o AgResearch Grasslands, Private Bag 11008, Palmerston
North, New Zealand

Abstract. *Trifolium occidentale* D.E. Coombe is a diploid, clonal perennial clover that is very closely related to white clover (*T. repens* L.). It has been previously reported to be self-pollinating and lacking in genetic diversity. However, new collections, especially in Spain and Portugal, have revealed that cross-pollinating populations with substantial genetic diversity do exist. This has led to *T. occidentale* being investigated as a potential genetic model species to facilitate the application of genomic methods for the improvement of white clover. Investigations have shown that *T. occidentale* has many attributes that make it suitable as a genetic model for white clover. It forms hybrids with white clover and the chromosomes of the two species pair and recombine at meiosis. Phylogenetic research shows that it is a very close relative, and probably an ancestor, of white clover. A framework linkage map based on SSR markers has shown it to be highly syntenic with white clover. A protocol for efficient transformation has been developed. An effective EMS mutagenesis method has been demonstrated by the induction of a high frequency of condensed tannin negative mutants. The clonal nature of *T. occidentale* is not shared by other dicotyledonous model species. It may, therefore, be useful for the genomic characterisation of traits associated with clonal growth and perenniality in this wider class of plants.

T. Yamada and G. Spangenberg (eds.), *Molecular Breeding of Forage and Turf,*
doi: 10.1007/978-0-387-79144-9_4, © Springer Science + Business Media, LLC 2009

Introduction

White clover (*Trifolium repens* L.) is a stoloniferous, clonal perennial plant species that has become the most important legume component of grazed pastures in temperate parts of the world. For the last 80 years, it has been continuously improved by conventional plant breeding, leading to hundreds of cultivars (Caradus and Woodfield 1997). However, white clover presents some difficulties for the application of genomic methods to enhance the efficiency of plant breeding. Such methods work best on species that are diploid, self-pollinated and homozygous. White clover is a cross-pollinated allotetraploid with amphidiploid inheritance and is consequently highly heterozygous. Inbred lines are scarce and difficult to create. Here, we show that *T. occidentale* D.E. Coombe, a diploid stoloniferous perennial clover, has the characteristics of a genetic model, not only for white clover, but also for clonal perennial dicotyledonous species, in general.

The required characteristics of an ideal genetic model species are that it must be closely related to the target species and be diploid and self-pollinated. It must also have high genetic diversity, a short generation time, available genetic maps with high synteny and collinearity with the target species and efficient transformation. A further desirable feature would be an efficient mutagenesis system.

T. occidentale: Morphology and Distribution

T. occidentale is a diploid stoloniferous perennial clover that closely resembles tetraploid white clover (*T. repens*). It is so similar that, until 1960, it was not recognised as a distinct species (Coombe 1961). Indeed, the main taxonomic treatise on the genus *Trifolium* (Zohary and Heller 1984) regarded it as a variety of white clover. It is small-leaved and is distinguished from white clover by hairy petioles and thick, opaque leaves with a shiny, waxy under-surface. Its flower heads are very similar to those of white clover. It is known only from the gulf-stream coasts of western Europe, from northern Portugal and Spain, Brittany and Normandy, Cornwall, Wales, and SE Ireland (Preston 1980). It is strictly coastal in distribution, on consolidated beach sand, dunes and cliff-tops, generally within 50–100 m of the sea. Usually it does not overlap in distribution with white clover (Kakes and Chardonnens 2000).

Phylogenetic Relationships

In a phylogenetic analysis of about 200 *Trifolium* species using chloroplast and ITS DNA sequences (Ellison et al. 2006), a group of European-Eurasian species closely related to white clover, including *T. occidentale*, was identified. The nuclear ITS regions of 18-26S rDNA are variable and strongly indicative of phylogenetic relationships among *Trifolium* species. *T. occidentale* and *T. repens* are virtually identical in ITS sequences, differing, at most, at 1 bp among 738 bp. A very close relationship was also indicated by evidence that shows that both species share a unique centromeric DNA repeat sequence (TrR350) (Ansari et al. 2004).

This evidence indicates that *T. occidentale* may have been one of the diploid ancestral species that hybridized with another species to form allotetraploid white clover. On the other hand, chloroplast *trnL* intron DNA sequences of all *T. occidentale* populations studied differ from those of *T. repens* by a single 5 bp deletion in 584 bp. Other species, including *T. pallescens*, show closer sequence similarity with this *T. repens* chloroplast sequence, strongly suggesting that *T. occidentale* was not the female parent of white clover (Ellison et al. 2006).

T. occidentale forms interspecific hybrids with white clover and several other closely related species. Artificial hybrids with white clover were reported by Gibson and Beinhart (1969). These were fertile and showed chromosome pairing and the potential for introgression between the species (Chen and Gibson 1970). Hybridization and introgression were also demonstrated with two subspecies of *T. nigrescens* (Williams et al. 2008). A hybrid with *T. pallescens* produced by embryo rescue was reported by Williams et al. (2006).

Collections and Genetic Diversity

The species was originally described, on the basis of the first discovered populations from Cornwall, Ireland, France, the Channel Islands, and northern Spain, as being self-pollinated and having limited variability (Coombe 1961). Later, more populations were discovered in Portugal (Géhu 1972) and Spain (Rivas-Martínez 1976) and proved to be substantially more variable than any of the northern ones.

T. occidentale populations from 42 different sites were grown in two common garden experiments in comparison with Kent, a small leaved white clover, in Palmerston North, New Zealand. Populations from north of Brittany looked relatively uniform – almost all having small leaves without any leaf markings, while leaves in populations from Portugal and Spain were highly polymorphic for leaf markings. *T. occidentale* populations showed a marked geographic distribution in the genetic diversity of white V markings. We observed 960 plants from Brittany northwards to Ireland (80 from each of 12 populations) and found no plant with a white V. By contrast, among 600 plants from Portugal and Spain (20 from each of 30 populations) the majority were V marked and 26 populations were totally marked (9) or had marked and unmarked plants (17). Similarly, clovers frequently have purple anthocyanin markings, often as a fleck in the leaf. Among 960 plants from Brittany northward three had purple markings, and all were from the southern-most population in Brittany. By contrast, 24 of the 30 southern populations had some flecked plants and in three populations all plants were flecked.

Similar comparisons between northern and southern populations can be made with other traits. For example, in leaf size, all northern populations were significantly much smaller ($P < 0.05$) than Kent white clover. On the other hand, among southern populations, two from Spain had significantly larger ($P < 0.05$) leaves than Kent, several were similar in leaf size to Kent and some were significantly smaller. The species is apparently almost monomorphic for cyanogenesis (Kakes and Chardonnens 2000). In our study, every plant checked, to date, from 42 populations was linamarase enzyme negative and all but three plants from one population were positive for cyanogenic glucosides.

The greater genetic diversity of southern populations was also clearly shown by SSR analyses (Griffiths et al. unpublished data). SSR variation among 16 plants from 10 northern populations was surveyed and it was found that two of the French populations were most distinct and, among the others, the four Irish populations formed a group that was distinct from the UK and other French populations. Plants from the two most genetically distant northern groups (Ireland and France) were crossed and an F_2 mapping population was generated. The same Irish plant was also crossed with a Portuguese plant and a second mapping population based on a north x south cross was generated. A total of 1,100 SSR primer pairs from white clover and *Medicago truncatula* and which amplified in *T. occidentale* were assessed for polymorphism in both *T. occidentale* mapping populations.

In the northern population, only 110 SSR primer pairs were fully informative and 948 (85%) were monomorphic and uninformative. By contrast, in the north x south population, 375 were fully informative and only 425 (38%) were monomorphic. This result shows that the addition of the plant from Portugal increased the genetic diversity more than threefold over the maximum detected in our sample of northern populations. Combined with data on the other traits this shows that the genetic diversity in some populations from Portugal and Spain is greater than that among populations from France, the United Kingdom and Ireland.

Breeding System and Generation Time

Populations from Portugal, France, England, Wales and Ireland are fully self-fertile. By contrast, most populations from NW Spain are self-incompatible cross-pollinators. Self-fertile plants set some seeds autogamously. Full seed-set is easily achieved by rolling the seed heads between thumb and fingers or otherwise disturbing the heads.

Seedlings from several northern European populations grown from germination at 11°C initiated flower heads within 6 weeks and three generations can thus be achieved annually with such material. Populations from southern Europe require less exposure to low temperature and those from Spain and Portugal might therefore achieve four or five generations annually. Apart from exceptional genotypes from very far northern European populations, no flower head initiation occurred in warm long days without previous exposure to low temperatures.

Mapping and Synteny

Based on integration of data from the two F_2 mapping populations already mentioned, we have completed a consensus linkage map of *T. occidentale* (Griffiths et al. 2006, 2007). The map has a length of 610 cM with 362 SSR loci spread evenly at an average density of one locus per 1.7 cM. The map length is very similar to that of *Medicago truncatula* (600 cM) and is half the length of our white clover map (1,271 cM). There are, as expected, eight linkage groups averaging 76 cM in length – essentially similar to white clover (77 cM). Many of the mapped SSRs were white clover derived, enabling the alignment of the maps of the two species. As expected, there was a clear synteny of marker order and spacing between the two species.

This highlights not only their relatedness but also the potential of *T. occidentale* as a genetic model for white clover.

Transformation

Development of the transformation protocol was carried out with a population from Cornwall (Richardson et al. unpublished). The subsequent availability of the Portuguese and Spanish populations led to an assessment of 16 of these for transformability. As with many species, success of *Agrobacterium*-mediated transformation was highly genotype dependent. The binary vector pHZBar-intGUS based on pART27 (Gleave 1992) was used for transformation. The T-DNA of this plasmid contains a *bar* selectable marker gene which confers resistance to the herbicide ammonium glufosinate expressed from the cauliflower mosaic virus 35S promoter and the *uid*A gene (coding for β-glucuronidase, GUS) also expressed from a cauliflower mosaic 35S promoter. Up to 600 cotyledonary explants were transformed for each population. Putatively transformed plants were assayed by PCR and histochemical activity of GUS. Results were combined to calculate the transformation frequencies. Stable integration of the T-DNA was determined by Southern hybridisation.

The control population from Cornwall gave a transformation frequency of 0.3% as compared to a frequency of 2% routinely achieved with *T. repens*. By contrast two Spanish populations gave transformation frequencies of 5.5 and 7.5% which provides a significant improvement over that obtained in white clover. Further experiments have demonstrated repeatability of high transformation rates for both of the Spanish populations. Because both high transformation populations came from the NW Spain, they were self-incompatible and further work is needed to insert an S_f allele to make them useful for genomics work.

EMS Mutagenesis

An EMS treatment protocol to generate homozygous recessive mutants in *T. occidentale* has been developed and found to work very efficiently, as demonstrated with mutants that are defective in condensed tannin biosynthesis (Williams and Rasmussen unpublished data). The protocol involved the soaking of imbibed seeds in EMS solutions at 25°C. Only low concentrations of 0.1–0.25% (v/v) EMS were required. Treatment times of 6–16 h gave useful survival rates and workable numbers of M_2 families. Treated seeds were germinated and the resulting M_1 plants were self pollinated to produce M_2 seeds. M_2 seeds were germinated to generate M_2 families which were expected to have approximately 25% of plants homozygous for any recessive mutant alleles that had resulted from the mutagenesis treatment. An indication of the effectiveness of the treatments was obtained when the 16-h treatments produced approximately 70% of M_2 families with albino or pale green seedlings. These figures compare with only 34 pale green seedlings (2.4%) among the 1,440 M_1 seedlings (the best available control) and indicate very significant potential mutagenic effectiveness of the treatments.

Condensed Tannin Mutants

A very desirable objective for white clover breeding is the development of plants with significant amounts of condensed tannins in the leaves. The presence of condensed tannins would greatly improve the efficiency of protein digestion in the gut of grazing animals (Aerts et al. 1999) and also serve as an anti-foaming agent to prevent animal deaths caused by bloat (Kendall 1966). Currently, the species has condensed tannins in the flower petals and in the leaf trichomes, but the overall concentration is too low to fulfil the desired nutritional functions. The presence of the biosynthetic pathway tempts speculation that a regulatory mutation in the genes controlling tannin biosynthesis could bring about a change in the tissues in which white clover accumulates tannins.

White clover is not likely to be amenable to mutagenesis because it is allotetraploid and out-crossing. However, *T. occidentale*, which has an identical condensed tannin pattern to white clover, is amenable to mutagenesis. In all, 1,075 M_2 families were generated and over 8,000 M_2 plants were screened for leaf condensed tannin variants. No mutants were found among the 1,080 M_1 plants, indicating that CT mutants were not present before mutagenesis treatment. Twenty four heritable condensed tannin mutants were obtained. This mutation rate (2.4%) is high and confirms the value of *T. occidentale* for mutagenesis research. The initial variants were categorised into three classes:

1. Lacking trichomes
2. Trichomes lacking CT
3. Abnormal trichomes or trichomes with abnormal CT distribution

Most of the confirmed mutants were from class 2 – trichomes lacking CT. Among the 24 mutants, seven had seeds with transparent testas, i.e. they lacked CT in the seed coat as well as the trichomes. Virtually all of the mutants also had modified flower morphology, indicating possible pleiotropism of the gene(s) causing CT changes. The flowers of the mutants were of three types: (1) pink petals, normal morphology, (2) white petals, normal morphology, (3) white petals, reduced morphology. Detailed characterisation of accumulating foliar flavonoids by LC-MS showed that all mutants accumulated the same types of flavonols as wild-type plants but with varying concentrations.

Acknowledgements

The study was funded by AgResearch Ltd and Pastoral Genomics, a research consortium jointly funded by Fonterra, Meat & Wool New Zealand, AgResearch, Deer Industry New Zealand, Dairy InSight and the Foundation for Research Science and Technology (FRST).

References

Aerts RJ, Barry TN, McNabb WC (1999) Polyphenols and agriculture: Beneficial effects of proanthocyanidins in forages. Agric Ecosyst Environ 75:1–12

Ansari HA, Ellison NW, Griffiths, AG, Williams WM (2004) A lineage-specific centromeric satellite sequence in the genus *Trifolium*. Chromosome Res 12:1–11

Caradus JR, Woodfield DR (1997) World checklist of white clover varieties II. N Z J Agric Res 40:115–206

Chen C-C, Gibson PB (1970) Meiosis in two species of *Trifolium* and their hybrids. Crop Sci 10:188–189

Coombe DE (1961) *Trifolium occidentale*, a new species related to *T. repens* L. Watsonia 5:68–87

Ellison NW, Liston A, Steiner JJ, Williams WM, Taylor NL (2006) Molecular phylogenetics of the clover genus (*Trifolium* – Leguminosae). Mol Phylogenet Evol 39:688–705

Géhu J-M (1972) *Trifolium occidentale* D.E. Coombe espece noevelle pour le littoral du nord du Portugal. Agron Lusitana 34:197–204

Gibson PB, Beinhart G (1969) Hybridization of *Trifolium occidentale* with two other species of clover. J Hered 60:93–96

Gleave A (1992) A versatile binary vector system with a T-DNA organisational structure conducive to cloned DNA into the plant genome. Plant Mol Biol 20:1203–1207

Griffiths AG, Bickerstaff P, Anderson CB, Franzmayr B (2006) Threshing the white clover genome for gene-associated molecular markers. In: Mercer CF (ed) Breeding for success: diversity in action. Proc 13th Australasian Plant Breeding Conference, Christchurch, New Zealand, 18–21 April 2006, pp. 817–821

Griffiths AG, Anderson CB, Williams WM, Hay MJM, Williamson ML, Tilbrook JC (2007) A microsatellite map of *Trifolium occidentale*, a model stoloniferous perennial legume. 5th International Symposium on the Molecular Breeding of Forage and Turf (MBFT2007), Sapporo, Japan, 1–6 July 2007

Kakes P, Chardonnens AN (2000) Cyanotypic frequencies in adjacent and mixed populations of *Trifolium occidentale* Coombe and *Trifolium repens* L. are regulated by different mechanisms. Biochem Syst Ecol 28:633–649

Kendall WA (1966) Factors affecting foams with forage legumes. Crop Sci 6:487–489

Preston CD (1980) *Trifolium occidentale* D.E. Coombe, new to Ireland. Ir Nat J 20:37–40

Rivas-Martínez S (1976) Notes systématiques, chorologiques et écologiques sur des plantes d'Espagne, 1. Candollea 31:111–117

Williams WM, Verry IM, Ellison NW (2006) A phylogenetic approach to germplasm use in clover breeding. In: Mercer CF (ed) Breeding for success: diversity in action. Proc 13th Australasian Plant Breeding Conference, Christchurch, New Zealand, 18–21 April 2006, pp. 966–971

Williams WM, Ansari HA, Hussain SW, Ellison NW, Williamson ML, Verry IM (2008) Hybridisation and introgression between two diploid wild relatives of white clover, *Trifolium nigrescens* Viv. and *T. occidentale* Coombe. Crop Sci 48:139–148

Zohary M, Heller D (1984) The genus *Trifolium*. The Israel Academy of Sciences and Humanities, Jerusalem

Characterization and Utilization of Genetic Resources for Improvement and Management of Grassland Species

Roland Kölliker[1,3], Beat Boller[1], Mahdi Majidi[2], Madlaina K.I. Peter-Schmid[1], Seraina Bassin[1] and Franco Widmer[1]

[1]Agroscope Reckenholz-Tänikon ART, Zurich, Switzerland
[2]Isfahan University of Technology (IUT), Isfahan, Iran
[3]Corresponding author, roland.koelliker@art.admin.ch

Abstract. Characterization and targeted utilisation of genetic diversity are crucial for any successful breeding program and may help to better understand ecological processes in complex multi-species grasslands. Molecular genetic markers allow for a rapid assessment of diversity at the genome level and highly informative, sequence-specific markers have become available for several grassland species. Although many grassland species are outbreeding and therefore require a large number of plants to be analysed per population, technical and statistical developments have enabled the molecular genetic characterisation of diversity for a broad range of purposes. Molecular markers allow for a targeted selection of parental plants in polycross breeding programs, which may lead to improved performance of the resulting progenies. Detailed characterisation of plant genetic resources existing in collections of ecotype populations, landraces or cultivars may enable breeders to specifically complement their breeding material. In addition, the identification and subsequent elimination of duplicate or highly similar accessions helps to reduce costs involved in maintaining large germplasm collections. The analysis of genetic diversity also allows to investigate the influence of management and environment on population structure or to determine the effect of *in situ* and *ex situ* conservation on genetic composition at the species or population level. Furthermore, molecular markers may be used to characterise the effect of environmental pollutants on genetic diversity within species present in permanent grassland. For several grassland species including ryegrasses, fescues and clovers, various studies have highlighted the benefit of detailed molecular genetic characterisation for a targeted utilisation of genetic resources.

T. Yamada and G. Spangenberg (eds.), *Molecular Breeding of Forage and Turf,*   55
doi: 10.1007/978-0-387-79144-9_5, © Springer Science + Business Media, LLC 2009

Introduction

Diversity at various levels may substantially influence composition, functioning, productivity and sustainability of managed and non-managed ecosystems (Meyers and Bull 2002; Vellend and Geber 2005). Different levels of diversity are closely connected and interact through a variety of processes such as immigration, drift, selection or competition (Fig. 1).

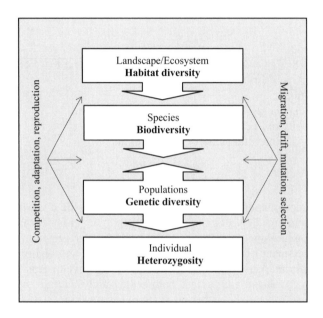

Fig. 1 Potential connections and factors influencing diversity at different levels of organization. (Modified from Vellend and Gerber (2005))

Heterogeneity, connectivity and fragmentation of regional landscapes as well as identity, abundance and diversity of habitats directly influence the abundance of individual species present in certain areas and are therefore important determinants of biodiversity (Noss 1990). In addition to harboring a broad range of species, diverse landscapes are also the basis for a broad range of human land use activities.

With increasing awareness that the Earth's biodiversity is declining (Chapin et al. 2000), the relationship between species diversity and ecosystem stability has been intensively investigated. Although diversity has been generally shown to increase ecosystem stability, the presence of functional groups capable of differential response, rather than persistence per se, may be responsible for this relationship (McCann 2000). In a decade-long grassland experiment, the number of species has been shown to directly influence ecosystem stability expressed as variation of biomass yield over time (Tilman et al. 2006). Average biomass yield from plots sown with 16 perennial grassland species from five functional groups showed significantly lower variation over time when compared to yield from plots sown with only 1, 2, 4 or 8 species, respectively. In contrast to the increased ecosystem stability, yield stability of individual species declined with increasing number of species sown per sward (Tilman et al. 2006). Increased species diversity may not only increase ecosystem stability but also influence productivity of grassland ecosystems. In a large scale experiment conducted at 28 European sites, Kirwan et al. (2007) were able to show that productivity of communities consisting of four sown species consistently exceeded productivity of communities consisting of only one sown species. Based on their data, authors estimated average transgressive overyielding, i.e. increased productivity in mixture when compared to monoculture, to reach up to 16%, depending on the species group investigated. In addition, the percentage of weeds present in the sward was considerably lower in multispecies communities when compared to monocultures (Kirwan et al. 2007).

While the importance and the benefit of diversity at the species level is generally recognized, the role of diversity at the lowest level of organization, i.e. genetic diversity within populations and individuals, may be less apparent. Genetic diversity is indispensable for the response of species and populations to selection, either natural through environmental changes or human mediated through processes such as targeted selection (Reed and Frankham 2003). This has long been recognized by plant breeders who routinely screen large germplasm collections for variation in specific traits or use ecotype populations to broaden their breeding germplasm (Allard 1999). Genetic variation of quantitatively inherited, fitness related traits is essential for the adaptability of populations and may therefore contribute to ecosystem stability. Although population fitness is a very complex trait and a direct relationship between genetic diversity and the various fitness components can often not be established (Booy et al. 2000), preservation of diversity for genes controlling specific traits such as disease resistance has been shown to be of great importance for population survival (Foster Hünneke 1991). In addition, reduced genetic diversity is often the result of partial inbreeding, which can directly influence population fitness (Oostermeijer et al. 1995). With regard to the importance of genetic resources for the improvement of agronomically important traits in forage crops and their value as a reservoir of diversity at the lowest level (Fig. 1), a detailed characterization of genetic diversity within grassland species is indispensable. In addition, monitoring changes in genetic diversity may also allow for estimating the effect of environmental change and agricultural management on permanent grassland.

Methods for Characterizing Genetic Diversity

Genetic diversity within species and, therefore, the diversity among individuals and groups of individuals can be estimated in various ways. Although environmental effects may be largely excluded through adequate measures and methods, estimates always reflect only a proportion of the overall genetic diversity of an individual, population or species. The choice of method must therefore be carefully made based on the particular question to be answered and the resources available.

Morpho-Physiological Characteristics

Phenotypic characterization of morpho-physiological traits such as height, shape, yield, date of flowering, etc. allows for the most direct approach to variability of agronomically important traits. However, in order to separate phenotypic plasticity from genetic diversity i.e. to account for environmental influences, laborious field experiments replicated across different environments are inevitable. This has been successfully applied in breeding and decade-long targeted phenotypic selection has resulted in forage grass and legume cultivars significantly improved for various traits (Humphreys 2005). Consequently, many quantitative genetic studies have been conducted under controlled environments with cultivated plants and little work has been undertaken in permanent grassland under natural conditions (Primack and Kang 1989). However, morpho-physiological traits, such as flowering time or growth habit, have been successfully used to characterize genetic resources of forage grass species (Casler 1998; Fjellheim et al. 2007). In addition, reliable phenotypic characterization is essential for the molecular genetic dissection of complex traits (Herrmann et al. 2008; Studer et al. 2007) and may valuably complement estimates of genetic diversity based on molecular genetic markers (Franco et al. 2001; Kölliker et al. 1998).

Molecular Genetic Markers

Molecular genetic markers allow for a rapid assessment of genetic diversity at the genome level and have been extensively used to characterize genetic resources in numerous plant species. Due to the initial lack of genetic information, non species-specific marker systems such as randomly amplified polymorphic DNA (RAPD, Welsh and McClelland 1990; Williams et al. 1990), inter simple sequence repeats (ISSR, Zietkiewicz et al. 1994) and amplified fragment length polymorphism (AFLP, Vos et al. 1995) have long dominated research on genetic diversity of forage and grassland species. These methods yield multilocus genetic data from single polymerase chain reaction (PCR) assays and therefore allow to investigate a large number of loci at relatively low cost. Although improved laboratory and data analysis techniques have resulted in high reproducibility and comparability of these methods, they fail to distinguish heterozygous from homozygous individuals and therefore markers must be treated as dominantly inherited (Nybom 2004). Consequently, considerable effort was made to develop co-dominant marker systems such as simple sequence repeats (SSR) and a fair number of markers has become publicly available for several grassland species such as *Lolium perenne* and *L. multiflorum* (Hirata et al. 2006; Jensen et al. 2005; Studer et al. 2006), *Festuca arundinacea* (Saha et al. 2004), *Trifolium repens* and *T. pratense* (Kölliker et al. 2001, 2006; Sato et al. 2005). Despite the many advantages of SSR markers, such as the ability to detect multiple alleles at one specific locus and a high rate of polymorphism, reproducibility and transferability, genomic SSRs may often be located in non-coding regions and the variability detected is poorly correlated with the variability of traits relevant for survival and performance (Booy et al. 2000). Functional markers directly linked to specific traits or functions (Andersen and Lübberstedt 2003) would allow for a more targeted characterization of genetic resources. Clearly, the development of a comprehensive set of functional markers for forage and grassland species will require major research efforts. So far, several gene based markers such as expressed sequence tag (EST) derived cleaved amplified polymorphic sequence (CAPS) markers (Miura et al. 2007), sequence tag site (STS) markers for resistance gene analogs (Ikeda 2005) and single nucleotide polymorphism (SNP) markers (Cogan et al. 2006, 2007) have been developed for *Lolium* spp. and *T. repens*, but their function remains to be clarified.

Statistical Analyses

The molecular genetic techniques available generate large datasets containing information on genetic variation across many loci in a large number of individuals or populations. To examine the relationship among individuals and the structure within and between populations, multivariate techniques such as principal component analysis (PCA), discriminant analysis or cluster analysis using the unweighted pair group method with arithmetic means (UPGMA) are employed and various coefficients such as Dice, Jaccard or squared Euclidean distance are used for calculation of genetic similarity or dissimilarity. In addition, analysis of molecular variance (Excoffier et al. 1992) or model based clustering methods (Pritchard et al. 2000) may be used to partition variance among experimental units or to infer population structure. The choice of appropriate coefficients and methods depends on the marker system used, the species under investigation, as well as on the objectives of the experiment and has been extensively discussed (Bonin et al. 2007; Kosman and Leonard 2005; Mohammadi and Prasanna 2003). Descriptive multivariate procedures such as cluster analysis or PCA are useful to highlight interesting groups of individuals or populations. However, hypotheses have to be tested with appropriate statistical methods wherever possible. For example, bootstrapping may be effectively utilized for estimating the statistical support to nodes in a dendrogram (Hillis and Bull 1993). In grassland research, it is often desirable to quantify the influence of geographic or environmental factors on genetic diversity and to test the significance of groupings observed in multivariate analyses. Redundancy analysis (implemented in CANOCO, ter Braak and Smilauer 2002) offers a powerful tool to relate a set of dependent variables to a set of independent variables. Hartmann et al. (2005) showed that this method can efficiently be used to relate genetic fingerprints to experimental treatments. In a study using bacterial community profiles from heavy metal treated and control soils, redundancy analysis revealed a highly significant influence of the heavy metal treatment on profile diversity (48% of the variance explained) and thus confirmed groups of bacterial communities previously identified by cluster analysis. Redundancy analysis may also successfully be used to evaluate the influence of environmental factors on genetic diversity within plant species.

Applications in Grassland Research

Thanks to the effort of numerous researchers, the knowledge on phylogenetic relationship and origin of grassland species has considerably increased.

For example, Charmet et al. (1997) demonstrated the close relationship of *F. pratensis* to *Lolium* spp. using chloroplast DNA and nuclear rDNA markers. More recently, Ellison et al. (2006) identified *T. occidentale* and *T. pallescens* to be the diploid progenitors of the allotetraploid *T. repens*. Chloroplast DNA sequences have also successfully been used to show that natural conditions rather than human activity influenced the present distribution of *F. pratensis* (Fjellheim et al. 2006).

Due to its importance for plant breeding, the characterization of genetic diversity within germplasm collections and natural populations has gained particular attention. Various marker systems have been employed to characterize genetic diversity in species such as *Lolium* spp., *F. pratensis*, *Trifolium* spp. or *Chloris gayana* (Bolaric et al. 2005; George et al. 2006; Herrmann et al. 2005; Kölliker et al. 1998; Ubi et al. 2003; Van Treuren et al. 2005). Such studies not only yield valuable information for the targeted utilization of particular germplasm collections, they may also give insight into evolutionary and ecological processes important for the long-term management of genetic resources of grassland species. In this context, studies on *T. pratense* not only showed that Mattenklee, a distinct Swiss form of *T. pratense*, most likely has originated from germplasm introduced from Flanders and Brabant, but also that cross fertilization with indigenous wild clover accessions occurred (Herrmann et al. 2005; Kölliker et al. 2003).

Many forage crops are outbreeding species with a high degree of self-incompatibility. Consequently, breeding often relies on intercrossing several selected parents using the polycross method. Knowledge on genetic diversity may be used directly to select optimal parental combinations in order to maximize heterosis and to minimize inbreeding. Although a general, direct correlation between molecular marker diversity and heterosis may not exist (Cerna et al. 1997; Joyce et al. 1999), a study in *L. perenne* showed that selection of genetically diverse parents may lead to significantly improved agronomic performance of first and second generation progenies (Kölliker et al. 2005).

Conservation of plant genetic resources has always been a great concern not only to plant breeders who rely on diverse germplasm, but also to conservationists and ecologists who aim to protect diversity at all levels (Fig. 1). In this context, research concerning the management of germplasm collections, as well as the influence of agricultural practices and environmental factors on genetic resources may be of particular interest.

Management of Germplasm Collections

Genetic resources of many species are maintained in large *ex situ* germplasm collections. Molecular markers were found to be useful to characterize germplasm stored in these collections, to eliminate duplicates, to establish core collections and to complement phenotypic and pedigree data (Grenier et al. 2000a,b).

In forage grass breeding, ecotype populations particularly adapted to specific environments are routinely used for broadening breeding germplasm. Such germplasm collections are often conserved *ex situ* in gene banks rather than *in situ* at collection sites but little is known on the effect of *ex situ* conservation on genetic diversity of ecotype populations. Therefore, we used five populations of *F. pratensis* sampled in 1973–1979 from five permanent meadows across Switzerland and conserved *ex situ* and compared them with five populations sampled from the same meadows in 2003, considered to be conserved *in situ*. Meadows were cut or grazed at inter-mediate intensity with only limited possibility to flower and produce seed. Genetic diversity was assessed using 28 plants per population and 20 SSR markers, which were chosen to represent each linkage group according to the *Lolium* reference map (Jones et al. 2002). The average number of alleles detected per locus was 5.4 across all ten populations, average expected (H_E) and observed (H_O) heterozygosity was 0.42 and 0.40, respectively. Diversity within populations, expressed as H_E, ranged from 0.38 to 0.47. There was no consistent trend towards increased or reduced diversity within *ex situ* or *in situ* conserved populations (Fig. 2).

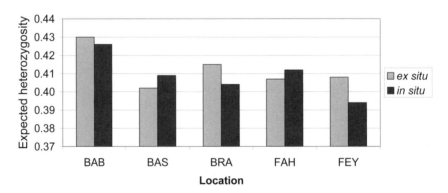

Fig. 2 Expected heterozygosity within *ex situ* and *in situ* conserved populations of *Festuca pratensis* sampled from five permanent meadows (location) across Switzerland determined on 28 individuals per population and 22 SSR markers

In addition, cluster analysis based on average Euclidean squared distance showed similar relationships among *ex situ* conserved populations when compared to *in situ* conserved populations (data not shown). In conclusion, we observed high levels of genetic diversity within populations but there was no clear influence of conservation method on within population diversity and no change of genetic relationship among populations was observed. Therefore, the two conservation methods seem not to affect genetic diversity of *F. pratensis* ecotype populations.

Effect of Habitat and Management on Genetic Diversity

In order to evaluate the influence of habitat and management on genetic diversity of forage grass ecotype populations, two valuable forage grass species, the highly abundant *L. multiflorum* and the only moderately abundant *F. pratensis* were studied (Peter-Schmid et al. 2008). Twelve ecotype populations per species were sampled from permanent pastures and meadows across Switzerland and compared to four reference cultivars each. Genetic diversity was determined using 23 plants per population and 24 SSR markers.

Analysis of molecular variance revealed 92.6% of the variation within *F. pratensis* to be due to variation within populations, 4.8% to be due to variation among populations and only 2.6% to be due to variation between ecotypes and cultivars (Peter-Schmid et al. 2008). This proportion even dropped to 0.6% in *L. multiflorum*, where 2.3% of the variation was due to variation among populations and 97.1% to variation within populations. Cluster analysis based on coancestry coefficients revealed a clear separation of cultivars and ecotypes for *F. pratensis*. In addition, *F. pratensis* ecotype populations were further subdivided into three groups, which corresponded to the altitude and to the management intensity prevailing at the respective habitats (data not shown). Redundancy analysis revealed a highly significant influence of all three factors (cultivar vs. ecotype, altitude and management intensity), but the variance explained by each factor was low (~1%). In contrast, for *L. multiflorum* no distinct grouping of ecotype populations or cultivars was observed (Peter-Schmid et al. 2008).

In general, the diversity within populations of both species was high and the level of population differentiation was low. The significant influence of environment and management on genetic structure of *F. pratensis* may reflect the limited and specialized distribution of this species. *L. multiflorum* ecotypes and cultivars on the other hand seem to form one large gene pool.

Effect of Environmental Pollutants on Grassland Species

Although environmental pollutants such as ozone (O_3) have been shown to negatively affect quality and yield of fruit, legume and cereal crops (Velissariou 1999) and to change the species composition of grassland (Volk et al. 2006), little is known on the effect of elevated O_3 on the genetic composition of grassland species. In a long-term fumigation experiment established in an old semi-natural grassland, the effect of 5 years treatment with ambient or elevated O_3 on genetic diversity of *Plantago lanceolata*, a representative dicotyledonous species of permanent grassland was investigated using AFLP and SSR markers (Kölliker et al. 2008). A total of 198 individual plants sampled along equally spaced transects through three fumigated (1.5 times elevated levels of O_3) and three control (ambient O_3) plots were analyzed using 87 polymorphic AFLP and 4 polymorphic SSR markers. Genetic diversity based on AFLP markers and expressed as percentage of polymorphic loci (PL) and H_E was significantly higher within fumigated plots (PL = 89.66%, H_E = 0.20) when compared to control plots (PL = 77.01%, H_E = 0.17). The respective values determined with SSR markers were not significantly different between the two

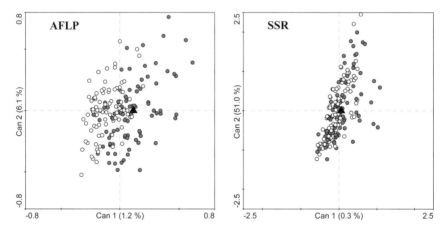

Fig. 3 Redundancy analysis for 198 *Plantago lanceolata* plants sampled from within and around six experimental plots with ambient or elevated O_3 levels (three plots per treatment) based on 87 AFLP and 4 SSR markers. Ozone treatment was used as environmental factor (*black triangle*), *white circles* indicate plants sampled from control plots and *grey circles* indicate plants sampled from plots treated with elevated O_3. Percentage of variance explained by the first two canonical axes (Can 1, Can 2) is given in *parentheses* (Kölliker et al. 2008, © 2008 Elsevier Ltd)

treatments, but H_O was again significantly higher for fumigated plots (Kölliker et al. 2008). With 1.2% of the variance explained, redundancy analysis based on AFLP markers showed a small but significant influence of elevated O_3 on genetic diversity, while no significant effect of O_3 was observed for SSR markers (Fig. 3).

Thus, genetic diversity within populations from elevated O_3 was slightly higher when compared to populations from ambient O_3 indicating no negative effect of O_3 in this experimental system. However, O_3 had a small but significant effect on population differentiation, indicating micro-evolutionary processes as a response to elevated O_3.

Conclusions

Molecular genetic markers provide a powerful tool to characterize genetic resources with a broad range of applications in breeding of forage crops and grassland ecology. However, estimates of genetic diversity based on anonymous genetic markers often show low correlation to diversity of agronomically important traits (Dias et al. 2008). Therefore, for efficient application, the link between neutral genetic diversity and functional diversity needs to be established. In the future, markers closely linked to traits of agronomic importance will become increasingly available and high throughput fingerprinting methods such as diversity array technology (Wenzl et al. 2004) will facilitate genotyping. However, there will be an increasing need for high throughput phenotyping facilities, bioinformatic solutions to integrate phenotypic and genetic data and carefully designed experiments established in systems relevant for the agricultural or ecological questions under investigation.

Acknowledgements

We would like to acknowledge Yvonne Häfele, David Schneider, Simone Günter and Mirjam Bleiker for technical assistance and Jürg Fuhrer for scientific support. The original research was partiall funded by the Swiss Federal Office for Agriculture and the Swiss Federal Office for the Environment.

References

Allard RW (1999) Principles of plant breeding, Wiley, New York

Andersen JR, Lübberstedt T (2003) Functional markers in plants. Trends Plant Sci 8:554–560

Bolaric S, Barth S, Melchinger AE, Posselt UK (2005) Molecular genetic diversity within and among German ecotypes in comparison to European perennial ryegrass cultivars. Plant Breed 124:257–262

Bonin A, Ehrich D, Manel S (2007) Statistical analysis of amplified fragment length polymorphism data: a toolbox for molecular ecologists and evolutionists. Mol Ecol 16:3737–3758

Booy G, Hendriks RJJ, Smulders MJM, van Groenendael JM, Vosman B (2000) Genetic diversity and the survival of populations. Plant Biol 2:379–395

Casler MD (1998) Genetic variation within eight populations of perennial forage grasses. Plant Breed 117:243–249

Cerna FJ, Cianzio SR, Rafalski A, Tingey S, Dyer D (1997) Relationship between seed yield heterosis and molecular marker heterozygosity in soybean. Theor Appl Genet 95:460–467

Chapin FS III, Zavaleta ES, Eviner VT, Naylor RL, Vitousek PM, Reynolds HL, Hooper DU, Lavorel S, Sala OE, Hobbie SE, Mack MC, Diaz S (2000) Consequences of changing biodiversity. Nature 405:234–242

Charmet G, Ravel C, Balfourier F (1997) Phylogenetic analysis in the *Festuca–Lolium* complex using molecular markers and ITS rDNA. Theor Appl Genet 94:1038–1046

Cogan NOI, Ponting RC, Vecchies AC, Drayton MC, George J, Dracatos PM, Dobrowolski MP, Sawbridge TI, Smith KF, Spangenberg GC, Forster JW (2006) Gene-associated single nucleotide polymorphism discovery in perennial ryegrass (*Lolium perenne* L.). Mol Genet Genomics 276:101–112

Cogan NOI, Drayton MC, Ponting RC, Vecchies AC, Bannan NR, Sawbridge TI, Smith KF, Spangenberg GC, Forster JW (2007) Validation of in silico-predicted genic SNPs in white clover (*Trifolium repens* L.), an outbreeding allopolyploid species. Mol Genet Genomics 277:413–425

Dias PMB, Julier B, Sampoux JP, Barre P, Dall'Agnol M (2008) Genetic diversity in red clover (*Trifolium pratense* L.) revealed by morphological and microsatellite (SSR) markers. Euphytica 160:189–205

Ellison NW, Liston A, Steiner JJ, Williams WM, Taylor NL (2006) Molecular phylogenetics of the clover genus (*Trifolium* – Leguminosae). Mol Phylogenet Evol 39:688–705

Excoffier L, Smouse PE, Quattro JM (1992) Analysis of molecular variance inferred from metric distances among DNA haplotypes: application to human mitochondrial DNA restriction data. Genetics 131:479–491

Fjellheim S, Rognli OA, Fosnes K, Brochmann C (2006) Phylogeographical history of the widespread meadow fescue (*Festuca pratensis* Huds.) inferred from chloroplast DNA sequences. J Biogeogr 33:1470–1478

Fjellheim S, Blomlie AB, Marum P, Rognli OA (2007) Phenotypic variation in local populations and cultivars of meadow fescue – potential for improving cultivars by utilizing wild germplasm. Plant Breed 126:279–286

Foster Hünneke L (1991) Ecological implications of genetic variation in plant populations. In: Falk DA, Holsinger KE (eds) Genetics and conservation of rare plants. Oxford University Press, New York, pp 31–44

Franco J, Crossa J, Ribaut JM, Betran J, Warburton ML, Khairallah M (2001) A method for combining molecular markers and phenotypic attributes for classifying plant genotypes. Theor Appl Genet 103:944–952

George J, Dobrowolski MP, de Jong EV, Cogan NOI, Smith KF, Forster JW (2006) Assessment of genetic diversity in cultivars of white clover (*Trifolium repens* L.) detected by SSR polymorphisms. Genome 49:919–930

Grenier C, Bramel-Cox PJ, Noirot M, Rao KEP, Hamon P (2000a) Assessment of genetic diversity in three subsets constituted from the ICRISAT sorghum collection using random vs non-random sampling procedures A. Using morpho-agronomical and passport data. Theor Appl Genet 101:190–196

Grenier C, Deu M, Kresovich S, Bramel-Cox PJ, Hamon P (2000b) Assessment of genetic diversity in three subsets constituted from the ICRISAT sorghum collection using random vs non-random sampling procedures. B. Using molecular markers. Theor Appl Genet 101:197–202

Hartmann M, Frey B, Kölliker R, Widmer F (2005) Semi-automated genetic analyses of soil microbial communities: comparison of T-RFLP and RISA based on descriptive and discriminative statistical approaches. J Microbiol Methods 61:349–360

Herrmann D, Boller B, Widmer F, Kölliker R (2005) Optimization of bulked AFLP analysis and its application for exploring diversity of natural and cultivated populations of red clover. Genome 48:474–486

Herrmann D, Boller B, Studer B, Widmer F, Kölliker R (2008) Improving persistence in red clover – insight from QTL analysis and comparative phenotypic evaluation. Crop Sci 48:269–277

Hillis DM, Bull JJ (1993) An empirical test of bootstrapping as a method for assessing confidence in phylogenetic analysis. Syst Biol 42:182–192

Hirata M, Cai H, Inoue M, Yuyama N, Miura Y, Komatsu T, Takamizo T, Fujimori M (2006) Development of simple sequence repeat (SSR) markers and construction of an SSR-based linkage map in Italian ryegrass (*Lolium multiflorum* Lam.). Theor Appl Genet 113:270–279

Humphreys MO (2005) Genetic improvement of forage crops – past, present and future. J Agric Sci 143:441–448

Ikeda S (2005) Isolation of disease resistance gene analogs from Italian Ryegrass (*Lolium multiflorum* Lam.). Grassl Sci 51:63–70

Jensen LB, Muylle H, Arens P, Andersen CH, Holm PB, Ghesquiere M, Julier B, Lubberstedt T, Nielsen KK, Riek JD, Roldan-Ruiz I, Roulund N, et al. (2005) Development and mapping of a public reference set of SSR markers in *Lolium perenne* L. Mol Ecol Notes 5:951–957

Jones ES, Dupal MP, Dumsday JL, Hughes LJ, Forster JW (2002) An SSR-based genetic linkage map for perennial ryegrass (*Lolium perenne* L.). Theor Appl Genet 105:577–584

Joyce TA, Abberton MT, Michaelson-Yeates TPT, Forster JW (1999) Relationships between genetic distance measured by RAPD-PCR and heterosis in inbred lines of white clover (*Trifolium repens* L.). Euphytica 107:159–165

Kirwan L, Lüscher A, Sebastia MT, Finn JA, Collins RP, Porqueddu C, Helgadottir A, Baadshaug OH, Brophy C, Coran C, Dalmannsdottir S, Delgado I, et al. (2007) Evenness drives consistent diversity effects in intensive grassland systems across 28 European sites. J Ecol 95:530–539

Kölliker R, Stadelmann FJ, Reidy B, Nösberger J (1998) Fertilization and defoliation frequency affect genetic variability of *Festuca pratensis* Huds. in permanent grasslands. Mol Ecol 7:1757–1768

Kölliker R, Jones ES, Drayton MC, Dupal MP, Forster JW (2001) Development and characterisation of simple sequence repeat (SSR) markers for white clover (*Trifolium repens* L.). Theor Appl Genet 102:416–424

Kölliker R, Herrmann D, Boller B, Widmer F (2003) Swiss Mattenklee landraces, a distinct and diverse genetic resource of red clover (*Trifolium pratense* L.). Theor Appl Genet 107:306–315

Kölliker R, Boller B, Widmer F (2005) Marker assisted polycross breeding to increase diversity and yield in perennial ryegrass (*Lolium perenne* L.). Euphytica 146:55–65

Kölliker R, Enkerli J, Widmer F (2006) Characterization of novel microsatellite loci for red clover (*Trifolium pratense* L.) from enriched genomic libraries. Mol Ecol Notes 6:50–53

Kölliker R, Bassin S, Schneider D, Widmer F, Fuhrer J (2008) Elevated ozone affects the genetic composition of *Plantago lanceolata* L. populations. Environ Pollut 152:380–386

Kosman E, Leonard KJ (2005) Similarity coefficients for molecular markers in studies of genetic relationships between individuals for haploid, diploid, and polyploid species. Mol Ecol 14:415–424

McCann KS (2000) The diversity-stability debate. Nature 405:228–233

Meyers LA, Bull JJ (2002) Fighting change with change: adaptive variation in an uncertain world. Trends Ecol Evol 17:551–557

Miura Y, Hirata M, Fujimori M (2007) Mapping of EST-derived CAPS markers in Italian ryegrass (*Lolium multiflorum* Lam.). Plant Breed 126:353–360

Mohammadi SA, Prasanna BM (2003) Analysis of genetic diversity in crop plants – salient statistical tools and considerations. Crop Sci 43:1235–1248

Noss RF (1990) Indicators for monitoring biodiversity: a hierarchical approach. Conserv Biol 4:356–364

Nybom H (2004) Comparison of different nuclear DNA markers for estimating intraspecific genetic diversity in plants. Mol Ecol 13:1143–1155

Oostermeijer JGB, Van Eijck MW, Van Leeuwen N, Den Nijs JCM (1995) Analysis of the relationship between allozyme heterozygosity and fitness in the rare *Gentiana pneumonanthe* L. J Evol Biol 8:739–759

Peter-Schmid MKI, Boller B, Kölliker R (2008) Habitat and management affect genetic structure of Festuca pratensis but not Lolium multiflorum ecotype populations. Plant Breed doi:10.1111/j.1439-0523.2007.01478.x

Primack RB, Kang H (1989) Measuring fitness and natural selection in wild plant populations. Annu Rev Ecol Syst 20:367–396

Pritchard JK, Stephens M, Donnelly P (2000) Inference of population structure using multilocus genotype data. Genetics 155:945–959

Reed DH, Frankham R (2003) Correlation between fitness and genetic diversity. Conserv Biol 17:230–237

Saha MC, Mian RMA, Eujayl I, Zwonitzer JC, Wang L, May GD (2004) Tall fescue EST-SSR markers with transferability across several grass species. Theor Appl Genet 109:783–791

Sato S, Isobe S, Asamizu E, Ohmido N, Kataoka R, Nakamura Y, Kaneko T, Sakurai N, Okumura K, Klimenko I, Sasamoto S, Wada T, et al. (2005) Comprehensive structural analysis of the genome of red clover (*Trifolium pratense* L.). DNA Res 12:301–364

Studer B, Widmer F, Enkerli J, Kölliker R (2006) Development of novel microsatellite markers for the grassland species *Lolium multiflorum*, *L. perenne* and *Festuca pratensis*. Mol Ecol Notes 6:1108–1110

Studer B, Boller B, Bauer E, Posselt UK, Widmer F, Kölliker R (2007) Consistent detection of QTLs for crown rust resistance in Italian ryegrass (*Lolium multiflorum* Lam.) across environments and phenotyping methods. Theor Appl Genet 115:9–17

ter Braak CJF, Smilauer P (2002) CANOCO reference manual and CanoDraw for Windows user's guide: software for canonical community ordination (version 4.5), Microcomputer power, Ithaca, New York, USA

Tilman D, Reich PB, Knops JMH (2006) Biodiversity and ecosystem stability in a decadelong grassland experiment. Nature 441:629–632

Ubi BE, Kölliker R, Fujimori M, Komatsu T (2003) Genetic diversity in diploid cultivars of rhodesgrass determined on the basis of amplified fragment length polymorphism markers. Crop Sci 43:1516–1522

Van Treuren R, Bas N, Goossens PJ, Jansen J, Van Soest LJM (2005) Genetic diversity in perennial ryegrass and white clover among old Dutch grasslands as compared to cultivars and nature reserves. Mol Ecol 14:39–52

Velissariou D (1999) Toxic effects and losses of commercial value of lettuce and other vegetables due to photochemical air pollution in agricultural areas of Attica, Greece. In: Fuhrer J, Achermann B (eds) Critical Levels for Ozonoe – Level II. Swiss Agency for Environment, Forest and Landscape, Berne, Switzerland, pp 253–256

Vellend M, Geber MA (2005) Connection between species diversity and genetic diversity. Ecol Lett 8:767–781

Volk M, Bungener P, Contat F, Montani M, Fuhrer J (2006) Grassland yield declined by a quarter in 5 years of free-air ozone fumigation. Global Change Biol 12:74–83

Vos P, Hogers R, Bleeker M, Reijans M, Vandelee T, Hornes M, Frijters A, Pot J, Peleman J, Kuiper M, Zabeau M (1995) AFLP – a new technique for DNA fingerprinting. Nucleic Acids Res 23:4407–4414

Welsh J, McClelland M (1990) Fingerprinting genomes using PCR with arbitrary primers. Nucleic Acids Res 18:7213–7218

Wenzl P, Carling J, Kudrna D, Jaccoud D, Huttner E, Kleinhofs A, Kilian A (2004) Diversity Arrays Technology (DArT) for whole-genome profiling of barley. Proc Natl Acad Sci USA 101:9915–9920

Williams JGK, Kubelik AR, Livak KJ, Rafalski JA, Tingey SV (1990) DNA polymorphisms amplified by arbitrary primers are useful as genetic-markers. Nucleic Acids Res 18:6531–6535

Zietkiewicz E, Rafalski A, Labuda D (1994) Genome fingerprinting by simple sequence repeat (SSR)-anchored polymerase chain-reaction amplification. Genomics 20:176–183

Genomic and Geographic Origins of Timothy (*Phleum* sp.) Based on ITS and Chloroplast Sequences

Alan V. Stewart[1,4], Andrzej Joachimiak[2] and Nick Ellison[3]

[1]PGG Wrightson Seeds, PO Box 175, Lincoln, New Zealand
[2]Institute of Botany, Jagiellonian University, Kraków, Poland
[3]AgResearch, Palmerston North, New Zealand
[4]Corresponding author, astewart@pggwrightsonseeds.co.nz

Abstract. The relationship among members of the subgenus *Phleum* was determined using nuclear ribosomal ITS and chloroplast *trnL* intron DNA sequences. This subgenus is derived from a progenitor of the diploid *Phleum alpinum* subsp. *rhaeticum*. The relationships provide evidence of migration, hybridization, polyploidy and speciation associated with historical glaciations.

The subgenus *Phleum* represents one enormous germplasm pool for breeders and it should now be possible to re-synthesize hexaploid *pratense* from a wider range of diploid forms than occurred historically. This requires the urgent collection of genetic resources from the centers of diversity within glacial refugia as these resources are almost entirely absent from genebanks.

Introduction

The agricultural grass Timothy, *Phleum pratense* and other members of the subgenus *Phleum* have provided taxonomists many challenges over the years. Today eight different "entities" in three species are recognized (Joachimiak 2005). The alpine *P. alpinum* L. includes forms with ciliate or glabrous awns and diploid or tetraploid cytotypes. These include a widespread tetraploid with glabrous awns known as *P. alpinum* L, syn. *P. commutatum* Gaudin, a ciliate awned diploid known as *Phleum alpinum* subsp. *rhaeticum* Humphries, syn. *P. rhaeticum* (Humphries) Rauschert and a glabrous awned diploid. As this latter form is unable to be differentiated morphologically

T. Yamada and G. Spangenberg (eds.), *Molecular Breeding of Forage and Turf*,
doi: 10.1007/978-0-387-79144-9_6, © Springer Science + Business Media, LLC 2009

from the widespread tetraploid we will refer to it by the informal name "commutatum" following Joachimiak and Kula (1993). The lowland species *P. pratense* L. consists of polyploid series from diploid to octoploid. Diploid forms occurring throughout much of Europe and parts of North Africa are known as *P. pratense* subsp. *bertolonii* (DC.) Bornm., syn. *P. bertolonii* DC. Tetraploid forms in southern Europe, the widespread agricultural hexaploid, and an octoploid restricted to southern Italy are all known as *P. pratense* subsp. *pratense*. The most unusual species within this subgenus is *P. echinatum* (Host), a winter-active annual grass of eastern Mediterranean mountains with 2n = 10 instead of the normal 2n = 14.

Materials and Methods

A wide range of *Phleum* populations (159) was obtained as plants, seed or herbarium specimens for molecular analysis. The samples were studied using the sequences of the *trnL* (UAA) gene intron of chloroplast DNA and the internal transcribed spacer regions ITS1 and ITS2 of nuclear ribosomal DNA following the methods of Ellison et al. (2006).

Results and Discussion

The results show that an ancestor of *P. alpinum* subsp. *rhaeticum* is the progenitor of the *Phleum* subgenus. The widespread allotetraploid *P. alpinum* is a hybrid of this early form with an unknown genome. The three other diploids, *P. alpinum* form "commutatum", *P. echinatum*, and *P. pratense* subsp. *bertolonii*, are all derived from *P. alpinum* subsp. *rhaeticum*. In each case these have undergone molecular, cytological and morphological changes as well as changes in adaptation to environmental conditions (Fig. 1).

Phleum alpinum

Phleum alpinum L. occurs in most mountains of Europe, northern Asia and North and South America. At 30° latitudes it occurs at altitudes over 4,000 m but this reduces to sea level at 60°. Tetraploids are widespread while diploids are restricted to Europe.

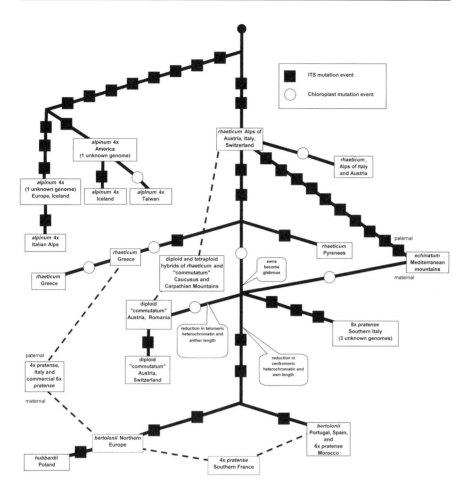

Fig. 1 The development of *Phleum* species from an Asian ancestor of diploid *P. alpinum* subsp. *rhaeticum* showing ITS and cpDNA mutations, hybridizations and polyploidizations

Tetraploid Euro-American *Phleum alpinum*

This has the widest distribution of any of the *Phleum* species. Not only is it present in many of the mountains of Europe and north Asia, it is the only species of the genus to successfully migrate to both North and South America (Conert 1998).

The molecular data reveal a central Asian origin with a divergent migration east to coastal Asia, Japan and to the Americas, and west into Europe. Once in America it migrated rapidly to South America with no additional mutations. Its circumpolar migration was completed in Iceland where derivatives of both the European and American molecular forms occur. The recent divergence of the European and American forms is supported by the lack of significant karyological differences between them (Kula et al. 2006) and only minor ecotypic differences (Heide and Solhaug 2001).

The formation of its allotetraploid ancestor is likely to have been in Asia over 300,000 years B.P., before it diverged into the two forms. The diploid behavior of the species and the difficulty in crossing with *P. pratense* also suggest an ancient formation of this allotetraploid (Nordenskiold 1945).

Diploid *Phleum alpinum*

On the basis of morphology and molecular form diploid *P. alpinum* can be divided into three groups, subsp. *rhaeticum*, "commutatum", and their hybrids.

Diploid Subsp. *rhaeticum*

This diploid subspecies is recognized by the presence of ciliate awns, a feature where intermediate forms can make this distinction difficult. Subspecies *rhaeticum* is the dominant species in Switzerland but it occurs throughout the European Alps and in most central and southern European mountain ranges including the Pyrenees and the Balkans. It usually occurs in the sub-alpine and alpine belt from 1,000 to 2,500 m in fertile and humid habitats (Zernig 2005).

From this study it is apparent that *rhaeticum* has migrated out from a base population in the Alps along adjacent mountain chains to the Pyrenees, Apennines and the Balkans. Associated with these migrations are small changes in the genome as typified by the reduction in centromeric heterochromatin observed in the Greek molecular form of *rhaeticum* (Kula 2005).

The *rhaeticum* molecular form present in the Alps is the progenitor of all the forms within subgenus *Phleum*: "commutatum", *bertolonii*, and polyploid *pratense*, with the notable exception of the widespread tetraploid *Phleum alpinum*. In particular the molecular form of *rhaeticum* from the eastern Alps and Greece identifies itself as an ancestor of all agricultural hexaploid *pratense* sharing an identical ITS sequence.

Diploid Form "Commutatum"

This diploid identified by its glabrous awns is found in the Alps and north into Germany and the Czech Republic, as well as the Carpathian mountains from Poland to Romania and Sweden (Joachimiak and Kula 1996). Diploid "commutatum" originates from *rhaeticum* of the Alps and is characterized by a cpDNA insertion. It differs from its *rhaeticum* ancestor by having glabrous awns, reduced telomeric heterochromatin (Joachimiak 2005), reduced anther length and by growing at higher altitudes among the snow-bed vegetation (Zernig 2005). From the geographic location of molecular derivatives it is apparent that there has been a general northern and eastern radiation outwards from the Alps towards Germany and the Carpathian mountains of Poland and Romania.

Hybrids Between *rhaeticum* and "Commutatum"

Of the 47 European diploids tested, there was approximately a third each of *rhaeticum*, "commutatum", and their hybrids. A high frequency of hybridization has occurred over time. We also report tetraploids based on these hybrids, something not previously reported, which adds further complexity to the many *alpinum* forms. Of the 19 hybrids discovered 3 were tetraploid, 2 of these occurring in the Caucasus Mountains.

Phleum pratense

P. pratense is a lowland species distributed naturally throughout Europe, parts of North Africa and Asia. It is now used for agricultural purposes in all cool temperate regions of the world.

Diploid Subsp. *bertolonii*

Subsp. *bertolonii* occurs throughout Europe but compared to the hexaploid it is less common in northern areas and more common in the south (Humphries 1980), except for Italy where it is uncommon (Cenci et al. 1984). It also occurs in the mountains of North Africa (Maire 1953).

Subsp. *bertolonii* is derived from *P. alpinum* subsp. *rhaeticum* and associated with this derivation is the loss of awns, loss of centromeric heterochromatin (Joachimiak 2005) and most importantly, a change in adaptation from sub-alpine to lowland conditions.

Two major molecular forms exist within *bertolonii*, one restricted to Spain and Portugal, a second widespread across northern Europe. It is likely that these have diverged as a result of glaciation events in Europe. The northern European molecular form is likely to have reinvaded northern Europe after the last glaciation from a refuge in Italy or the Balkans.

Tetraploid *P. pratense*

Tetraploid *P. pratense* has been reported to occur in central and northern Italy, France, Belgium, Spain and Poland (Joachimiak 2005). We have found four molecular forms of tetraploid *pratense*, two different allotetraploids and two autotetraploid. Two allotetraploid *pratense* were discovered in the Alps, one a hybrid of *bertolonii* with hexaploid *pratense* and the other a hybrid of *bertolonii* with *rhaeticum*. One autotetraploid *pratense* from southern France has molecular characteristics of both northern European and Spanish *bertolonii*, while the second French population exhibits only northern European *bertolonii* molecular characteristics.

Hexaploid *P. pratense*

Hexaploid *pratense* can be divided into three molecular forms. The most common is the agricultural hexaploid with the cytoplasmic molecular pattern of *bertolonii* and an ITS molecular pattern from the *rhaeticum* found from the eastern alps to Greece. Although it is not possible from our results to determine the origin of the third genome there is cytological and molecular evidence that this may also be a *bertolonii* genome (Nordenskiold 1945; Cai and Bullen 1994; Joachimiak 2005). More recent GISH studies also suggest the presence of two *bertolonii* genomes (Joachimiak unpublished) and on the basis of geographic origin these two *bertolonii* genomes are most probably both the northern European form. Chromosome pairing in triploid plants of hexaploid *pratense* (7II + 7I) (Nordenskiold 1945) suggest at least minor differences in the structure of *bertolonii* and *rhaeticum* genomes. The small difference between these two genomes readily allows a synthesized auto-hexaploid *bertolonii* to cross with natural hexaploid *P. pratense* (Nordenskiold 1957). The uniformity of the molecular profile in 30 accessions of agricultural *P. pratense* suggests that the formation of these hexaploid *pratense* is probably post-glacial. Its distribution suggests it has expanded throughout

Europe from a glacial refuge, most likely from the Balkans glacial refuge (Hewitt 1999). Two different hexaploid molecular forms were also found in Southern Italy and Morocco. Our results show that hexaploid *pratense* is polyphyletic in origin, having formed at least three times from different diploid ancestors, a situation very common in polyploid species (Soltis and Soltis 2000).

Octoploid *P. pratense*

Octoploid *pratense* is reported only from southern Italy (Cenci et al. 1984). Two samples with the very short stature of diploid *bertolonii* have the maternal genome derived from *rhaeticum* but it is not possible to determine from our results the origin of the other three genomes.

Phleum echinatum

The annual *P. echinatum* occurs in eastern Mediterranean mountain ranges from Sicily to Crete (Humphries 1980). Our results show that this species has developed from hybridization between two different derivatives of *rhaeticum*. Furthermore it is likely it originated from a single hybridization event as this species has undergone genetic reconstruction from the genus norm of 14 chromosomes to 10 (Ellestrom and Tijo 1950). One short chromosome has the centromere at the end, suggestive of half the chromosome being lost. This species exhibits some features of *P. rhaeticum*, but has a longer awn and reduced centromeric heterochromatin (Joachimiak 2005).

Genomic Formula

Genomic formulae have been assigned to 22 genomic forms within the subgenus:

P. pratense

diploid subsp. *bertolonii* in northern Europe	$B^N B^N$
diploid subsp. *bertolonii* in Spain and Portugal	$B^S B^S$
diploid subsp. *bertolonii* in Greece and the Balkans	$B^G B^G$
autotetraploid in France	$B^S B^S B^N B^N$
allotetraploid in the Italian Alps	$B^N B^N R^G R^G$
tetraploid hybrid of *bertolonii* and hexaploid *pratense*	$B^N B^N B^N R^G$
common agricultural hexaploid *pratense*	$B^N B^N B^N B^N R^G R^G$

hexaploid *pratense* in southern Italy	$R^G R^G XXXX$
hexaploid *pratense* in Morocco	$B^S B^S XXXX$
octoploid *pratense* in southern Italy	$R^8 R^8 XXXXXX$
P. alpinum	
ancestral diploid *rhaeticum* in Asia	$R^A R^A$
diploid *rhaeticum* in the Alps	$R^S R^S$
diploid *rhaeticum* in the Pyrenees	$R^P R^P$
diploid *rhaeticum* in Italy	$R^I R^I$
diploid *rhaeticum* in Greece	$R^G R^G$
diploid "commutatum" in the Carpathian mountains	CC
diverse diploid hybrids of *rhaeticum* and "commutatum"	RC
tetraploid *rhaeticum* "commutatum" hybrids, Italy	$R^S R^S CC$
tetraploid *rhaeticum* "commutatum" hybrids, Caucasus	$CCR^G R^G$
tetraploid *alpinum* of Europe across to Iceland	$R^E R^E XX$
tetraploid *alpinum* East Asia, the Americas, to Iceland	$R^W R^W XX$
P. echinatum	EE

Migration History in Relation to Glaciation Events

The molecular results show an Asian origin for the subgenus *Phleum* and identify two separate migrations into Europe.

The first migration into Europe was of an ancestor of diploid *P. alpinum* subsp. *rhaeticum* **RR**. The penultimate Riss glaciation 130,000–150,000 years B.P. provided ample opportunity for this alpine species to migrate vast distances through lowland areas to eventually become isolated on the Alps during subsequent warmer interglacial periods. Subsequent migration along mountain ranges has occurred so that today *rhaeticum* occurs in the Alps, Pyrenees, Apennines, and the Balkans. Migration also occurred onto the colder mountain ranges to the north Germany and to the Carpathian mountains of Poland and Romania but was associated with micro-evolutionary changes in morphology and cytology to develop into diploid "commutatum" **CC**. The overlap of the range of *rhaeticum* and "commutatum" has since allowed considerable hybridization so that a swarm of hybrids **RC** overlaps the range of "commutatum" and part of the *rhaeticum* range. Occasional tetraploid hybrids **RRCC** or **CCRR** have developed and have migrated east at least as far as the Caucasus mountains, if not further.

Migration of *rhaeticum* populations back into the lowlands as a result of climate cooling eventually resulted in the first lowland species of this group, *P. pratense* subsp. *bertolonii* **BB**. This was also accompanied by micro-evolutionary changes in cytology, morphology and adaptation. As the climate cooled during the last glaciation (the Würm 22,000–13,000 years B.P.) this lowland species retreated into southern European glacial refugia. Upon warming these subsequently reinvaded northern Europe from the Balkan/Italy refugia as molecular form $B^N B^N$, with a second molecular form, $B^S B^S$, remaining restricted to the Spanish glacial refuge. Hybridization occurred when these forms met at the interface in France resulting in a recent autotetraploid $B^S B^S B^N B^N$.

Hybrids formed in the Italian Alps where subsp. *bertolonii* and Greek *rhaeticum* overlapped resulting in an allotetraploid pratense $B^N B^N R^G R^G$. It is probable that a further hybridization with the adjacent northern European subsp. *bertolonii* $B^N B^N$ lead to the formation of agricultural hexaploid *pratense* $B^N B^N B^N B^N R^G R^G$. Upon warming in the holocene these subsequently reinvaded northern Europe from the Balkan/Italy refugia, a refugia common to a wide range of European biota (Hewitt 1996, 1999).

Two further hexaploid forms and an octoploid occur within glacial refugia, all based on local diploids but in all cases the remaining genomes are unknown, a hexaploid in southern Italy $R^G R^G XXXX$, another hexaploid in Morocco $B^S B^S XXXX$ and an octoploid in southern Italy $R^8 R^8 XXXXXX$.

Hybridization among *rhaeticum* forms lead to the Mediterranean mountain annual *P. echinatum* **EE** with a reconstructed genome of only ten chromosomes. This event probably took place prior to the last glaciation enabling it to spread throughout the Balkan/Italy glacial refuge and to subsequently re-colonize the mountains of the eastern Mediterranean as the climate warmed.

The very widespread allotetraploid *P. alpinum* formed over 300,000 years B.P. in Asia from hybridization of an ancestral *rhaeticum* with another unknown genome, $R^A R^A XX$. This form remained in Asia until eventually migrating into Europe during the last glaciation (the Würm 22,000–13,000 years B.P.) when conditions were suitable. At the same time many species including this one were able to migrate into the Americas via the Bering/Aleutian route, although probably not completing their entry into America until this route became open around 8,000 years ago (Hong et al. 1999; Weber 2003). This divergent migration has lead to a divergence in molecular forms, one in northern Europe $R^E R^E XX$ and the other in Japan and the Americas $\mathbf{R^WR^WXX}$. The circumpolar migration was completed in Iceland where derivatives of both forms occur.

Conclusion

The understanding of the genomic constitution of entities within subgenus *Phleum* gained from this study should allow breeders to utilize the broader genepool more effectively than previously. It should now be possible to re-synthesize hexaploid *pratense* from a wider range of diploid forms than occurred historically. This requires the collection of genetic resources from the centers of diversity within glacial refugia as these are almost entirely absent from genebanks. As climatic temperatures increase in these regions a large proportion of this germplasm is vulnerable to extinction and collection must be considered urgent.

Acknowledgments

The authors wish to thank the numerous people and organizations who kindly provided samples for analysis.

References

Cai Q, Bullen MR (1994) Analysis of genome specific sequences in *Phleum* species: identification and use for study of genomic relationships. *Theoretical and Applied Genetics* 88: 831–837.

Cenci CA, Pegiati MT, Falistocco E (1984) *Phleum pratense* (Gramineae): chromosomal and biometric analysis of Italian populations. *Willdenowii* 14: 343–353

Conert HJ (1998) *Phleum*. In: Hegi (Ed.) Illustrierte Flora von Mitteleuropa. I/3, pp. 190–206, Verlag Paul Parey, Berlin-Hamburg

Ellestrom S, Tijo JH (1950) Note on the chromosomes of *Phleum echinatum*. *Botanical Notes* 463–465

Ellison NW, Liston A, Steiner JJ, Williams WM, Taylor NL (2006) Molecular phylogenetics of the clover genus (Trifolium – Leguminosae). *Molecular Phylogenetics and Evolution* 39: 688–705

Heide OM, Solhaug KA (2001) Growth and reproduction capacities of two bipolar *Phleum alpinum* populations from Norway and South Georgia. *Artic, Antarctic, and Alpine Research* 33: 173–180

Hewitt GM (1996) Some genetic consequences of ice ages, and their role in divergence and speciation. *Biological Journal of the Linnean Society* 58: 247–276

Hewitt GM (1999) Post glacial recolonisation of European biota. *Biological Journal of the Linnaean Society* 68: 87–112

Hong Q, White P, Klinka K, Chourmouzis C (1999) Phytogeographical and community similarities of alpine tundras of Changbaishan Summit, and Indian Peaks, USA. *Journal of Vegetation Science* 10: 869–882

Humphries CJ (1980) *Phleum*. In Tutin TC et al. (Eds) Flora Europeaea 5. Alismataceae to Orchidaceae (Monocotyledones), pp. 239–241, Cambridge University Press, Cambridge

Joachimiak A (2005) Heterochromatin and microevolution in *Phleum*. In Sharma AK, Sharma A (Eds) Plant Genome: Biodiversity and Evolution. Vol. 1, Part B: Phanerogams, chapter 4, pp. 89–117, Science Publishers., Enfield

Joachimiak A, Kula A (1993) Cytotaxonomy and karyotype evolution in *Phleum* sect. *Phleum* (Poaceae) in Poland. *Plant Systematics and Evolution* 188: 11–25

Joachimiak A, Kula A (1996) Karyosystematics of the *Phleum alpinum* polyploid complex (Poaceae). *Plant Systematics and Evolution* 203: 11–25

Kula A (2005) Searching for a Primeval *Phleum* karyotype. In Ludwick F (Ed) Biology of Grasses, Polish Academy of Sciences: Krakow, Poland

Kula A, Dudziak B, Śliwińska E, Grabowska-Joachimiak A, Stewart AV, Golczyk H, Joachimiak A (2006) Cytomorphological studies on American and European *Phleum* commutatum Gaud. (Poaceae). *Acta Biologica Cracoviensia* 48: 99–108

Maire RD (1953) Flore de L'Afrique du Nord. Fl. Afrique N.: 367

Nordenskiold H (1945) Cyto-genetic studies in the genus *Phleum*. *Acta Agriculturae Suecana* 1: 1–138

Nordenskiold H (1957) Segregation ratios in progenies of hybrids between natural and synthesized *Phleum pretense*. *Hereditas* 43: 525–540

Soltis PS, Soltis DE (2000) The role of genetic and genomic attributes in the success of polyploids. *Proceedings of the National Academy of Sciences of the United States of America* 97: 7051–7057

Weber WA (2003) The Middle Asian Element in the Southern Rocky Mountain Flora of the western United States: a critical biogeographical review. *Journal of Biogeography* 30: 649–688

Zernig K (2005) *Phleum* commutatum and *Phleum rhaeticum* (Poaceae) in the Eastern Alps: characteristics and distribution. *Phyton* 45: 65–79

Diploid *Brachypodium distachyon* of Turkey: Molecular and Morphologic Analysis

Ertugrul Filiz[1], Bahar Sogutmaz Ozdemir[1], Metin Tuna[2] and Hikmet Budak[1,3]

[1]Sabanci University, Biological Science and Bioengineering Program, Istanbul, Turkey
[2]Namik Kemal University, Department of Field Crops, Tekirdag, Turkey
[3]Corresponding author, budak@sabanciuniv.edu

Abstract. Brachypodium distachyon is a model species for the grass family, Poaceae, which includes major cereal crops such as wheat and barley. The aim of this study were to assess morphological and phylogenetic relationships among diploid accessions of Brachypodium representing diverse geographic regions of Turkey based on Sequence related Amplified Polymorphism (SRAP) analyses. The similarity matrix indicated close relation among species used in the section using SRAP primer combinations, produced 156 fragment bands, of which 120 were polymorphic. Genetic distance ranged from 0.03 to 0.62. Plant genotypes were grouped into two major clusters based on SRAP analysis. There was a high level of diversity among the native diploid Brachypodium genotypes. These genotypes can be used for a better understanding of grass genomics.

Introduction

The genome sequences of the *Arabidopsis* (The Arabidopsis Genome Initiative 2000), and rice (Khan and Stace 1999) are already available and are a major resource for functional genomics. However, neither of these species serve as model for temperate grasses. *Arabidopsis* as a dicot species, it does not share with grass crops most of the biological features related to agricultural traits. Although rice would be a better alternative, the rice plant itself does not fulfill requirements such as the short size,

T. Yamada and G. Spangenberg (eds.), *Molecular Breeding of Forage and Turf,*
doi: 10.1007/978-0-387-79144-9_7, © Springer Science + Business Media, LLC 2009

rapid life cycle or ease of transformation. As a tropical species, it does not display all the agronomic traits that are relevant for temperate grasses, as resistance to temperate grass pathogens, freezing tolerance, vernalization or postharvest biochemistry of silage. Rice is also phylogenetically distant from the *Pooidae* subfamily that includes wheat, barley and temperate grasses (Catalan et al. 1995). Additionally molecular phylogenetic analysis has demonstrated that the genus *Brachypodium* diverged from the ancestral *Pooidae* clade immediately prior to the radiation of the modern 'core pooids' (*Triticeae*, *Bromeae* and *Avenae*) which includes the majority of important temperate cereals (Catalan and Olmstead 2000).

Brachypodium accessions have chromosome numbers ranging from 10 to 30 (Martín A and Sánchez-Monge-Laguna 1980). The reported size of the diploid *Brachypodium* ($2n = 2x = 10$) genome varies from 172 to 355 Mbp (Anamthawat-Jónsson et al. 1997; Draper et al. 2001), and, given that the former value may be an underestimate (Anamthawat-Jónsson et al. 1997), it is assumed to be approximately 355 Mbp. Its genome size is between the sizes of *Arabidopsis thaliana* with 157 Mbp, (Bennett and Leitch 2004) and rice with 490 Mbp (Bennett and Leitch 2004). GISH analysis of somatic chromosomes has shown the preponderance of repetitive DNA in the pericentromeric regions, reflecting the compactness of its genome (Sharma and Gill 1983).

The aim of this study was to assess morphological and phylogenetic relationships of diploid *Brachypodium* accessions sampled diverse geographic regions of Turkey based on sequence-related amplified Polymorphism (SRAP).

Materials and Methods

Plant Materials

A total of 500 *Brachypodium* individuals collected from diverse geographic regions of Turkey were firstly stratified at 4°C for 7–10 days in dark between the wetted filter papers in the petri plates. After cold treatment, they were put under light at room temperature. The germinated seeds were first transferred to peat-soil mixture in the viols and then grown in soil-pots under a 16/8-h (light/dark) photoperiod at the greenhouse.

DNA Extraction

DNA was extracted from leaves of 2-week-old seedlings using 1 ml of extraction buffer (50 mM Tris–HCl, 25 mM EDTA, 1 M NaCl, 1% CTAB, 1 mM 1, 10-phenathroline, and 0.15% 2-mercaptoethanol). The extract was incubated at 60°C for 1 h, and then mixed with equal volume of chloroform: isoamyl alcohol (24:1). After centrifuging at 12,000 rpm, the supernatant was transferred to a new tube and isopropanol was added and then incubated for 30 min at room temperature to precipitate the DNA. The pellet was dried, resuspended in 200 μl of TE buffer (10 mM Tris–HCl, 0.1 mM EDTA, pH 8.0) plus 20 μg of RNase, and then incubated at room temperature overnight. The DNA concentration was quantified by spectrophotometry (TKO100 Fluorometer, Hoefer Scientific Instruments, San Francisco).

Genetic relationships were evaluated using a combination of SRAP markers as reported by Budak et al. (2004a,b). Nuclear genome amplifications were carried out as follows: for 32 cycles of 1 min at 94°C; 1 min at 47°C and 37°C for SRAP analyses; 1 min at 72°C; followed by a final extension at 72°C for 5 min before cooling to 24°C. The PCR products (25 μl) were fractionated on 12% polyacrylamide gel using a Hoefer vertical-gel apparatus (SE600). Amplified fragments were visualized using ethidium bromide staining and photographed using a Gel Doc 2000 (Bio-Rad) (Hercules, CA).

Scoring Gels and Data Analysis

Presence or absence of each fragment was coded as "1" and "0", where "1" indicated the presence of a specific allele and "0" indicated its absence. The distance matrix and dendrogram were constructed using the Population Genetic Analysis (POPGEN32) version 1.32 software package. Nei's gene diversity (He) was used to compute Nei's standard genetic distance coefficients (Nei and Li 1979). NTSYS-pc version 2.1 software package (Rohlf 2000) was used for PCA analysis. Additionally, regression analysis using PROC REG (SAS, Cary, NC) was performed to determine associations between pair wise genetic distance from nuclear DNA data sets.

Flow Cytometry Analysis

Flow cytometry analysis was performed as described by Arumuganathan and Earl (1991) to identify the ploidy levels of the accessions sampled from Turkey. Mean DNA content was based on analysis of 1,000 nuclei. Each genotype was analyzed by four separate extractions and flow cytometric runs.

Results and Discussion

Morphological Characterization of Sampled Accessions

Approximately 500 diploid genotypes of *Brachypodium distachyon* were grown in the greenhouse. Phenotypic characterization of the genotypes was identified and depicted in Table 1. Average height differed from 21 to 52 cm. Feathery leaf structure was determined and classified according to its degree from 1 to 5 in ascending order. Leaf color was grouped as 1 (light green), 2 (green) and 3 (dark green). Stem structure also varied from one ecotype to other as fairly erect to branchy. First seed production was recorded at earliest as 7 weeks and at latest as 22 weeks. Like all other characteristics, seed yield has shown differences among ecotypes. While Bd TR-4 had the highest yield (average of 793 seeds/plant), Bd TR-12 exhibited a very low yield (average of 4 seeds/plant). Since this was a huge difference, another experiment was performed and the seeds vernalized for 6 weeks at 4°C in order to see if vernalization requirement changes among the ecotypes resulting in an effect on seed production. On the other hand,

Table 1 Morphological characteristics of some *Brachypodium* ecotypes collected from different locations of Turkey

ID	†Average Height (cm)	Stem Structure	†Seed size (width/length) (cm)	***Seed Production	††Seed yield (seed #/plant)
Bd TR-1	29.2 - 40.0	Fairly erect	0,121 / 0,645	8 weeks after	63
Bd TR-2	29.7 - 40.6	Fairly erect	0,115 / 0,599	10 weeks after	183
Bd TR-3	30.0 - 38.3	Erect	0,104 / 0,658	12 weeks after	34
Bd TR-4	38.3 - 49.3	Erect	0,127 / 0,678	7 weeks after	793
Bd TR-5	29.2 - 37.4	Fairly erect	0,108 / 0,682	14 weeks after	20
Bd TR-6	35.0 - 51.6	Erect	0,141 / 0,784	10 weeks after	443
Bd TR-7	26.0 - 36.0	Erect	0,112 / 0,633	22 weeks after	49
Bd TR-8	25.0 - 35.0	Erect	0,127 / 0,647	22 weeks after	67
Bd TR-9	30.0 - 45.0	Branchy	0,123 / 0,611	9 weeks after	14
Bd TR-10	30.0 - 40.5	Erect	0,123 / 0,621	20 weeks after	17
Bd TR-11	28.6 - 38.3	Erect	0,208 / 0,617	17.5 weeks after	100
Bd TR-12	25.0 - 32.3	Erect	0,112 / 0,578	20 weeks after	4
Bd TR-13	30.7 - 42.1	Branchy	0,104 / 0,663	19.5 weeks after	214

- ***   The time of first seed production observed.
- †   Values are the means of 15 replicates (individual plant).
- ††   Values are the means of 7 replicates (individual plant).

viability of the harvested seeds was checked and the germination percentages ranged from 60 to 100%. Most of the ecotypes showed a high germination rate.

Flow Cytometry Analysis

Mean nuclear DNA content of the 500 native *Brachypodium* genotypes ranged from 0.67 to 1.35 pg/2C. Based on their nuclear DNA content, genotypes were grouped into three different ploidy levels as diploid, and tetraploid. Only the diploid genotypes (Fig. 1) were used in this study. The results of the study indicate that three different ploidy levels exist within this species growing naturally in Turkey. The survey indicated the most prevalent ploidy level in Turkish *Brachypodium* is tetraploid. The results showed the presence of a high genetic source of variation among Turkish populations and can be used in genetics and genomics program.

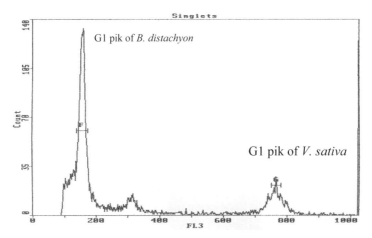

Fig. 1 Flow histogram of diploid *Brachypodium*

Nuclear DNA Analysis

Twenty-three SRAPD markers (Budak et al. 2004a) were used in the study. A total of 120 polymorphic fragment bands were obtained. The size of the markers ranged from 245 to 2,000 bp. The UPGMA dendrogram includes two main clusters and Principal Component analysis was depicted in Fig. 2. Genetic distance ranged from 0.03 to 0.62. The genotypes representing various geographic regions did not differ for SRAP indicating

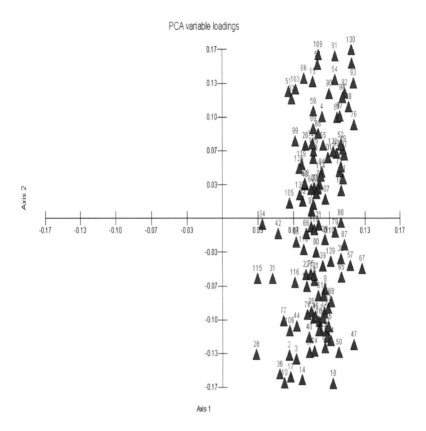

Fig. 2 Principal component analysis of diploid *Brachypodium* of Turkey

that germplasm from different geographical regions grouped together. The SRAP polymorphism detected in this study can be used to genetically classify *Brachypodium* genotypes However; morphological characteristics coupled with the nuclear DNA variation helped a better understanding of the classification and evolution of genotypes.

References

The Arabidopsis Genome Initiative (2000) Analysis of the genome sequence of the flowering plant *Arabidopsis thaliana*. Nature 408: 796–815

Anamthawat-Jónsson K, Bödvarsdóttir S, Bragason B, Gudmundsson J, Martin PK, Koebner RMD (1997) Wide hybridization between wheat (*Triticum* L.) and lymegrass (*Leymus* Hochst.). Euphytica 93: 293–300

Arumuganathan K, Earl ED (1991) Estimation of nuclear DNA contents of plants by flow cytometry. Plant Mol Biol Rep 9: 229–241

Bennett MD, Leitch IJ (2004) Plant DNA C-values database (release 3.0, December 2004). <http://www.rbgkew.org.uk/cval/homepage.html> (Accessed 20 May, 2007)

Budak H, Shearman RC, Parmaksiz I, Gaussoin RE, Riordan TP, Dweikat I (2004a) Molecular characterization of buffalograss germplasm using sequence related amplified polymorphism markers. Theor Appl Genet 108: 328–334

Budak H, Shearman RC, Parmaksiz I, Dweikat I (2004b) Comparative analysis of seeded and vegetative buffalograsses based on phylogenetic relationship using ISSR, SSR, RAPD and SRAP. Theor Appl Genet 109: 280–288

Catalan P, Olmstead R (2000) Phylogenetic reconstruction of the genus *Brachypodium* Beauv. (Poaceae) from combined sequences of chloroplast gene and nuclear ITS. Plant Syst Evol 220: 1–19

Catalan P, Shi Y, Armstrong L, Draper J, Stace CA (1995) Molecular phylogeny of the grass genus *Brachypodium* P-Beauv based on RFLP and RAPD analysis. Bot J Linn Soc 117: 263–280

Draper J, Mur LAJ, Jenkins G, Ghosh-Biswas GC, Bablak P, Hasterok R, Routledge APM (2001) *Brachypodium distachyon*. A new model system for functional genomics in grasses. Plant Physiol 127: 1539–1555

Khan M, Stace C (1999) Breeding relationships in the genus *Brachypodium* (Poaceae: Pooideae). Nord J Bot 19: 257–269

Martín A, Sánchez-Monge-Laguna E (1980) A hybrid between *Hordeum chilense* and *Triticum turgidum*. Cereal Res Comm 8: 349–353

Nei M, Li WH (1979) Mathematical model for studying genetic variation in terms of restriction endonucleases. Proc Natl Acad Sci USA 76: 5269–5273

Rohlf JF (2000) NTSYSpc: numerical taxonomy and multivariate analysis system. Exeter Software, Setauket, New York, USA

Sharma HC, Gill BS (1983) Current status of wide hybridization in wheat. Euphytica 32: 17–31

Remnant Oak Savanna Acts as Refugium for Meadow Fescue Introduced During Nineteenth Century Human Migrations in the USA

M.D. Casler[1], E. van Santen[2], M.W. Humphreys[3], T. Yamada[4], K. Tamura[5], N.W. Ellison[6], R.D. Jackson[7], D.J. Undersander[7], R. Gildersleeve[8], C. Stendal[7], M. Reiter[7] and C. Opitz[9]

[1]USDA-ARS, U.S. Dairy Forage Res. Ctr., 1925 Linden Dr., Madison, WI 53706, USA, mdcasler@wisc.edu
[2]Department of Agronomy and Soils, Auburn University, Auburn, AL, USA
[3]Institute for Grassland and Environmental Research, Aberystwyth, Wales, UK
[4]Field Science Center for Northern Biosphere, Hokkaido University, Sapporo, Japan
[5]National Agricultural Research Center for Hokkaido Region, Sapporo, Japan
[6]Grasslands Research Centre, AgResearch, Palmerston North, New Zealand
[7]Department of Agronomy, University of Wisconsin, Madison, WI, USA
[8]Iowa County UW Extension, University of Wisconsin, Dodgeville, WI, USA
[9]Hidden Valley Farms, Mineral Point, WI, USA

Abstract. In 1990, an unknown forage grass was discovered growing in the shade of a remnant oak savanna in southwestern Wisconsin. Over 12 years, the practice of feeding mature hay on winter pastures spread this grass onto over 500 ha via seedling recruitment. Analysis of amplified fragment length polymorphic (AFLP) markers on 561 plants, compared to a diverse sample of wild European collections of perennial ryegrass (*Lolium perenne* L.), Italian ryegrass (*L. multiflorum* Lam.), meadow fescue (*Festuca pratensis* Huds. = *L. pratense* (Huds.) Darbysh.), and tall fescue (*F. arundinacea* Schreb.), identified a highly diverse population that was more closely allied with *F. pratensis* than the other species, based on genetic distances. Genomic in situ hybridization (GISH), using both *Lolium*- and *Festuca*-specific probes, led to effective hybridizations by only the *Festuca*-specific probes and gave indications of close homology to the *F. pratensis* genome. Similarly, genetic distance analysis using PCR-based *Lolium* expressed sequence tag (EST) markers on a subset of genotypes, compared to the four control species, clearly identified *F. pratensis* as the closest relative. Sequence analysis of the *trnL* intron of cpDNA distinguished the unknown plants from *F. arundinacea*, but not from

T. Yamada and G. Spangenberg (eds.), *Molecular Breeding of Forage and Turf*, doi: 10.1007/978-0-387-79144-9_8, © Springer Science + Business Media, LLC 2009

Lolium. Additional survey work has identified this grass on 12 other farms within an area of about 20,000 ha. Soil samples accompanying plant samples indicated no seed banks and most farm records indicate no commercially introduced seeds during the twentieth century. We hypothesize that seeds of meadow fescue may have arrived with some of the earliest European immigrants to Wisconsin and spread along the historic Military Ridge Trail, a network of frontier U.S. Army forts connected by a major thoroughfare.

Introduction

Prior to European settlement of the USA, the tallgrass prairie ecosystem occupied a large region, extending from the Dakotas and western Minnesota south to parts of Oklahoma and Texas. Oak savanna formed the transition zone between the tallgrass prairie and the eastern deciduous forests that formed the dominant ecosystems in the eastern USA (Bailey 1998). Bur oak (*Quercus macrocarpa* Mich.) is one of the dominant oak species of this ecosystem and many small patches of bur oaks can be found throughout the region from Minnesota and Wisconsin to Texas, Arkansas, and Louisiana. Bur oak is extremely fire resistant and one of the more long-lived oaks (Abrams 1990, 1992).

European settlement of Wisconsin began during the early 1800s (Current 1977). In particular, settlement in the driftless (unglaciated) region of southwestern Wisconsin (Paleozoic Plateau), largely dominated by oak savanna, focused on lead and zinc mining and the agriculture required to feed miners and their families. As cattle were brought into the region, grazing soon removed most of the oak savanna understory, which was gradually replaced by grasses introduced from Europe, one of which was meadow fescue.

Meadow fescue was commonly used for pasture and hay production in the USA during the nineteenth century (Buckner et al. 1979) and thought to have been introduced from Great Britain before 1800 (Kennedy 1900). As agronomic research gained momentum during the early twentieth century, trials of new forage grasses soon identified tall fescue (*Festuca arundinacea* Schreb. = *Lolium arundinaceum* (Schreb.) Darbysh.) as having considerably higher forage yield and better disease resistance than meadow fescue, particularly in the southeastern USA (Buckner et al. 1979). By the 1940s, USDA seed production statistics indicate that tall fescue had completely replaced meadow fescue in the livestock industry of the USA. Meadow fescue did not appear again, to any significant degree, until the grazing movement of the 1980s, when it was recognized for superior forage production, livestock acceptance and utilization, and desirable grazing characteristics (Casler et al. 1998).

During the 1990s, Charles Opitz discovered an unusual and unknown grass growing on a small part of Hidden Valley Farms near Mineral Point, WI. Recognizing that this grass was spreading across his farm, most likely from seeds ingested by grazing livestock, he began to bale hay from areas in which seed had been allowed to ripen. By feeding these bales of hay on other pastures during winter, he soon had established this grass onto approximately 500 ha during the 1990s. The objectives of this study were to (1) identify the species of grass on this farm, (2) determine if pasture longevity is due to survival of individual plants or to seed production and seedling recruitment, and (3) identify potential habitat differentiation within the population of plants on this farm.

Materials and Methods

In September 2002, we sampled 17 sites on Hidden Valley Farms. Sites were chosen to represent a range of habitats on the farm, ranging from stream bottoms to hilltops, with a maximum elevation range of 30 m, including hillsides with north or south aspects, and one site in the deep shade of a bur oak grove. Plants were sampled using a spoke-and-wheel design in which one center plant was identified and eight equidistant spokes were sampled at intervals of 0.3, 0.6, 1.2, and 2.4 m from the center for a total of 33 plants. One tiller per plant was sampled within a radius of 5 cm of the each pre-determined sampling point. A small number of sampling points did not have a plant within 5 cm, so these points were represented by missing data.

Three soil cores, 10 cm in diameter and 10 cm deep, were collected from each sampling site. Each soil core was spread out in a flat in the glasshouse and watered to encourage germination of seeds. All germinated meadow fescue seeds were counted after 4 weeks. Following the germination test, soil samples were washed and all seeds were collected and inspected for presence of meadow fescue seeds.

Plants were moved to a glasshouse, where they were maintained for several months. Control plants representing ten geographically diverse accessions of perennial ryegrass, Italian ryegrass, meadow fescue, and tall fescue were established at the same time (five plants per accession). Nuclear

DNA was extracted from leaves using the methods of Stendal et al. (2006). Amplified fragment length polymorphism (AFLP) DNA markers were developed by testing a large number of primer pairs on a subset of 32 plants representing all 17 sampling sites. Three primer pairs were selected based on a large number of polymorphic markers, providing a total of 223 polymorphic markers for the entire population. Preselective amplification was conducted with *Mse*I+C and *Eco*RI+A primers. Selective amplification was conducted with two primer pairs: *Mse*I+CGA in combination with *Eco*RI+AAG and *Eco*RI+AGC primers. All AFLP marker reactions and capillary electrophoresis were conducted under the conditions described by Johnson et al. (2003). Polymorphic AFLP markers were analyzed by AMOVA and multidimensional scale plots. Data from AFLP markers were also used to generate autocorrelograms for each of the 17 sampling sites and autocorrelations were tested using permutation tests (Smouse and Peakall 1999).

Eleven plants from Hidden Valley Farms were chosen for additional analyses, based on divergence in the first two dimensions of the multi-dimensional scale plot of AFLP marker data (Fig. 2), along with one random plant from each of five accessions within each of the four control species. Chromosome counts were made on root-tip squashes of the 11 Hidden Valley Farms plants. Genomic in situ hybridization (GISH) was performed on each of the 11 selected plants using both *Lolium* and *Festuca* probes and methods as described by Humphreys et al. (1995). The cpDNA *trnL* (UAA) intron was PCR-amplified and sequenced for each plant, ranging in size from 483 to 586 bp, using the primers "c" and "d" described by Taberlet et al. (1991) and the methods described by Ellison et al. (2006). Only those cpDNA haplotypes observed in two or more plants are presented. Finally, 11 PCR-based *Lolium* EST primers were analyzed and scored for each of the 11 unknown and 20 control plants, resulting in a total of 35 alleles. Ten of the 11 primers were developed by T. Yamada and K. Tamura (unpublished data): LpEST21, LpEST62, LpEST64, LpEST107, LpEST122, LpEST163, LpEST453, LpEST774, 7-49320, and 9-03610. Primer S2359 was developed by Armstead et al. (2005). PCR reaction conditions were 94°C (120 s); 94°C (30 s), 65°C (60 s), and 72°C (60 s) for ten cycles; 94°C (30 s), 55°C (60 s), and 72°C (60 s) for 25 cycles; and 72°C (630 s). Alleles were visualized on agaraose gel, scored 0/1 for absence/presence, and analyzed by UPGMA cluster analysis.

Results and Discussion

The germination test and the soil screen revealed no meadow fescue seeds or seedlings, indicating that there was no seed bank of meadow fescue at Hidden Valley Farms at the time of sampling in September 2003. Auto-correlation analyses revealed no relationships between autocorrelation and distance between sampled plants within any of the 17 sampling sites (one of 17 sites shown in Fig. 1). Autocorrelations analyses were sufficiently precise to detect some significant correlation coefficients, but no relationships were detected. These results all indicate that this population of plants has not been propagated by additional sexual reproduction since its initial establishment by on-pasture feeding with bales of seed-ripe hay during the 1990s. Sexual reproduction, followed by seedling recruitment, would lead to mothers and daughters in very close proximity, which was not observed for any of the 17 sampling sites. Thus, individual plants sampled on our survey have survived many years of intermittent drought, freezing stress, and grazing pressure and appear to be reasonably long-lived.

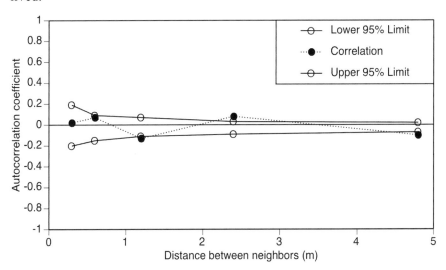

Fig. 1 Autocorrelogram, showing the relationship between the autocorrelation between neighbors, based on AFLP markers, as a function of distance between neighbors, for one of 17 sampling sites on Hidden Valley Farms

Ten of the 11 unknown plants were diploid with $2n = 2x = 14$ chromo-somes. One plant was a mixoploid with chromosome counts of $2n = 2x = 14$, $2n = 4x = 28$, and $2n = 6x = 42$. All plants had a similar physical appearance, with leaf and tiller morphology intermediate to meadow fescue and perennial ryegrass and a fescue-type panicle. We originally thought these plants to be very unusual hybrids between *Lolium* and *Festuca*. The predominance of diploids makes that hypothesis unlikely, because it would involve multiple rare events: interspecific hybridization, polyploidization, and (eventual) haploidization.

Analysis of molecular variance of 640 control and unknown plants revealed a small, but significant, amount of variation between the controls and unknowns (Table 1). Within the control group, there was a small amount of differentiation among species and among accessions within species, but most of the variability was within accessions. Within the unknowns, there was some variation associated with the 17 sampling sites, but this was not related to the four habitats from which they were sampled. As with the accessions, most of the variability was within sampling sites, as expected for a highly self-incompatible, cross-pollinated grass.

Table 1 Analysis of molecular variance (AMOVA) of 640 control and unknown plants analyzed for 223 amplified fragment length polymorphic (AFLP) DNA markers

Source of variation	df	Variance component	Sum of squares (%)	P-value
Control vs. unknown	1	1.83	15.8	<0.0001
Control species	3	0.15	1.3	<0.0001
Accessions/species	26	0.20	1.7	<0.0001
Plants/Acc/species	87	1.14	9.8	<0.0001
Habitats	3	0.00	0.0	0.6892
Sites/habitats	13	1.01	8.7	<0.0001
Plants/sites/habitats	506	7.25	62.6	<0.0001

A multidimensional scale plot revealed a large amount of marker diversity within the population of unknown plants, relative to that found among plants of the four control species (Fig. 2). Most plants within the population appear to have highly unusual haplotypes that are not found within any of the accessions of the control species. The dominant AFLP markers failed to provide sufficient clear discrimination to identify the unknown species.

Land records, dating back approximately 80 years, reveal that no seed of commercial forage grasses was introduced onto this farm. The farm was cropped for about 50 years, before the current owner allowed it to revert back to grassland by allowing plants to grow from the existing seedbank and providing intermittent grazing pressure. We hypothesize that this population was present on the farm, or neighboring farms prior to the long history of row cropping. The remnant oak savanna ecosystem appears to have acted as a refugium for these plants and their progeny, allowing meadow fescue to survive in close proximity to land that was plowed and row-cropped each year. Prior to the deforestation and spread of agriculture across Europe, meadow fescue was restricted to forest margins, suggesting that is possesses some degree of shade tolerance (Scholz 1975). Genetic bottlenecks that may have occurred as a result of immigration from Europe to North America may have been overcome by subsequent hybridizations, mutations, and natural selection under a different set of environmental conditions relative to the native habitats of this European grass.

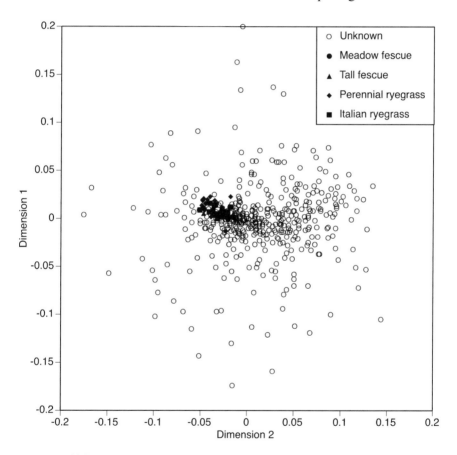

Fig. 2 Multidimensional-scale plot of 640 unknown or control plants evaluated for 223 amplified fragment length polymorphic (AFLP) DNA markers

Cluster analysis, based on 11 PCR-based *Lolium* EST markers provided clarification of the identity of the unknown grass (Fig. 3). The cluster analysis provided three nearly discrete clusters: *Lolium* (including both species), *F. arundinacea*, and *F. pratensis* (including all 11 of the unknown plants). The integration of the five known *F. pratensis* plants with the 11 unknown plants, combined with the clear separation of the three clusters is strong evidence that the unknown plants belong to the *F. pratensis* taxon. These results were supported by the GISH results, which revealed that *F. pratensis* probes hybridized with chromosomes from all 11 Hidden Valley Farms plants. *F. arundinacea* probes hybridized to a limited extent with six of 11 Hidden Valley Farms plants. *L. perenne* probes hybridized only to a limited number of chromosomes and only in two of 11 Hidden Valley Farms plants.

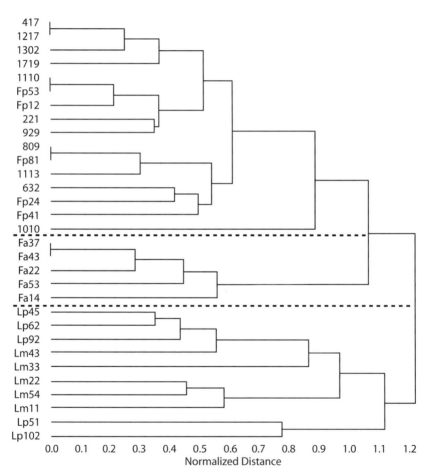

Fig. 3 Cluster dendrogram of ten *Lolium* (Lp or Lm) plants, five *F. arundinacea* (Fa) plants, five *F. pratensis* (Fp) plants, and 11 unknown plants, based on 35 alleles at 11 *Lolium* EST loci

Finally, analysis of cpDNA revealed a clear discrimination between the Hidden Valley Farms meadow fescue plants and the control tall fescue plants, in the presence of a guanine unit at position 395 of the *trnL* intron (Table 2). There were two haplotypes of meadow fescue observed among the 11 Hidden Valley Farms plants and all three haplotypes observed among the five control plants of meadow fescue, with zero, one, or two thymine units at positions 393 and 394, two of which were identical to the two *Lolium* haplotypes found within the control plants.

Table 2 Six haplotypes of the *trnL* intron, from position 380 to 399 (*bolded* nucleotides were discriminatory between *F. arundinacea* and the other species; *hyphens* indicate deleted base pairs)

Species	Nucleotide sequence (positions 380–399)
F. arundinacea	CAGAATTTTTTTT**TG**GAATT
F. pratensis	CAGAATTTTTTTT**TT**GAATT
F. pratensis	CAGAATTTTTTTT**T**-GAATT
F. pratensis	CAGAATTTTTTTT--GAATT
L. perenne	CAGAATTTTTTTT**T**-GAATT
L. perenne	CAGAATTTTTTTT--GAATT

The presence of two haplotypes of *trnL* cpDNA suggests the possibility of multiple introduction events of meadow fescue onto this farm, although this hypothesis must be further investigated before any conclusions can be drawn. We have since discovered meadow fescue populations on over 150 farms in the Iowa, Grant, and Lafayette Counties of southwestern Wisconsin. Most of these populations are closely or proximally associated with remnant oak savanna. Grazing pressure on these farms ranges from intensive, with set stocking and extremely sparse populations of meadow fescue, to nil, with dense monocultures that are often harvested for hay at fairly advanced stages of seed maturation. Multiple introduction events, the presence of a large regional population, frequent pollen migration, and a diverse array of natural selection pressures would all contribute to expansion of genetic diversity following a series of bottleneck-inducing introduction events. These forces could all work to expand the genetic diversity of meadow fescue deriving from a single oak savanna refuge, such as that found at Hidden Valley Farms (Fig. 2).

We have three possible hypotheses regarding the origin and introduction of meadow fescue into this region of Wisconsin – the hypotheses are not mutually exclusive. (1) The primary immigration hypothesis involves direct immigration of Europeans to Wisconsin, including meadow fescue seed from their homeland, largely northern Europe or higher altitudes of southern Europe and southwestern Asia. (2) The secondary immigration hypothesis involves immigration of descendants from the original European immigrants, including meadow fescue populations that resided in the USA since the original immigration. Consistent with both of these hypotheses, our survey has determined that the highest concentration of remnant meadow fescue populations occurs in the region associated with the historic Military Ridge Trial, a network of frontier U.S. Army forts connected by a major thoroughfare. (3) The summer/winter pasture hypothesis involves immigration of meadow fescue to the mid-South of the USA (Buckner et al. 1979) for autumn-winter-spring grazing, followed by shipment of cattle on railroads to southwestern Wisconsin for summer grazing. This practice was very common in the late nineteenth and early twentieth centuries, leading to the possibility of multiple introduction events of meadow fescue seed ingested by cattle just prior to their journey to northern pastures. Further work on meadow fescue collections will be required to identify the more likely of these hypotheses and to identify the potential European origin of these meadow fescue populations.

References

Abrams MD (1990) Adaptations and responses to drought in *Quercus* species of North America. Tree Physiol 7:227–238

Abrams MD (1992) Fire and the development of oak forests. Bioscience 42:246–253

Armstead IP, Skøt L, Turner LB, Skøt K, Donnison IS, Humphreys MO, King IP (2005) Identification of perennial ryegrass (*Lolium perenne* (L.)) and meadow fescue (*Festuca pratensis* (Huds.)) candidate orthologous sequences to the rice *Hd1* (*Se1*) and barley *HvCO1 CONSTANS*-like genes through comparative mapping and microsynteny. New Phytol 167:239–247

Bailey RG (1998) Ecoregions: the ecosystem geography of the oceans and continents. Springer, Berlin Heidelberg New York

Buckner RC, Powell JB, Frakes RV (1979) Historical development. In: Buckner RC, Bush LC (eds) Tall fescue. ASA-CSSA-SSSA, Monograph No. 20, pp 1–8

Casler MD, Undersander DJ, Fredericks C, Combs DK, Reed JD (1998) An on-farm test of perennial forage grass varieties under management intensive grazing. J Prod Agric 11:92–99

Current RN (1977) Wisconsin: a history. WW Norton & Co, New York

Ellison NW, Liston A, Steiner JJ, Williams WM, Taylor NL (2006) Molecular phylogenetics of the clover genus (*Trifolium* – Leguminosae). Mol Phylogenet Evol 39:688–705

Humphreys MW, Thomas HM, Morgan WG, Meredith MR, Harper JA, Thomas H, Zwierzykowski Z, Ghesquière M (1995) Discriminating the ancestral progenitors of hexaploid *Festuca arundinacea* using genomic in situ hybridisation. Heredity 75:171–174

Johnson EL, Saunders JA, Mischke S, Helling CS, Emche SD (2003) Identification of *Erythroxylum* taxa by AFLP DNA analysis. Phytochemistry 64:187–197

Kennedy PB (1900) Cooperative experiments with grasses and forage plants. USDA Bull No 22. US Govt Printing Office, Washington, DC

Scholz H (1975) Grassland evolution in Europe. Taxon 24:81–90

Smouse PE, Peakall R (1999) Spatial autocorrelation analysis of individual multiallele and multilocus genetic structure. Heredity 82:561–573

Stendal C, Casler MD, Jung G (2006) Marker-assisted selection for neutral detergent fiber in smooth bromegrass. Crop Sci 46:303–311

Taberlet L, Gielly L, Pautou G, Bouvet J (1991) Universal primers for amplification of three non-coding regions of chloroplast DNA. Plant Mol Biol 17:1105–1109

Development of Microsatellite Markers for *Brachiaria brizantha* and Germplasm Diversity Characterization of this Tropical Forage Grass

Letícia Jungmann[1,2], Patrícia M. Francisco[2], Adna C.B. Sousa[2], Jussara Paiva[2], Cacilda B. do Valle[1] and Anete P. de Souza[2]

[1]Embrapa Beef Cattle, CP 154, 79002-970, Campo Grande, MS, Brazil, jungmann@cnpgc.embrapa.br
[2]State University of Campinas, CP 6010, 13083-875, Campinas, SP, Brazil

Abstract. Grasses of the African genus *Brachiaria* are the most widely planted forages in Brazil. We previously reported the construction of microsatellite-enriched libraries for five *Brachiaria* species. Now the development of micro-satellite markers for *B. brizantha* and their use for the genetic characterization of morphologically divergent accessions of the germplasm collection of Embrapa are presented. Fifteen pairs of primers were designed and assayed on 23 genotypes. From the fifteen loci evaluated, ten were polymorphic. The divergence between genotypes was estimated using Jaccard's coefficient of similarity and UPGMA method was used for clustering genotypes. The results showed that microsatellites are powerful tools for characterizing genetic diversity of *Brachiaria* species.

Historical Background, Importance and Manipulation of *Brachiaria*

The genus *Brachiaria* includes about 100 species, which occur in the tropical and subtropical regions of both eastern and western hemispheres, but mostly in Africa (Renvoize et al. 1996). Germplasm diversity in America was scarce until the introduction of almost 700 accessions by the International Center for Tropical Agriculture (CIAT) in the middle of the 1980s

T. Yamada and G. Spangenberg (eds.), *Molecular Breeding of Forage and Turf,*
doi: 10.1007/978-0-387-79144-9_9, © Springer Science + Business Media, LLC 2009

supported by the International Plant Genetic Resources Institute (IPGRI) and the International Livestock Center for Africa (ILCA). Subsequently, the Brazilian Agricultural Research Corporation (Embrapa) imported to Brazil more than 400 accessions, which have been preserved and used in a breeding program.

Over the last years, Brazil has been classified as one of the main beef exporters in the world. A key factor that has enormously contributed to this scene is the fact that cattle business is strongly based on animals fed with pastures. Cattle activities in Brazil comprise a herd of 185 million animals on about 180 million hectares of native and cultivated pastures. In general, cultivated forage is represented by exotic species.

Brachiaria reached great economic importance as forage in Brazil in the 1930s, by making feasible the expansion of cow activities on the weak and acid soils of Central Brazil (Pereira et al. 2001). In the last decades, few cultivars has become the basis for cattle production in tropical America, corresponding to 80% of tropical forage seeds produced (Valle, personal communication). Ten cultivars are commercialized by a vital seed industry and nine of them originated as direct selections from germplasm collections of four African species: *Brachiaria brizantha* (A. Rich) Stap; *B. decumbens* Stap; *B. humidicola* (Rendle) Schweick and *B. ruziziensis* Germain & Ervard.

Understanding the genetics and cytogenetics of *Brachiaria* have opened the way for controlled genetic manipulation (Miles and Valle 1996). In addition, a lot of effort has been made in order to raise the basic knowledge about *Brachiaria* in terms of taxonomy, morphology, physiology, nutritional quality, diversity and mode of reproduction. An initiative of CIAT and Embrapa resulted in a book (Miles et al. 1996) covering the biology, agronomy and the improvement aspects of *Brachiaria* grasses.

The major results in Biotechnology of *Brachiaria* were achieved only in the last decade: Pagliarini et al. (2005, 2006, 2007) studied cytogenetic aspects of species and hybrids; Rodrigues et al. (2003) and Alves et al. (2007) have clarified some features of the gene expression associated with aposporous apomixis development in *B. brizantha*; Carneiro et al. (2000) and Silveira et al. (2003) worked on *B. brizantha* genetic transformation. Genetic diversity in the germplasm collection of Embrapa has been studied using Random Amplified Polymorphic DNA (RAPD) markers by our Bio-technology group at Embrapa (unpublished data). However, to date, not too many molecular techniques have been incorporated as auxiliary tools for the breeding programs. Although Chiari (personal communication) has used some selected RAPD marks to early discriminate hybrids in intra- and interspecific progenies in crosses, no great advances have been reached, in part due to the unspecific nature of this molecular marker.

Microsatellite Markers for Performing Genetic Studies on *Brachiaria*

The main goals pursued by the breeding program at Embrapa include improvement on spittlebug resistance; tolerance to acid soils containing aluminum (which prevail in the most important regions for cattle production in Brazil); seed yield and quality; forage yield and nutritive value for animal production. Barriers for crossing imposed by the predominance of the apomictic mode of reproduction also represent a limit to be bypassed. Mapping genes controlling apomixis and desired agronomic traits in segregating progenies would represent a vast progress towards launching new cultivars with more aggregated value.

Attempting to generate powerful tools for performing genetic studies within this material our group has been working in the generation of microsatellite markers (hereinafter also mentioned as SSRs, as an abbreviation for Simple Sequence Repeats) for the five species of *Brachiaria*, which constitute the major source of agronomic valuable traits pursued: *B. brizantha* (e.g. spittlebug resistance), *B. decumbens* (e.g. aluminum tolerance), *B. ruziziensis* (e.g. sexuality), *B. humidicola* (e.g. adaptation to poorly drained soils) and *B. dictyoneura*. In this manuscript, the development of SSRs for *B. brizantha* and their use for characterization the genetic diversity present in morphologically divergent accessions of the germplasm collection of Embrapa is presented. Because the data are still being analyzed, we describe here only the characterization of 23 out of the 226 accessions available in the germplasm collection of *B. brizantha* held in Embrapa.

Material and Methods

Clones of a genomic microsatellite-enriched library previously constructed (Jungmann et al. 2005) were sequenced and used as source for SSRs development. Searches for simple perfect repetitive motifs were performed using the Simple Sequence Repeat Identification Tool-SSRIT (Temnykh et al. 2001). Primers were designed for SSR-flanking regions using the Primer Select/DNAStar software package. Conditions for amplification were determined using 23 morphologically divergent genotypes. The ability of primers for amplifying the same loci in four other *Brachiaria* species was also evaluated. Amplification products were resolved by electrophoresis in 3% agarose and 6% denaturing polyacrylamide gels. The Jaccard's coefficient of similarity was calculated and the UPGMA method was used for clustering the genotypes, within the NTNSYSpc2.1 software.

Results and Discussion

Sequences of 384 clones showed that about 80% of them presented 360 different repetitive motifs. From the repetitive motif found, 160 were considered SSRs for having a minimum size of 10 nucleotides and at least 3 repetitions. Dinucleotides SSRs were the most frequent (51%), followed by the trinucleotide (17%) and tetranucleotide (22%). Penta- and hexanucleotide together accounted for the 10% of the perfect motifs found.

Microsatellites formed by AC/GT and CA/TG were more frequent than those formed by AG/CT and GA/TC. Motifs containing AT/AT and CG/CG were the less frequent. According to McCouch et al. (1997), this type of genomic analysis can be biased in enriched libraries due to the formation of secondary structures in the selection of fragments.

A total of 15 pairs of primers were designed for regions flanking the predicted SSRs. Conditions for amplification were determined using 23 morphologically divergent accessions of *B. brizantha*. Moreover, one genotype of each *B. decumbens*, *B. dictyoneura*, *B. humidicola* and *B. ruziziensis* species were included to evaluate the power of these SSRs to be transferred to other species of the genus. Ten SSRs revealed polymorphic loci between *B. brizantha* accessions. Most of the polymorphic loci amplified also in both *B. ruziziensis* and *B. decumbens*, but not in *B. humidicola* and *B. dictyoneura*.

This result was previously expected, since *B. decumbens* and *B. ruziziensis* are closer related to *B. brizantha* than *B. humidicola* and *B. dictyoneura*, according to Renvoize et al. (1996), who sorted the species of *Brachiaria* into nine groups based on morphological characters.

In fact, *B. ruziziensis* and *B. decumbens* were first regarded as closely related to each other by Bogdan (1977) and *B. decumbens* was introduced into Brazil in 1952 under the name *B. brizantha* (Serrão and Simão Neto 1971), what demonstrates the difficulty to establish a defined boundary between these three species. This genetic proximity is confirmed by a practical approach used in the *Brachiaria* breeding program, in which *B. brizantha*, *B. decumbens* and *B. ruziziensis* form a reproduction complex, with an induced tetraploid ecotype of *B. ruziziensis* being used as a sexual genitor in crosses performed with apomictic accessions of the other two species.

The ten polymorphic loci were used to measure the genetic diversity in the tested plants. The divergence between genotypes was estimated using the Jaccard's coefficient of similarity, since SSRs were treated as dominant markers in this study because the polyploidy in *Brachiaria* avoids determination of the correct frequencies of each allele per locus and, consequently, the heterozygosity required to calculate genetic distances. Similarity varied from 0.22 to 1. Figure 1 shows the grouping pattern observed after the UPGMA method was used for clustering the genotypes. Results revealed that the analysis with this small number of loci does not distinguish between the genotypes B095 and B123, which was previously located in different groups by a principal components analysis of morphological descriptors. In order to confirm the high level of identity between these

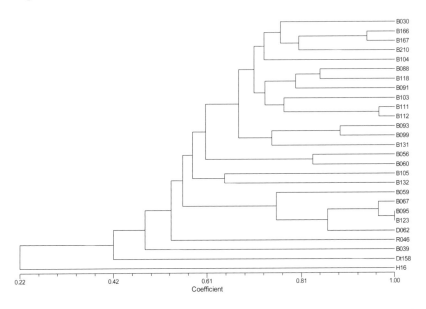

Fig. 1 Jaccard's coefficient of similarity estimated for 23 accessions of *B. brizantha* (beginning with the letter *B*, identification given by Embrapa) and one genotype of the species *B. decumbens* (*D062*), *B. dictyoneura* (*Dt158*), *B. humidicola* (*H16*) and *B. ruziziensis* (*R046*)

two genotypes, a higher number of loci should be used, for higher genome coverage.

As expected for the other species analyzed, *B. decumbens* showed the lower divergence from *B. brizantha*, followed by *B. ruziziensis*. *B. dictyoneura* and *B. humidicola* presented small similarity with *B. brizantha*. It can be explained by the mentioned biological aspects resulting in misamplification of most loci in these two accessions of these species. Although it is a preliminary analysis, this work showed that SSRs, are powerful markers to characterize genetic diversity in germplasm of *Brachiaria* species.

At the moment, a higher number of loci are being used to characterize the genetic diversity in the whole germplasm collection of *B. brizantha* in Embrapa Beef Cattle.

Acknowledgments

We thank Fernanda Megda for English review of this manuscript. This work is supported by grants from Embrapa, CNPq, Fapesp, Fundect and Unipasto.

References

Alves ER, Carneiro VTC, Dusi DMA (2007) In situ localization of three cDNA sequences associated to the later stages of aposporic embryo sac development of *Brachiaria brizantha*. Protoplasma 231:161–171

Bogdan AV (1977) Tropical pasture and fodder plants. Longman, London

Carneiro VTC, Araújo ACG, Lenis-Manzano SJ, Rodrigues JCM, Cabral GB, Leite JA, Silveira ED, Pereira RFA (2000) Genetic transformation of *Brachiaria* sp. by microprojectile bombardment. In: XVI International Congress on Sexual Plant Reproduction. Program and Abstracts, Banff, p 44

Jungmann L, Valle CB do, Laborda PR, Resende RMS, Jank L, Souza AP (2005) Construction of microsatellite-enriched libraries for tropical forage species and characterization of repetitive sequences found in *Brachiaria brizantha*. In: Humphreys MO (ed) Molecular breeding for the genetic improvement of forage crops and turf. Wageningen Academic Publishers, Wageningen, p 128

McCouch SR, Chen XL, Panaud O, Temnykh S, Xu Y, Cho YG, Huang N, Ishii T, Blair M (1997) Microsatellite marker development, mapping and applications in rice genetics and breeding. Plant Mol Biol 35:89–99

Miles JW, Valle CB do (1996) Manipulation of apomixis in *Brachiaria* breeding. In: Miles JW, Maass BL, Valle CB do (eds) *Brachiaria*: biology, agronomy and improvement. Embrapa/CIAT, Cali, pp 164–177

Miles JW, Maass BL, Valle CB do (1996) *Brachiaria*: biology, agronmy and improvement. Embrapa/CIAT, Cali

Pagliarini MS, Pascoto CR, Valle CB do (2005) Meiotic behavior in interspecific hybrids between *Brachiaria ruziziensis* and *Brachiaria brizantha* (Poaceae). Euphytica 42:155–159

Pagliarini MS, Mendes DV, Boldrini KR, Bonato ABM, Valle CB do (2006) Cytological evidence of natural hybridization in *Brachiaria brizantha* Stapf (Gramineae). Biol J Linnean Soc 150:441–446

Pagliarini MS, Valle CB do, Gallo PH, Lonardoni PM, Boldrini KR, Pascotto CR (2007) 2n gamete formation in the genus *Brachiaria* (Poaceae: Paniceae). Euphytica 154:255–260

Pereira AV, Valle CB do, Ferreira RP, Miles JW (2001) Melhoramento de forrageiras tropicais. In: Nass LL, Valois ACC, Valadares-Inglis MC, de Melo IS (eds) Recursos genéticos & melhoramento – Plantas. Fundação MT, Rondonópolis, pp 549–601

Renvoize SA, Clayton WD, Habuye CHS (1996) Morphology, taxonomy, and natural distribution of *Brachiaria* (Trin.) Griseb. In: Miles JW, Maass BL, Valle CB do (eds) *Brachiaria*: biology, agronmy and improvement. Embrapa/CIAT, Cali, pp 1–15

Rodrigues JCM, Cabral GB, Dusi DM de A, Mello LV, Rigden D, Carneiro VTC (2003) Identification of differentially expressed cDNA sequences in ovaries of sexual and apomictic plants of *Brachiaria brizantha*. Plant Mol Biol 53:745–757

Serrão EAS, Simão Neto MS (1971) Informações sobre duas espécies de gramíneas forrageiras do gênero *Brachiaria* na Amazônia: *B. decumbens* Stapf e *B. ruziziensis* Germain et Evrard. Instituto de Pesquisa e Experimentação Agropecuária do Norte, Belém

Silveira ED, Cabral GB, Rodrigues JCM, Costa SS, Carneiro VTC (2003) Evaluation of exogenous promoters for use in *Brachiaria brizantha* transformation. J Plant Biotech 5:87–93

Temnykh S, DeClerck G, Lukashova A, Lipovich L, Cartinhour S, McCouch S (2001) Computational and experimental analysis of microsatellites in rice (*Oryza sativa* L.): frequency, length variation, transposon associations, and genetic marker potential. Genome Res 11:1441–1452

Functional Genomics of Forage and Bioenergy Quality Traits in the Grasses

Iain S. Donnison[1,2], Kerrie Farrar[1], Gordon G. Allison[1], Edward Hodgson[1], Jessic Adams[1], Robert Hatch[1], Joe A. Gallagher[1], Paul R. Robson[1], John C. Clifton-Brown[1] and Phillip Morris[1]

[1]Institute of Grassland and Environmental Research, Plas Gogerddan, Aberystwyth, SY23 3EB, UK
[2]Corresponding author, isd@aber.ac.uk

Abstract. Biomass from forage and energy crops can provide a renewable source of meat, milk, and wool, or power, heat, transport fuels and platform chemicals, respectively. Whilst in forage grasses some improvements have been made, the potential of energy grasses is limited because plant varieties have not yet been selected for this purpose. There are distinct challenges to determine and improve quality traits which increase ultimate energy yield but experience from forage crops can help. Energy grasses offer the potential to be utilised through either thermal or biological conversion methods with the route chosen being largely determined by the calorific value, moisture content and the ratio of soluble to structural carbohydrates. Plant chemical composition underlies these characteristics, for example whichever way grass feedstocks are converted the major determinates of energy are lignin, cell wall phenolics and the soluble and cell wall carbohydrates. These components affect the efficiency of the energy conversion process to meat, milk, wool, energy, platform chemicals and the end quality of certain liquid fuels such as pyrolysis oils. To associate phenotype to genotype for such underlying chemical composition, it is necessary to develop both DNA based molecular markers and high throughput methods for compositional analysis. The genetic resources available in forage and energy grasses are limited in comparison with several model grasses including maize and for some traits it may be appropriate to work initially on such a model and then translate this research back to the forage or bioenergy crop. However not all traits will be present in the model, and so genetic and genomic resources are and will have to be developed in the crops themselves. As part of the EU project GRASP, SNP based markers have been developed in carbohydrate associated genes which map to soluble carbohydrate QTL in *Lolium perenne* (perennial ryegrass) and these have been used in

T. Yamada and G. Spangenberg (eds.), *Molecular Breeding of Forage and Turf,*
doi: 10.1007/978-0-387-79144-9_10, © Springer Science + Business Media, LLC 2009

association studies in a synthetic population of *L. perenne* to measure allele shifts. High throughput calibration models have been developed using near infrared reflectance spectroscopy (NIRS) and Fourier transform infrared spectroscopy (FTIR) in the mid-infrared spectral range which allow accurate predictions of a number of composition traits including lignin, cellulose and hemicellulose contents in several forage and energy grasses including *Miscanthus*, *L. perenne* and related species. These calibrations have allowed a comparison of chemical composition from different grass genotypes, species and environments. Both tools and genetic resources for the optimisation of biomass as forage and energy feedstocks are therefore being developed to enable association of phenotype with genotype.

Introduction

There are major strategic drivers for the further development of forage and bioenergy crops. For example, increasing human population size, urbanisation, industrialisation and meat consumption are increasing the demand for forage crops; whilst climate change, fuel security, and the economic impact of oil price, the effects of climate change and potential benefits to the rural economy are driving the adoption of targets set by many governments to reduce CO_2 emissions and increase renewable sources of energy. Meeting these targets will require a large increase in the use of energy crops, particularly in the case of transport fuels where bioenergy is the only realistic option at short/middle-long term. A limitation for many countries will be access to sufficient land to provide both food and energy needs. Sugarcane is potentially a highly productive source of biomass but its geographical range is restricted to around the tropics. Moreover current 'first generation' annual food crops such as oilseed rape, maize and wheat being used for energy production in temperate regions have a poor energy balance when the entire lifecycle is considered. Indeed, it has been argued that biofuels from annual crops will lead to more expensive food and fuel for very little gain in terms of reducing CO_2 emissions. It is plain that a more sustainable and effective approach must be taken if bioenergy is to feature as a realistic alternative and this will be dependent on the availability of non-food 'second generation' energy crops. These will need to be higher yielding, reducing the land take from food crops, and require lower inputs thereby providing a sustainable method of generating energy and greatly reducing CO_2 emissions. For example dedicated perennial energy crops such as giant grasses (*Miscanthus*, switchgrass, reed canary grass) and woody species (willow, *Eucalyptus*, poplar) generate high yields over successive generations with minimal requirements for energy-demanding nitrogen fertilizers. The three main breeding targets for forage and energy crops are therefore to maximise yield, increase crop quality to increase conversion efficiency and increase or maintain sustainability.

Dedicated bioenergy crops are already being planted for use in heat and power applications and in the case of energy grasses, there is also significant interest in conversion, in biorefineries, to transport fuels. Forage and energy crop quality is determined by chemical composition and therefore the association of biomass feedstock chemistry with the underlying genetics, to determine the genes influencing chemical composition and the efficiency with which the biomass can be converted to food, energy or other biorenewable products will be an important resource to future plant breeding. Moreover an important, if not the main quality trait for forage and energy grasses is the cell wall composition, i.e. lignin, cellulose, hemicellulose and the cell wall phenolics. However the relative concentrations of these components will influence the suitability of the crop for specific conversion routes. Plant biomass with a high lignin content has a higher calorific value and is more likely suited to thermal conversion, whilst biomass with more available sugars is more likely suited as forage and to fermentation.

Functional Genomics

The association of genotype to phenotype can be determined by a number of methods. Candidate genes may be identified informatically using public or private data sets such as sequences derived from expressed sequence tag (EST) or GeneThresher™ libraries. Alternatively candidate genes can be identified experimentally by PCR through homology to genes from model and related grass species such as rice, *Brachypodium*, oats, barley, wheat, *Sorghum*, and maize, or by comparative expression studies such as microarray analysis using cDNA or custom oligo-arrays generated using, for example, a Nimblegen format. In forage and energy grasses where whole genome sequences do not yet exist, candidate genes can be genetically mapped and related to quantitative trait loci (QTL) for traits of interest including for example soluble sugars and cell wall components (Cogan et al. 2005; Turner et al. 2006). If genes map close to a QTL, this can be followed up by fine mapping in a larger population, by association mapping or functionally testing using reverse genetic approaches such as transgenesis, transposon mutagenesis, or targeted induced local lesions in genomes (TILLING). Where QTL exist but no candidate genes map, it is more likely that a map based cloning approach will be necessary, for example by use of a physical map anchored to a closely related model species. If the QTL also exists in a closely related model it may be easier to identify the gene in the model first and then translate back to the crop. There is therefore often a choice of whether to develop a particular resource in the crop of interest or whether to exploit a model or related crop where genetic and genomic resources are more extensive. For forage and energy grasses this could be rice and *Brachypodium* for C3 crops, and *Sorghum* and maize for C4 crops.

Identification and Characterisation of Senescence Enhanced Genes in Maize

Senescence and the delayed senescence phenotype, staygreen, have been studied in maize with the intention of subsequently applying the information back to other forage and energy grasses. The staygreen phenotype can provide a number of advantages to a forage or energy crop including higher biomass yields, higher nutritional quality, increased drought resistance, post harvest stability, reduced lodging and higher stem sugars. Senescence has been studied in maize lines contrasting in onset of senescence (Smart et al. 1995; Martin et al. 2005). In such lines differences in the onset of visible chlorophyll loss as measured by a SPAD meter (Minolta), are underlied by the breakdown of photosynthesis associated proteins (e.g. rbcL, PEP carboxylase, glutamine synthase) as measured by western analyses and the detection of greater proteolytic activity (Smart et al. 1995). Genes which exhibit senescence enhanced expression (See) have been identified in maize by a progression of methods including by a differential screen, differential display, cDNA-AFLP, subtractive hybridisation and microarray analysis (Smart et al. 1995; Griffiths et al. 1997; Thomas et al. 1997; Martin et al. 2005). The See genes identified to date display a range of expression patterns, some genes being more associated with chlorophyll loss and others more associated with age. This work has been translated to temperate forage grasses through the identification of homologous genes in the forage grass *Lolium multiflorum* which are also senescence enhanced (Li et al. 2000).

To functionally test these senescence associated genes, a maize *Mutator* transposon population (Robertson 1978) was exploited to identify transposon insertions in candidate See genes. DNA from pools of mutagenised maize plants was screened using nested primer pairs complementary to the candidate gene and the *Mutator* transposon. Once a positive DNA pool was identified and the PCR fragment validated by DNA sequencing, the screen was repeated on plants within the pool to identify an individual parent plant. Seeds from this parental line were grown and leaf material from the resulting plants harvested for DNA extraction, PCR and DNA sequencing. By exploiting this procedure, mutants in two related senescence enhanced genes, *See2a* and *See2b*, both putative cysteine proteases were identified and characterised. Mutations in *See2b* and the double mutant exhibited an altered phenotype compared to sibling controls from a segregating population of plants. This altered phenotype included early tassel emergence and enhanced mobilisation (including of nitrogen) efficiency when the plants were grown at low nitrogen concentrations (Donnison et al. 2007). Mutator populations of maize also offer the opportunity to perform forward genetic studies, by screening for forage and bioenergy associated traits either directly by visual scoring. For example, brown mid-rib type mutants are likely to predict an altered cell wall composition. In addition, phenotype can be assessed directly by biochemical analysis or indirectly by virtual phenotyping methods such as infrared spectroscopy coupled with chemometric data analysis. Mutants identified from a transposon mutagenised population have the advantage that, as long as the transposon has not excised, they can be rapidly characterised at the molecular level using a technique such as transposon display (Frey et al. 1998; Edwards et al. 2002).

Biotechnological Application of Senescence Enhanced Promoters

The upstream regulatory regions of senescence enhanced genes have been identified in maize and *L. multiflorum* and despite being less than 50% identical both are capable of driving senescence specific expression of GUS (Robson et al. 2004; Li et al. 2004). The maize *See1* promoter has been used to drive the expression of the cytokinin-biosynthetic gene isopententyltransferase (IPT) to successfully delay senescence (Robson et al. 2004; Li et al. 2004) in a similar way to that demonstrated in tobacco using a promoter from a senescence associated gene *SAG12* from *Arabidopsis* (Gan and Amasino 1995). Such extreme staygreen plant material is also an ideal experimental system to study the difference between genes associated with degreening and remobilisation of resources, and genes associated with age. In addition the maize *See2* promoter has been used to express a fungal derived ferulic acid esterase (FAE) to increase the fermentability of grasses (Buanafina et al. 2006) which has applications in both forage and energy crops to increase conversion efficiency and create self processing plants.

Forage Grasses as Energy Crops

Forage grasses such as *Lolium* species in addition to being forage crops have enormous potential as feedstocks for the production of bioenergy and platform chemicals. Forage grasses have a high moisture content and are therefore more probably suited to biological fermentation conversion methods, for example to make ethanol, butanol, methane (Mähnert et al. 2005) or hydrogen (Martínez-Pérez et al. 2007), rather than combustion or gasification. Moreover forage grasses have already been selected and bred for improved digestibility and fermentability, and often contain high concentrations of water soluble carbohydrate (WSC) and low concentrations of lignin (typically 2–6%). However while stem digestibility is correlated with lignin concentration, leaf digestibility is correlated with cell wall phenolic concentration. The opportunity to exploit forage grasses as an energy or industrial crop, is important because dedicated energy crops will not cover the whole landscape in many countries because of restrictions of winter cold, water availability, soil type and social factors. For example studies by Clifton-Brown et al. (2004) and Powlson et al. (2005) envisage up to 12% of UK electricity may be generated from dedicated energy crops and up to one third of this could be obtained by converting 10% of existing grassland. However since 70% of UK agricultural land is grassland, it will still remain the dominant land use and therefore potentially a significant industrial as well as feed crop. Grassland species such as *L. perenne* have many other factors which will be advantageous to the production of biomass destined for bioenergy production. For example, they are perennial and require low annual energy inputs, especially when grown with nitrogen fixing clovers, and because they are native these species can be grown on marginal land at low temperatures. In addition farmers are already experienced in growing the crop, the current biodiversity of grasslands will be maintained and the public and tourist industry will not be concerned with a change to environmentally sensitive landscapes. *L. perenne* is also available to be harvested fresh during a long growing season or can be harvested and stored by ensiling for winter use. Not all grassland will be suitable for biomass intended for bioenergy production because of slopes or site inaccessibility, and grassland is still needed for livestock production. However, in parts of the world including Europe, dairy and livestock stocking levels are falling and farmers are already looking for a new use for grass. For example a reduction in production to more sustainable winter stocking levels is providing an excess of grass during the summer. Such grassland could be considered dual use providing a feedstock for both livestock and bioenergy. However other grassland could be grown as a dedicated energy (Martínez-Pérez et al. 2007) or industrial crop.

Alongside the progress in model plants, genetic and genomic resources have also been developed in the forage grasses themselves which enables a more rapid validation of candidate genes from model organisms in these species. Such resources include EST, GeneThresher™ and bacterial artificial chromosome (BAC) libraries, cDNA microarrays, mapping populations, genetic and physical maps. Recent collaborations in the perennial ryegrass community have increased the number of molecular markers and genetic and trait maps available to plant breeders. For example as a result of the International *Lolium* Genome Initiative (ILGI), a reference linkage map of perennial ryegrass was produced (Jones et al. 2002). The map contains heterologous anchor probes from wheat, barley, oat and rice, allowing comparative relationships to be investigated between *L. perenne* and other Poaceae. The genetic maps of perennial ryegrass and the Triticeae cereals are highly conserved in terms of synteny and colinearity (Jones et al. 2002; King et al. 2002; Alm et al. 2003). An additional linkage map of *L. perenne* based on an F_2 mapping population has also been produced. This map, generated for the genetic analysis of WSC accumulation (Armstead et al. 2002), has been used to identify quantitative trait loci (QTL) associated with WSC in the leaves and leaf sheaths of ryegrass (Turner et al. 2001, 2006). Interestingly, some QTL are already known to overlie or fall close to the location of genes with known function. These maps have also been aligned with maps constructed by other researchers using publicly available markers (Jensen et al. 2005). Recently tools developed in the EU Framework IV project Development of ryegrass allele specific markers for sustainable grassland improvement (GRASP; http://www.grasp-euv.dk; Lübberstedt et al. 2003; Farrar et al. 2007) are enabling a more rapid identification of genes important for tailoring plants for optimised forage or bioenergy production.

Chemical Phenotyping of Biomass

To understand biomass quality, it is important to establish the relationship between biomass quality, conversion efficiency and end product quality characteristics. In the UK, the Engineering and Physical Sciences Research Council (EPSRC) fund the Sustainable Power Generation Initiative (SUPERGEN; http://www.supergen-bioenergy.net) in which biologists at IGER and Rothamsted Research have been working with chemical engineers and engineers at the Universities of Aston, Cranfield, Leeds, and Sheffield to work at the biomass/conversion interface (for example: Bridgeman et al. 2007; Fahmi et al. 2007a,b). *Lolium* and related forage grasses are already known to exhibit a wide range of chemical composition making this ideal material for optimization of biomass for a range of bioenergy applications (Mähnert et al. 2005; Martínez-Pérez et al. 2007; Fahmi et al. 2007a).

Grasses have a complex cell wall structure that is composed largely of polysaccharides which are extensively cross linked to lignin by p-hydroxycinnamic acids (p-coumaric and ferulic acids) in complex ester linkages. In grasses, lignins are largely composed of two types of subunit, guaiacyl (G) and syringyl (S) subunits with comparatively lower concentrations of hydroxyphenyl (H) subunits (Fahmi et al. 2007b) and the technique of analytical pyrolysis in the presence of tetramethylammonium hydroxide (thermochemolysis) is a useful approach for determining not only the amount of lignin present in biomass but also the relative proportions of the lignin subunits and the amounts of the grass cell wall phenolic components, which are converted to methylated derivatives in contrast to those derived from lignin (Marques et al. 2007). To enable the screening of large populations of plants, for example in germplasm collections, large agronomic experiments with many treatments and replicates, mapping or breeding populations, for cell wall components it is desirable to develop high throughput screening methods for biomass samples. Virtual phenotyping methods based on Fourier Transform infrared (FTIR), Raman and near infrared reflectance spectroscopy (NIRS) methods (Stewart 1997; Chen et al. 1998; Kacurakova et al. 2000; Mouille et al. 2003; Landau et al. 2006) are well suited for high-throughput sample analysis and spectral data can be calibrated to wet chemistry (Van Soest 1963, 1974) determined values for lignin, cellulose and hemicellulose using standard chemometric procedures. Depending on the similarity between species and sample preparations such methods can sometimes be applied more generally to other grasses (global) or may only be applicable to specific data sets (local).

Use of Molecular Markers

Even though high throughput infrared spectroscopy methods may be applicable to a plant breeding programme, they are still dependent on the phenotype being present. In the case of large perennial grasses this may take several years to become apparent or stable and therefore it is also desirable to identify DNA based molecular markers for these traits. As described above, some QTL for high WSC content are known to map close to the location of genes with known function, several of which are invertases and fructosyltransferases (Gallagher et al. 2004). A BAC library of *L. perenne* (Farrar et al. 2007) has allowed the isolation of upstream and other non-coding regions in candidate genes of interest for the identification of single nucleotide polymorphism (SNPs) by using a PCR based screen (Farrar and Donnison 2007). Primers designed to the full length gene sequences were used to amplify the comparative allelic regions from 20 diverse *L. perenne* genotypes. Alignment of these sequences revealed molecular markers including SNPs which have been used to trace shifts in allele frequency in an *L. perenne* population undergoing selection pressure for high and low WSC concentration. Comparison of allele frequencies in these contrastingly selected populations indicates some candidate gene alleles may contribute to high or low WSC phenotypes.

Dedicated Energy Grasses

A number of dedicated energy grasses have been proposed including *Miscanthus*, switchgrass and reed canary grass. These grasses differ from current first generation bioenergy crops such as wheat and maize in that they are high yielding with low inputs. However most of the energy is locked up in complex polymers such as cellulose, hemicellulose and lignin. Currently such crops are largely being used for conversion to power and heat by combustion, including by co-firing with coal, but in the future as technologies are developed this is likely to expand to transport fuels, such as ethanol, and platform chemicals. *Miscanthus* is particularly interesting as an energy crop because it combines the fast growth rate of a tropical grass, such as sugarcane, with a tolerance to growth under temperate conditions. Furthermore *Miscanthus* is a very 'eco-friendly' crop since it requires herbicide treatment only during establishment, produces a high yield of biomass each year thereafter and highly effective nutrient recycling at the end of the year reduces the need for cultivation and fertiliser inputs.

The resources available for dedicated energy crops are currently limited as there has been little historical breeding or underpinning research into these crops. However some quality traits, including potassium and chloride content, which cause corrosion in engineering plant, have been studied (Atienza et al. 2003a,b). In another study, the European *Miscanthus* Improvement (EMI) project (Clifton-Brown and Lewandowski 2000; Clifton-Brown et al. 2001), 15 different *Miscanthus* genotypes were grown in five countries across Europe and sampled on two harvest dates. The experiment included the most commonly planted genotypes, *M. ×giganteus*, together with representatives of the two parental species of this sterile triploid hybrid, *M. sinensis* and *M. sacchariflorus*. Using NIRS calibrations to wet chemistry (Hodgson et al. 2007), lignin, cellulose and hemicellulose contents were determined for the 366 *Miscanthus* samples. The biggest differences detected in chemical composition were between *M. sinensis* and *M. sacchariflorus*, whilst *M.* × *giganteus* was more similar in chemical composition to *M. sacchariflorus*.

Conclusions

There are major social, political and environmental drivers for the improvement of forage and bioenergy grasses. The use of high throughput phenotyping methods and associated DNA based molecular markers can play a significant role in helping plant breeders to select grasses with enhanced forage and bioenergy characteristics such as higher conversion efficiency. Such techniques can enable the more rapid development of varieties, which is important as the environmental and societal need for these crops is immediate. Moreover as new dedicated bioenergy crops become developed, concomitant with chemical engineering and microbiology improvements, biomass conversion will move from combustion based technologies to more sophisticated processes such as in biorefineries where maximum use of the crop constituents is made and a range of higher value products including liquid transport fuels, platform and speciality chemicals are produced.

Acknowledgements

The Institute of Grassland and Environmental Research (IGER) is sponsored by the Biotechnology and Biological Sciences Research Council of the United Kingdom.

References

Alm V, Fang C, Busso CS, Devos KM, Vollan K, Grieg Z, Rognli OA (2003) A linkage map of meadow fescue (*Festuca pratensis* Huds.) and comparative mapping with other Poaceae species. Theor Appl Genet 108: 25–40

Armstead IP, Turner LB, King IP, Cairns AJ, Humphreys MO (2002) Comparison and integration of genetic maps generated from F_2 and BC_1-type mapping populations in perennial ryegrass (*Lolium perenne* L.). Plant Breed 121: 501–507

Atienza SG, Satovic Z, Petersen KK, Dolstra O, Martin A (2003a) Influencing combustion quality in *Miscanthus sinensis* Anderss.: identification of QTLs for calcium, phosphorus and sulphur content. Plant Breed 122: 141–145

Atienza SG, Satovic Z, Petersen KK, Dolstra O, Martin A (2003b) Identification of QTLs influencing combustion quality in *Miscanthus sinensis* Anderss. II. Chlorine and potassium content. Theor Appl Genet 107: 857–863

Bridgeman TG, Darvell LI, Jones JM, Williams PT, Fahmi R, Bridgwater AV, Barraclough T, Shield I, Thain SC, Donnison IS (2007) Influence of particle size on the analytical and chemical properties of two energy crops. Fuel 86: 60–72

Buanafina MMde O, Langdon T, Hauck BD, Dalton SJ, Morris P (2006) Manipulating the phenolic acid content and digestibility of Italian ryegrass (*Lolium multiflorum*) by vacuolar targeted expression of a fungal ferulic acid esterase. Appl Biochem Biotech 130: 415–426

Chen LM, Carpita NC, Reiter WD, Wilson RH, Jeffries C, McCann MC (1998) A rapid method to screen for cell-wall mutants using discriminant analysis of Fourier transform infrared spectra. Plant J 16: 385–392

Clifton-Brown JC, Lewandowski I (2000) European *Miscanthus* improvement (FAIR3 CT- 96-1392). Final report, Chapter 9. Mapping the most suitable climatic zones for different *Miscanthus* genotypes in Europe, University of Hohenheim, Germany

Clifton-Brown JC, Lewandowski I, Andersson B, Basch G, Christian DC, Kjeldsen JB, Jorgensen U, Mortensen JV, Riche AB, Schwarz KU, Tayebi K, Teixeira F (2001) Performance of 15 *Miscanthus* genotypes at five sites in Europe. Agron J 93: 1013–1019

Clifton-Brown JC, Stampfl P, Jones MB (2004) *Miscanthus* biomass production for energy in Europe and its potential contribution to decreasing fossil fuel carbon emissions. Global Change Biol 10: 509–518

Cogan NOI, Smith KF, Yamada T, Francki MG, Vecchies AC, Jones ES, Spangenberg GC, Forster JW (2005) QTL analysis and comparative genomics of herbage traits in perennial ryegrass (*Lolium perenne* L.). Theor Appl Genet 110: 364–380

Donnison IS, Gay AP, Thomas H, Edwards KJ, Edwards D, James CL, Thomas AM, Ougham HJ (2007) Modification of nitrogen remobilisation, grain fill and leaf senescence in maize (*Zea mays* L.) by transposon insertional mutagenesis in a protease gene. New Phytol 173: 481–494

Edwards D, Coghill J, Batley J, Holdsworth M, Edwards KJ (2002) Amplification and detection of transposon insertion flanking sequences using fluorescent MuAFLP. Biotechniques 32: 1090

Fahmi R, Bridgwater AV, Darvell LI, Jones JM, Yates N, Thain S, Donnison I (2007a) The effect of alkali metals on combustion and pyrolysis of *Lolium* and *Festuca* grasses, switchgrass and willow. Fuel 86: 1560–1569

Fahmi R, Bridgwater AV, Thain SC, Donnison IS, Morris PM, Yates N (2007b) Prediction of lignin and lignin thermal degradation products by py-gcms in a collection of *Lolium* and *Festuca* grasses. J Anal Appl Pyrol 80: 16–23

Farrar K, Donnison IS (2007) Construction and screening of BAC libraries made from *Brachypodium* genomic DNA. Nat Protocols 2: 1661–1674

Farrar K, Asp T, Lübberstedt T, Xu M, Thomas A, Christiansen C, Humphreys M, Donnison I (2007) Construction of two *Lolium perenne* BAC libraries and identification of BACs containing candidate genes for disease resistance and forage quality. Mol Breed 19: 15–23

Frey M, Stettner C, Gierl A (1998) A general method for gene isolation in tagging approaches: amplification of insertion mutagenised sites (AIMS). Plant J 13: 717–721

Gallagher JA, Cairns AJ, Pollock CJ (2004) Cloning and characterization of a putative fructosyltransferase and two putative invertase genes from the temperate grass *Lolium temulentum* L. J Exp Bot 55: 557–569

Gan S, Amasino RM (1995) Inhibition of leaf senescence by autoregulated production of cytokinin. Science 270: 1986–1988

Griffiths CM, Hosken SE, Oliver D, Chojecki J, Thomas H (1997) Sequencing, expression pattern and RFLP mapping of a senescence-enhanced cDNA from *Zea mays* with high homology to oryzain and aleurain. Plant Mol Biol 34: 815–821

Hodgson EM, Clifton-Brown J, Lister S, Donnison I (2007) Development of a near-infrared reflectance spectroscopy calibration (NIRS) for the determination of cell wall composition of *Miscanthus*. Proceedings of the 15th European Biomass Conference and Exhibition, Berlin 7–11 May, 2007

Jensen LB, Aarens P, Andersen CH, Holm PB, Ghesquiere M, Julier B, Lübberstedt T, Muylle H, Nielsen KK, de Riek J, Roldán-Ruiz I, Roulund N, Taylor C, Vosman B, Barre P (2005) Development and mapping of a public reference set of SSR markers in *Lolium perenne* L. Mol Eco Notes 5: 951–957

Jones ES, Mahoney NL, Hayward MD, Armstead IP, Jones JG, Humphreys MO, King IP, Kishida T, Yamada T, Balfourier F, Charmet G, Forster JW (2002) An enhanced molecular marker based genetic map of perennial ryegrass (*Lolium perenne*) reveals comparative relationships with other Poaceae genomes. Genome 45: 282–295

Kacurakova M, Capek P, Sasinkova V, Wellner N, Ebringerova A (2000) FT-IR study of plant cell wall model compounds: pectic polysaccharides and hemicelluloses. Carbohydr Polym 43: 195–203

King J, Armstead IP, Donnison IS, Thomas HM, Jones RN, Kearsey MJ, Roberts LA, Thomas A, Morgan WG, King IP (2002) Physical and genetic mapping in the grasses *Lolium perenne* and *Festuca pratensis*. Genetics 161: 315–324

Landau S, Glasser T, Dvash L (2006) Monitoring nutrition in small ruminants with the aid of near infrared reflectance spectroscopy (NIRS) technology: a review. Small Rumin Res 61: 1–11

Li Q, Bettany AJE, Donnison I, Griffiths CM, Thomas H, Scott IM (2000) Characterisation of a cysteine protease cDNA from *Lolium multiflorum* leaves and its expression during senescence and cytokinin treatment. Biochim Biophys Acta 1492: 233–236

Li Q, Robson PRH, Bettany AJE, Donnison IS, Thomas H, Scott IM (2004) Modification of senescence in ryegrass transformed with IPT under the control of a monocot senescence-enhanced promoter. Plant Cell Rep 22: 816–821

Lübberstedt T, Andreasen BS, Holm PB (2003) Development of ryegrass allele-specific (GRASP) markers for sustainable grassland improvement – a new framework V project. Czech J Genet Plant Breed 39: 125–128

Mähnert P, Heiermann M, Linke B (2005) Batch- and semi-continuous biogas production from different grass species. Agricultural Engineering International: the CIGR Ejournal. Manuscript EE 05 010, vol. VII

Marques G, Gutierrez A, del Rio JC (2007) Chemical characterization of lignin and lipophilic fractions from leaf fibers of curaua (*Ananas erectifolius*). J Agric Food Chem 55: 1327–1336

Martin A, Belastegui-Macadam X, Quilleré I, Floriot M, Valadier M-H, Pommel B, Andrieu B, Donnison I, Hirel B (2005) Physiological and molecular characterization of the stay-green phenotype in a maize hybrid. New Phytol 167: 483–492

Martínez-Pérez N, Cherryman SJ, Premier GC, Dinsdale RM, Hawkes DL, Hawkes FR, Kyazze G (2007) The potential for hydrogen-enriched biogas production from crops: Scenarios in the UK. Biomass Bioenergy 31: 95–104

Mouille G, Robin S, Lecomte M, Pagant S, Hofte H (2003) Classification and identification of *Arabidopsis* cell wall mutants using Fourier-transform infrared (FT-IR) microspectrocopy. Plant J 35: 393–404

Powlson DS, Riche AB, Shield I (2005) Biofuels and other approaches for decreasing fossil fuel emissions from agriculture. Ann Appl Biol 146: 193–201

Robertson DS (1978) Characterization of a *Mutator* system in maize. Mutat Res 51: 21–28

Robson PRH, Donnison IS, Wang K, Frame B, Pegg SE, Thomas A, Thomas H (2004) Leaf senescence is delayed in maize expressing the *Agrobacterium* IPT gene under the control of a novel maize senescence-enhanced promoter. Plant Biotech J 2: 101–112

Smart CM, Hosken SE, Thomas H, Greaves JA, Blair BG, Schuch W (1995) The timing of maize leaf senescence and characterization of senescence-related cDNAs. Physiol Plant 93: 673–682

Stewart D (1997) Application of Fourier-transform infrared and Raman spectroscopies to plant science. Rec Adv Food Agric Chem 1: 171–193

Thomas H, Evans C, Thomas HM, Humphreys MW, Morgan G, Hauck B, Donnison IS (1997) Introgression, tagging and expression of a leaf senescence gene in *FestuLolium*. New Phytol 137: 29–34

Turner LB, Humphreys MO, Cairns AJ, Pollock CJ (2001) Comparison of growth and carbohydrate accumulation in seedlings of two varieties of *Lolium perenne*. J Plant Physiol 158: 891–897

Turner LB, Cairns AJ, Armstead IP, Ashton J, Skøt K, Whittaker D, Humphreys MO (2006) Dissecting the regulation of fructan metabolism in perennial ryegrass (*Lolium perenne*) with quantitative trait locus mapping. New Phytol 169: 45–58

Van Soest PJ (1963) Use of detergents in the analysis of fibrous feeds. J Assoc Off Agric Chem 46: 829–835

Van Soest PJ (1974) Composition and nutritive value of forages. In: Heath ME, Metcalfe DS, Barnes RF (eds) Forages (3rd edition), Iowa State University Press, Ames, IA, pp. 53–63

Proanthocyanidin Biosynthesis in Forage Legumes with Especial Reference to the Regulatory Role of R2R3MYB Transcription Factors and Their Analysis in *Lotus japonicus*

Mark P. Robbins[1,5], David Bryant[1], Samantha Gill[1], Phillip Morris[1], Paul Bailey[2], Tracey Welham[2], Cathie Martin[2], Trevor L. Wang[2], Takakazu Kaneko[3], Shusei Sato[3], Satoshi Tabata[3] and Francesco Paolocci[4]

[1]Biorenewables and Plant Cell Biology Team, Institute of Grassland and Environmental Research, Aberystwyth Research Centre, Plas Gogerddan, Aberystwyth, Ceredigion, SY23 3EB, UK
[2]John Innes Centre, Norwich Research Park, Colney Lane, Norwich, NR4 7UH, UK
[3]Kazusa DNA Research Institute, Kisarazu, 292-0818 Chiba, Japan
[4]Institute of Plant Genetics- CNR, Research Division of Perugia, Via Madonna Alta 130, 06128, Perugia, Italy
[5]Corresponding author, mpr.@aber.ac.uk

Abstract. Proanthocyanidins (condensed tannins) can play an important part in ruminant nutrition both by increasing ruminal efficiency and preventing pasture bloat. In this chapter we discuss the control of this pathway by transcription factors and focus particularly upon genes of the *bHLH* and *R2R3MYB* classes. Results from studies using transgenic approaches, TILLING and similar techniques are discussed here.

Introduction

The important role of proanthocyanidins and other polyphenolic polymers is discussed elsewhere in the literature. However, it is increasingly clear that the accumulation of natural biopolymers in forage crops can have a profound effect upon their value as feedstuffs for ruminant livestock or as source material for other bioindustrial processes (bioenergy and biofuels being two major examples). In the case of ruminants being fed with high

T. Yamada and G. Spangenberg (eds.), *Molecular Breeding of Forage and Turf,*
doi: 10.1007/978-0-387-79144-9_11, © Springer Science + Business Media, LLC 2009

protein forages a number of reviewers have noted that the presence of proanthocyanidins within legume species can result in an increase in protein protection within the rumen and a decrease in levels of nitrogenous excreta (Aerts et al. 1999). In order to address these important biological and ecological issues, researchers at IGER and elsewhere initially focused upon cloning genes of the proanthocyanidin (PA) pathway from forage legumes with the aim of analyzing aspects of tissue-specific expression as well as trying to identify some of the environmental factors which may modulate the molecular expression of this pathway. Subsequent experimental approaches involved using these genes, or potential orthologues from other higher plants, to modify levels and structures of PA polymers (e.g. Carron et al. 1994).

A limitation of approaches based upon the augmentation of metabolic pathways using genes that encode enzymes was that while phenotypes were generally predictable and gave information upon points of metabolic control; resultant plants showed relatively modest chemical phenotypes.

Subsequent approaches in forage legumes benefited from observations in Arabidopsis that a ternary complex made up of bHLH, R2R3MYB and a WD40 protein (encoded by *TTG1*) regulated the PA pathway in the testa of this non-leguminous model species (Baudry et al. 2004). Building upon this approach, genes of the *bHLH* class that had been implicated in the regulation of anthocyanin pathways were then deployed in experiments in *Lotus corniculatus* (birdsfoot trefoil), Lucerne and white clover (Robbins et al. 2003; Ray et al. 2003; de Majnik et al. 2000).

Some of the transgenic experiments employing genes encoding bHLH proteins significantly modified the PA pathway. We discuss some general conclusions from these experiments in this chapter and then go on to discuss approaches aimed at cloning and identifying genes for R2R3MYB transcription factors that are intimately associated with the tissue-specific expression of the PA pathway in forage legumes.

Ectopic Expression of Genes Encoding *bHLH* Transcription Factors that Transactivate the PA Pathway in Forage Legumes

We discuss two examples using this experimental approach. Firstly work resulting from the transformation of Lucerne with Maize *Lc*, a gene which regulates leaf colour in maize plants (Ray et al. 2003). Secondly we consider publications based upon the transformation of *L. corniculatus* with *Sn*, a maize gene which controls the accumulation of anthocyanin pigments in scutellar nodes (Robbins et al. 2003; Paolocci et al. 2007).

With reference to experiments based upon the expression of *Lc* in Alfalfa (Ray et al. 2003) as predicted, red or deep red-purple colouration was noted in leaves and stems of transgenics when plants were stressed by cold or strong light. This phenotype was accompanied by the rapid accumulation of transcripts corresponding to two early flavonoid pathway genes; *CHS* (chalcone synthase) and *F3H* (flavanone 3-hydroxylase).

More surprisingly, the introduction and expression of *Lc* induced PA accumulation in leaf tissues after 48 h of continuous light treatment. In one line (88-19) levels of over 0.3 mg/g fresh weight (FW) were reported while a control genotype had non-detectable levels of PA in leaves (Ray et al. 2003). An induction of the activity of leucoanthocyanidin 4-reductase (LAR), an enzyme specific to the terminal part of the PA pathway, accompanied this induction of PA polymer. Such observations implied that bHLH proteins may have redundant functions and that over-expression of these transcription factors may result in novel 'neomorphic' phenotypes (Zhang 2003).

At the Institute of Grassland and Environmental Research in collaboration with the Institute of Plant Genetics, CNR, we have focused on introducing anthocyanin regulators of the *bHLH* class into *L. corniculatus*, a model legume and minor forage crop which normally biosynthesises PA in leaf and stem tissues. The introduction of *Sn* (and the related sequence *B-Peru* also from maize) resulted in the accumulation of anthocyanins in petioles and leaf bases and also a dramatic induction of PA in leaf tissues, with phenotypes being restricted to spongy and palisade mesophyll layers. Phenotypes were only significant in transgenics produced from genotypes which accumulated low and medium levels of PA in foliar tissues (Robbins et al. 2003). No obvious changes were noted in other related flavonoid and phenolic pathways and up- and down-regulation of the PA pathway had no obvious effects upon levels of flavonols, lignins or inducible isoflavans.

Transcript analysis indicated that the ectopic expression of *Sn* increased the levels of transcripts corresponding to terminal enzymatic steps of the pathway in *L. corniculatus* (Paolocci et al. 2005); specifically dihydroflavonol 4-reductase (DFR), anthocyanidin synthase (ANS), anthocyanidin reductase (ANR) and leucoanthocyanidin reductase (LAR). Interestingly *Sn* transgenics phenocopied *L. corniculatus* genotypes which accumulated high levels of PA and the expression of *LAR1*, but not *LAR2*, correlated with levels of PA polymer *in planta* (Paolocci et al. 2007).

In summary, the expression of *Sn* resulted in the transactivation of anthocyanin and PA pathways and we have also noted an increase in trichome numbers in some transgenic lines. Such observations are similar to those made in Arabidopsis where a network of redundant bHLH proteins has been noted to function in *TTG1*-dependent pathways which regulate a range of chemical and developmental outcomes via a ternary transcriptional complex (Zhang et al. 2003).

In conclusion, transgenic experiments have yielded compelling evidence that *bHLH* transcription factors exert control over the tissue-specific expression of the PA pathway in forage legumes. We also note recent genetic evidence indicating the role of *bHLH* genes in controlling PA biosynthesis in *Ipomoea purpurea* and rice (Park et al. 2007; Furukawa et al. 2007).

The R2R3MYB Family in *Lotus japonicus*

R2R3MYB transcription factors play important roles in controlling aspects of plant development including those associated with the biosynthesis and accumulation of natural products. This topic has been extensively reviewed e.g. Martin and Paz-Ares (1997). As a background for this work on *L. japonicus*, previous work has shown that the Arabidopsis genome contains in the region of 125 family members (Stracke et al. 2001).

Recent work (Bailey et al. 2007) reported the cloning of 84 *R2R3MYB* sequences from *L. japonicus* using information derived from data mining of genomic sequences and combining this with the cloning of family members using cDNA prepared from a variety of developmental (and environmental) sources. When a tree was produced using the sequences spanning the DNA-binding motif (the R2R3 domain) groupings were noted which appear to correlate with function when co-plotted with better characterized genes from Arabidopsis and rice.

With reference to genes that may be involved in the regulation of PA biosynthesis; four full-length sequences have been cloned which have significant homology to *AtTT2*, an *R2R3MYB* gene that regulates the biosynthesis of PAs in the testa of Arabidopsis seed (Bryant et al. 2005). To further characterize these we have used three approaches for ascertaining function.

Expression Analysis

Preliminary analysis for tissue-specific expression of the four family members in *L. japonicus* has been performed using reverse transcriptase PCR (RT-PCR).

Sequence	Expression pattern
LjMYB38	Root and stem
LjMYB41	Root and stem (induced by low phosphate)
LjMYB71	Flower
LjMYB72	Flower

Transgenic Analysis

We have transformed *L. japonicus* with a construct over-expressing the full-length *LjMYB41* coding sequence. Even though transgene expression (up-regulation) of *LjMYB41* was noted in vegetative tissues no obvious PA phenotypes have been noted. By contrast, there is evidence that genes of the *R2R3MYB* class do regulate PA accumulation in *Lotus* and we note that the introduction and expression of *FaMYB1* decreases the levels of PAs in leaf tissues of *Lotus corniculatus* (Paolocci et al. 2006).

TILLING (Targeted Induced Local Lesions IN Genomes)

Using the JIC gene machine (Perry et al. 2004) we have TILLED for mutations in the DNA-binding site of *LjMYB38*. Analysis of homozygous mutants has yet to yield any obvious PA phenotype.

In conclusion, we have circumstantial evidence regarding four *TT2* orthologues from *L. japonicus* but still await confirmation using transgenics or mutants. One possible explanation is functional redundancy of family members, such that single mutations will not give rise to a mutant phenotype. Future expression analysis should throw more light on this topic but it is interesting to note that three of these sequences are adjacent on the *L. japonicus* genome (Sato and Tabata, unpublished). Other workers have commented on the presence of tandem arrays in the genomes of higher plants and attendant problems for functional genomics and mutation-based plant breeding (Jander and Barth 2007).

Acknowledgements

IGER acknowledges core funding from BBSRC. DB and PB received funding from BBSRC grants: P17923 and P17925 'Transcriptional regulation of phenylpropanoid metabolism in legumes'.

References

Aerts RJ, Barry TN, McNabb WC (1999) Polyphenols and agriculture: beneficial effects of proanthocyanidins in forages. Agric Ecosyst Environ 75: 1–12

Bailey P, Bryant DB, Welham T, Perry JA, Pike JM, Asamizu E, Tabata S, Sato S, Morris P, Robbins MP, Martin C, Wang TL (2007) Mining *myb* genes in *Lotus japonicus*. Proceedings of the 3rd International Conference on Legume Genomics and Genetics, Brisbane, Australia, 9–13 April 2007, p 70

Baudry A, Heim MA, Dubreucq B, Caboche M, Weisshaar B, Lepiniec L (2004) TT2, TT8 and TTG1 synergistically specify the expression of *BANYULS* and proanthocyanidins biosynthesis in *Arabidopsis thaliana*. Plant J 39: 366–380

Bryant DB, Bailey P, Morris P, Robbins MP, Wang TL, Martin C (2005) Identification of putative AtTT2 R2R3-MYB transcription factor orthologues in tanniferous tissues of *L. corniculatus* var. *japonicus* cv Gifu. In: Humphreys MO (ed) Molecular breeding for genetic improvement of forage crops and turf. Wageningen Academic Publishers, The Netherlands, p 166

Carron TR, Robbins MP, Morris P (1994) Genetic modification of condensed tannin biosynthesis in *Lotus corniculatus* 1. Heterologous antisense dihydroflavonol reductase down-regulates tannin accumulation in 'hairy root' cultures. Theor Appl Genet 76: 1006–1015

de Majnik J, Weinman JJ, Djordjevik MA, Rolfe BG, Tanner GJ, Joseph RG, Larkin PJ (2000) Anthocyanin regulatory gene expression in transgenic white clover can result in an altered pattern of pigmentation. Aust J Plant Physiol 27: 659–667

Furukawa T, Maekawa M, Oki T, Suda I, Iida S, Shimada H, Takamure I, Kadowaki K (2007) The *Rc* and *Rd* genes are involved in proanthocyanidin synthesis in rice pericarp. Plant J 49: 91–102

Jander G, Barth C (2007) Tandem gene arrays: a challenge for functional genomics. Trends Plant Sci 12: 203–210

Martin C, Paz-Ares J (1997) MYB transcription factors in plants. TIGS 13: 67–73

Paolocci F, Bovone T, Tosti N, Arcioni S, Damiani F (2005) Light and an exogenous transcription factor qualitatively and quantitatively affect the biosynthetic pathway of condensed tannins in *Lotus corniculatus*. J Exp Bot 56: 1093–1103

Paolocci F, Robbins MP, Madeo L, Arcioni S, Damiani F (2006) The genetic control of the two branches of the pathway leading to the biosynthesis of proanthocyanidins in legume leaves. Proceedings of the 15th Italian Society of Agricultural Genetics, 10–14 September, Ishia, Italy, 2:02

Paolocci F, Robbins MP, Madeo L, Arcioni S, Martens S, Damiani F (2007) Ectopic expression of a *basic helix-loop-helix* gene transactivates parallel pathways of proanthocyanidin biosynthesis. Structure, expression analysis, and genetic control of *leucoanthocyanidin 4-reductase* and *anthocyanidin reductase* genes in *Lotus corniculatus*. Plant Physiol 143: 504–516

Park K-I, Ishikawa N, Morita Y, Choi J-D, Hoshino A, Iida S (2007) A *bHLH* regulatory gene in the common morning glory, *Ipomoea purpurea*, controls anthocyanin biosynthesis in flowers, proanthocyanidins and phytomelanin pigmentation in seeds, and seed trichome formation. Plant J 49: 641–654

Perry JA, Wang TL, Welham TJ, Gardner S, Pike JM, Yoshida S, Parniske M (2004) A TILLING reverse genetics tool and a web-accessible collection of mutants of the legume *Lotus japonicus*. Plant Physiol 131: 866–871

Ray H, Yu M, Auser P, Blahut-Beatty L, McKersie B, Bowley S, Westcott N, Coulman B, Lloyd A, Gruber MY (2003) Expression of anthocyanins and proanthocyanidins after transformation of alfalfa with maize *Lc*. Plant Physiol 132: 1448–1463

Robbins MP, Paolocci F, Hughes JW, Turchetti V, Allison GG, Arcioni S, Morris P, Damiani F (2003) *Sn*, a maize bHLH gene, modulates anthocyanin and condensed tannin pathways in *Lotus corniculatus*. J Exp Bot 54: 239–248

Stracke R, Werber M, Weisshaar B (2001) The R2R3-MYB family in *Arabidopsis thaliana*. Curr Opin Plant Biol 4: 447–456

Zhang JL (2003) Overexpression analysis of plant transcription factors. Curr Opin Plant Biol 6: 430–440

Zhang F, Gonzales A, Zhao M, Payne CT, Lloyd A (2003) A network of redundant bHLH proteins functions in all TTG1-dependent pathways of *Arabidopsis*. Development 130: 4859–4869

Molecular Dissection of Proanthocyanidin and Anthocyanin Biosynthesis in White Clover (*Trifolium repens*)

Aidyn Mouradov[1,2], Stephen Panter[1], Shamila Abeynayake[1,2], Ross Chapman[1], Tracie Webster[1] and German Spangenberg[1,2,3]

[1]Biosciences Research Division, Department of Primary Industries, Victorian AgriBiosciences Centre, La Trobe R&D Park, 1 Park Drive, Bundoora, Victoria 3083, Australia
[2]Molecular Plant Breeding Cooperative Research Centre, Australia
[3]Corresponding author, german.spangenberg@dpi.vic.gov.au

Abstract. Proanthocyanidins (PAs) form the basis for bloat-safety in a number of forage legumes. An attractive strategy for increasing the level of PAs in the foliage of forage legumes, including white clover (*Trifolium repens*) and alfalfa (*Medicago sativa*), involves metabolic reprogramming to divert intermediates from the pre-existing anthocyanin (ANT) pathway to PA biosynthesis. The ANT and PA pathways show remarkable similarities at the molecular and biochemical levels. However, modification of flavonoid biosynthesis to produce an agronomically desirable level of PA in foliage (2–4% of dry weight) is still a formidable task. To meet this challenge, a deeper understanding of the spatial patterns of ANT and PA accumulation in different tissues and cells of white clover and changes associated with development and exposure of plants to stress is required. Improved knowledge of PA and ANT biosynthesis should enhance the ability to reprogram the flavonoid pathway to develop bloat-safe white clover plants with an elevated level of PA in the foliage.

Modification of Proanthocyanidin (PA) Biosynthesis in White Clover

White clover (*Trifolium repens* L.) is a major component of temperate improved pastures, worldwide, and a key forage plant in countries with intensive livestock production systems (Forster and Spangenberg 1999).

This species is commonly used for forage throughout mainland Europe, the United Kingdom, New Zealand, Australia, USA and Japan. White clover has many benefits for farmland and grazing animals, including symbiotic nitrogen fixation, a high protein content and the accumulation of many plant secondary metabolites, including flavonols, anthocyanins, proanthocyanidins (condensed tannins) and isoflavonoids. Flavonoid biosynthesis in plants is one of the most intensively studied secondary metabolite pathways. Most of the products of this pathway are known to have important functions in plants, including attraction of pollinators and agents of seed dispersal to flowers and fruit, pollen development, signaling associated with plant–microbe interactions, and protection of plants from ultraviolet radiation, herbivores and pathogens (Dixon and Sumner 2003). Flavonoid biosynthesis is regulated by a complex network of signals triggered by a number of internal metabolic cues and external signals. Exposure to light, ultraviolet radiation, pathogen attack, deficiencies of nitrogen, phosphorus, and iron, low temperature, and wounding all influence the biosynthesis of different compounds derived from the flavonoid pathway (Dixon and Steele 1999).

The regulation of the flavonoid pathway branch leading to the production of flavan-3-ols and polymeric PAs has been studied in many plants including *Arabidopsis thaliana*, forage legumes, grape and apple (reviewed in Xie and Dixon 2005; Bogs et al. 2007; Ubi et al. 2006; Paolocci et al. 2007). Much of our recent understanding of PA biosynthesis has arisen from the genetic and biochemical analysis of mutants in the model plant *A. thaliana* that fail to accumulate PAs in the seed coat and therefore exhibit a transparent testa (tt) phenotype due to a loss of brown pigmentation provided by oxidized PAs (Grotewold 2006). Seventeen genes have already been identified at the molecular level, of which nine encode pathway enzymes (CHS, CHI, F3H, F3′H, F3′5′H, DFR, ANS, LAR and ANR), seven encode regulatory proteins (*TT1*, *TT2*, *TT8*, *TT16*, *TTG1*, *TTG2*, and *PAP1*) and three encode proteins involved in flavonoid compartmentalization (*TT12*, *TT19* and *AHA10*) (reviewed by Grotewold 2005). Recent studies have identified key genes and enzymes of the PA branch in the two pathways controlling the biosynthesis of the monomeric subunits, the 2,3-*trans*-flavan-3-ols (afzelechin, catechin, and gallocatechin) and the 2,3-*cis*-flavan-3-ols (epiafzelechin, epicatechin, and epigallocatechin) from a common precursor, the flavan-3,4-diols (leucoanthocyanidins) (Winkel-Shirley 2002). The first pathway involves the direct reduction of 2,3-flavan-3,4-diols to 2,3-*trans*-flavan-3-ol by leucoanthocyanidin reductase (LAR). The gene encoding this enzyme has been identified in *Desmodium uncinatum* and *Medicago truncatula* and the enzyme has been characterized at the biochemical level in grape, apple, *D. uncinatum* and *Lotus* spp. The second pathway involves the conversion of 2,3-flavan-3,4-diols to anthocyanidin intermediates by anthocyanidin synthase (ANS), followed by their reduction to 2,3-*trans*-flavan-3-ols by anthocyanidin reductase (ANR) (Xie et al. 2003). Stereo-chemical analyses of PAs have shown that in all of these species except *Arabidopsis*, PA occurs as a mixture of *cis* and *trans* 2,3-flavan-3-ols. In *Arabidopsis*, both terminal and extender units have been shown to consist exclusively of 2,3-*cis*-flavan-3-ols (Xie and Dixon 2005).

The agronomic importance of PAs lies in their ability to suppress bloat-inducing characteristics of some forage legumes, including white clover and alfalfa, by binding to dietary plant proteins. The large protein component in the leaves of these plants is rapidly fermented by rumen micro-organisms, generating protein foams that can trap ruminal gases and lead to pasture bloat, a disease that is estimated to cost the Australian pastoral industry over \$AU100 million each year. The presence of a low level of PAs (2–4% of dry weight) in foliage can prevent bloat and improve the efficiency of protein uptake by ruminants, leading to increased milk, meat and wool production (Wang et al. 1996).

Accumulation of PA in White Clover Tissues

The forage value of white clover is compromised by a lack of bloat-safety caused by high protein content but an insignificant level of PAs in herbage (Aerts et al. 1999). 4-Dimethylaminocinnemaldehyde (DMACA) staining has shown that PAs and/or their monomers are not detectable in white clover foliage outside of glandular trichomes (Fig. 1).

In contrast, PAs are produced at a high level in the floral organs of these plants (Foo et al. 2000) (Fig. 1a–c). Since flowering is seasonal, the development of white clover germplasm that produces a sufficient level of foliar PAs to confer bloat safety is very desirable and at the same time, a challenge for biotechnology.

Accumulation of PA is developmentally regulated in white clover inflorescences. A higher level of PA accumulation was seen in the more mature florets at the base of immature inflorescences than in less mature terminal florets (Fig. 1a). In developing petals, the accumulation of a high level of PA was observed first in standard petals, then in wing and then in keel petals (Fig. 1c).

In all vegetative organs examined, including stolons, stipules, petioles as well as mature and immature leaves, PAs were detected only in multicellular trichomes. Figure 1d–f shows the accumulation of PAs in trichomes at different stages of leaf development. Strongly DMACA-stained trichomes were detected in the immature leaf at stage 0.2 where the leaf is emerging from the stipular sheath (not shown). At later stages of leaf development (e.g. stage 0.3) and in the fully developed leaf (stage 1.0), PA accumulation was still restricted to multicellular trichomes (Fig. 1d and e, respectively). Figure 1f shows the accumulation of PA within the structure of a multi-cellular trichome from a leaf under high magnification. The presence of PAs in a subset of mesophyll cells in leaves of *L. corniculatus* is shown (Fig. 1g and h).

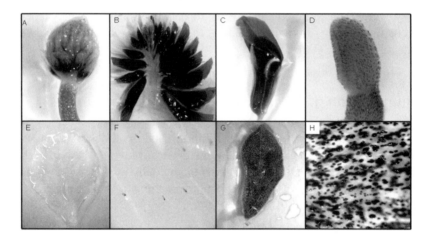

Fig. 1 Accumulation of PAs in white clover organs and tissues

Accumulation of Anthocyanins in White Clover Tissues

PAs and ANTs are produced by two branches of the flavonoid pathway. We have studied the accumulation of ANT in the same organs, tissues and group of cells as PAs in order to identify tissues and cells that accumulate both metabolites. ANT accumulation was seen in the epidermal cells of the petals in white clover flowers (Fig. 2a and b). A low level of ANTs was detected in carpels and stamens (not shown).

In vegetative organs, ANTs were found in epidermal and sub-epidermal groups of cells (Fig. 2c–l). In peduncles (Fig. 2c and d), stipules (Fig. 2e and f), petioles (Fig. 2g) and stolons (Fig. 2h and i), ANT accumulation was observed in sub-epidermal cells. In leaves ANT accumulation was seen in epidermal cells (Fig. 2j–l). ANTs were not detected in trichomes. ANTs were not visualised in unfolded petals during the early stages of floret development (not shown). These phenotypic analyses revealed that PAs and ANTs, and hence the underlying biosynthetic pathways, are co-localized only in a restricted number of epidermal cells in petals of mature florets. The ANT and CT pathways responded differently to abiotic stress treatment. A combination of low temperature and high light intensity induced the ANT pathway in most of the vegetative and reproductive organs and tissues of white clover plants. However, plants grown under conditions of low temperature and low light intensity showed a low level of ANT accumulation in vegetative and reproductive tissues. DMACA staining revealed that these conditions did not result in any gross changes in the accumulation of PA in the same organs and tissues (not shown).

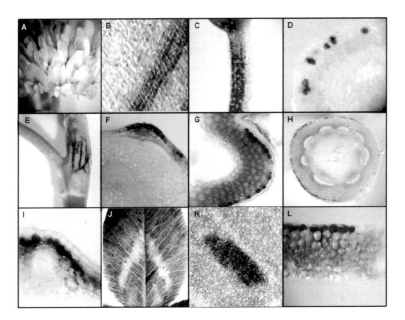

Fig. 2 Accumulation of ANTs in white clover organs and tissues

Future Strategies

PA levels could potentially be enhanced in white clover leaves by increasing the expression domain of flavonoid pathway enzymes that are normally active only in white clover flowers and trichomes. An alternative approach involves metabolic reprogramming to divert intermediates from another branch of the flavonoid pathway (e.g. ANT), to the PA-specific branch. However, both of these strategies are complicated by a lack of detailed information about molecular and biochemical basis of PA biosynthesis and its regulation in forage legumes, as well as factors that limit the rate of PA production in white clover foliage. For example, the molecular mechanism for the transport of PA monomers from the cytoplasm to the vacuole and their polymerisation is poorly understood. To complicate matters further, some flavonoid pathway enzymes are encoded by multigene families. This suggests that specific genes may encode specific enzyme isoforms that are involved in the biosynthesis of particular flavonoids.

A multidisciplinary approach involving plant transformation, marker-assisted plant breeding, plant functional and structural genomics, metabolomics and computational biology would greatly enhance our ability to convert genomics discoveries into improved white clover cultivars with benefits for the profitability and environmental sustainability of the pastoral industries.

References

Aerts RJ, Barry TN, McNabb WC (1999) Polyphenols and agriculture: beneficial effects of proanthocyanidins in forages. Agric Ecosyst Environ 75: 1–12

Bogs J, Jaffe FW, Takos AM, Walker AR, Robinson SP (2007) The grapevine transcription factor *VvMYBPA1* regulates proanthocyanidin synthesis during fruit development. Plant Physiol 143: 1347–1361

Dixon RA, Steele CL (1999) Flavonoids and isoflavonoids – a gold mine for metabolic engineering. Trends Plant Sci 4: 394–400

Dixon RA, Sumner LW (2003) Legume natural products: understanding and manipulating complex pathways for human and animal health. Plant Physiol 131: 878–885

Foo LY, Lu Y, Molan AL, Woodfield DR, McNabb WC (2000) The phenols and prodelphinidins of white clover flowers. Phytochemistry 54: 539–548

Forster JW, Spangenberg G (1999) Forage and turf grass biotechnology: principles, methods and prospects. In: Setlow JK (ed) Genetic Engineering: Principles and Methods, Vol 21, Kluwer/Plenum, New York, pp 191–237

Grotewold E (2005) Plant metabolic diversity: a regulatory perspective. Trends Plant Sci 10: 57–62

Grotewold E (2006) The genetics and biochemistry of floral pigments. Annu Rev Plant Biol 57: 761–780

Paolocci F, Robbins MP, Madeo L, Arcioni S, Martens S, Damiani F (2007) Ectopic expression of a basic helix-loop-helix gene transactivates parallel pathways of proanthocyanidin biosynthesis. Structure, expression analysis, and genetic control of leucoanthocyanidin 4-reductase and anthocyanidin reductase genes in *Lotus corniculatus*. Plant Physiol 143: 504–516

Ubi B, Honda C, Bessho H, Kondo S, Wada M, Kobayashi S, Moriguchi T (2006) Expression analysis of anthocyanin biosynthetic genes in apple skin: effect of UV-B and temperature, Plant Sci 170: 571–578

Wang Y, Douglas GB, Waghorn GC, Barry TN, Foote AG, Purchas RW (1996) Effect of condensed tannins on the performance of lambs grazing *Lotus corniculatus* and lucerne (*Medicago sativa*). J Agric Sci 126: 87–98

Winkel-Shirley B (2002) Biosynthesis of flavonoids and effects of stress. Curr Opin Plant Biol 5: 218–223

Xie DY, Dixon RA (2005) Proanthocyanidin biosynthesis – still more questions than answers? Phytochemistry 66: 2127–2144

Xie DY, Sharma SB, Paiva NL, Ferreira D, Dixon RA (2003) Role of anthocyanidin reductase, encoded by *BANYULS* in plant flavonoid biosynthesis. Science 299: 396–939

Transcript Profiling of Cold Responsive Genes in *Medicago falcata*

Chaoshu Pang[1], Congying Wang[1], Huiping Chen[1], Zhenfei Guo[1,3] and Cong Li[2]

[1]Biotechnology laboratory for Turfgrass and Forages, College of Life Science, South China Agricultural University, Guangzhou 510642, China
[2]Institute of Animal Science, CAAS, Beijing 100094, China
[3]Corresponding author, zhfguo@scau.edu.cn

Abstract. A cold-induced cDNA library containing 2,016 cDNA clones was constructed using suppression subtractive hybridization (SSH) from cold hardy *Medicago sativa* L. ssp. *falcata* (L.) Arcang. A total of 928 plasmids were initially screened using reverse Northern blot analysis, and 523 clones were sequenced. We identified 238 unique gene transcripts (84 repeats and 154 singlets). The EST sequences were deposited into GenBank and annotated. Eight representative clones, which encode a EREBP, phosphoinositide-specific phospholipase C (PLC), FtsH, sucrose phosphate synthase (SPS), sucrose synthase (SS), L-myo-inositol-1-phosphate synthase (MIPS), dehydrin-like protein, and a protein of unknown function, were selected as probes for Northern blot analysis. They were all induced at low temperature. The data indicate that SSH is an effective tool for identification of cold responsive genes and suggest that *M. falcata* rely on typical stress-inducible molecular mechanisms for cold acclimation.

Introduction

Alfalfa (*Medicago sativa*) is the most widely cultivated forage legume with a high feeding value (Michaud et al. 1988). Plant breeders have mostly focused on improvements of forage yield with however, limited success. Selecting for greater disease resistance or improved winter survival (cold tolerance) has not resulted in consistent improvement of stand persistence

T. Yamada and G. Spangenberg (eds.), *Molecular Breeding of Forage and Turf,* 141
doi: 10.1007/978-0-387-79144-9_13, © Springer Science + Business Media, LLC 2009

(Volenec et al. 2002). Large differences in winter hardiness exist among alfalfa germplasms and cultivars, but the physiological and molecular bases for these differences are not well understood. Although the mechanisms of the determination of winter hardiness of alfalfa have been investigated (Volenec et al. 2002; Castonguay and Nadeau 1998; Cunningham et al. 2003), our understanding is still poor and significant research efforts aimed at gene identification and characterization, protein function, and their impact on phenotype is needed.

Medicago sativa L. ssp. *falcata* (L.) Arcang. is a forage legume, closely related to *M. truncatula* and *M. sativa*. Wild and cultivated germplasms of *M. falcata* are widely distributed in the northernmost area of adaptation of alfalfa because of its resistance to cold environments (Michaud et al. 1988). It provides good genetic material for studies on mechanism of cold tolerance in *Medicago* species, but limited investigations on the cold tolerance of *M. falcata* have thus far been conducted. The objectives of this study were to generate a cDNA library for the discovery of cold responsive genes in *M. falcata* using suppression subtractive hybridization (SSH). We subsequently confirmed cold inducibility and assess the putative roles of a selected number of genes in the acquisition of cold tolerance. The preliminary investigation is reported here.

Materials and Methods

Plant Growth and Low Temperature Treatment

Seeds of *Medicago sativa* L. ssp. *falcata* (L.) Arcang. cv. Hulunbeir were sown in 15-cm diameter plastic pots containing mixture of peat and perlite (3:1, v/v) for germination. Seedlings were grown under natural light in a greenhouse with temperature ranging from 20 to 28°C from February to May, 2005. Plants were irrigated daily and fertilized once a week with 0.3% of N–P–K fertilizer (15–15–15). Eight-week old plants of similar size were moved to a growth chamber with a 12-h photoperiod at 200 μmol m^{-2} s^{-1} PPFD at 2°C for acclimation to low temperature.

Isolation of RNA

Leaves (0.1 g) were harvested separately at 0, 8, 16, 24, and 48 h after low temperature exposure, frozen in liquid nitrogen, and then stored at −80°C.

Frozen samples were ground in liquid nitrogen. Total RNA was isolated using TRIzol (Invitrogen) according to the manufacturer's protocol. RNA was quantified spectrophotometrically and was checked for quality by assessment on agarose gel. For suppression subtractive hybridization, equal amounts of total RNA from cold treated samples (8, 16, 24 and 48 h) were combined. The mRNA was purified from this pooled sample and a control (0 h), using PloyATtract® mRNA Isolation System I kit (Promega) according to the manufacturer's instruction.

Suppression Subtractive Hybridization (SSH)

SSH was performed by using PCR-Select™ cDNA Subtraction Kit (Clontech) starting with 2 μg of mRNA from the leaves of control versus low temperature-treated plants. The mRNA from cold-treated leaves was used as the 'tester', while the mRNA from the non-treated leaves was used as the 'driver'. The forward subtraction was performed following the manufacturer's instructions. The secondary PCR products were inserted into pGEM T-easy Vector (Promega) and transformed into *E. coli* DH10B competent cells.

Amplification of cDNA Inserts

The cDNA clones were randomly picked from the SSH library as the templates, and the cDNA inserts were amplified by PCR using nested PCR primers 1 and 2R provided in the PCR Select-cDNA Subtraction Kit, which were complementary to sequences flanking both ends of the cDNA insert. The colonies that contained the cDNA inserts were selected and stored in 60-well microtitre plates.

Reverse Northern Blot Analysis

For reverse Northern, equal amounts of PCR amplification products were spotted in duplicate onto Hybond N⁺ nylon membranes (Amersham). The membranes were immersed in 0.4 M NaOH solution for 5 min followed by neutralization in $2 \times$ SSC. The membranes were then hybridized with equivalent amounts of the probes from either tester or driver cDNAs. DNA probes were radioactively labeled with [α-^{32}P]dCTP using Random Primer DNA Labeling Kit (Takara). Blots were hybridized and washed according to standard procedures. Detection was carried out using Molecular Imager® FX (Bio-Rad).

cDNA Sequencing and Blasting

cDNA sequencing was conducted at Sangon Biological Engineering Technology & Services Co., Ltd (Shanghai, China). Once low quality regions, vector and adaptor sequences were removed, the resulting high quality sequences were then subjected to a BlastX or BlastN comparison to the GenBank database online (http://www.ncbi.nlm.nih.gov/). All the EST sequences generated under this study were deposited into GenBank.

Northern Blot Analysis

Total RNA (20 µg) was fractionated on a 1% agarose RNA gel and transferred to Hybond N$^+$ nylon membrane (Amersham). DNA probes specific to the identified cDNA clones were radioactively labeled by [α-^{32}P]dCTP with Random Primer DNA Labeling Kit (Takara). Blots were hybridized and washed according to standard procedures. Detection was carried out using Molecular Imager$^®$ FX (Bio-Rad).

Results

A Cold Up-Regulated cDNA Library Was Constructed Using SSH

A suppression subtractive hybridization was performed using mRNA from low temperature-treated plants as the 'tester', and mRNA from non-treated controls as the 'driver'. After two rounds of hybridization and PCR amplification, PCR products were inserted into pGEM T-easy vector and transformed into *E. coli* DH10B competent cells. A number of colonies were randomly picked and the presence of cDNA inserts was confirmed by PCR amplification. Figure 1 shows a typical distribution of PCR products,

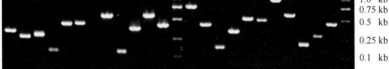

Fig. 1 PCR products from the differentially expressed cDNA clones. The colonies were used as templates, while the nested PCR primers 1 and 2R provided in the PCR Select-cDNA Subtraction Kit were used as primers. M indicates DNA ladder; the number 1–23 indicates randomly-picked cDNA clones

which ranged from 0.1 to 1 kb in size. Except for lane 7, the other 22 lanes were shown to have inserts, indicating a high efficiency of ligation between cDNA inserts and the cloning plasmid. The colonies that contained the cDNA inserts were then selected and stored to generate a SSH library of 2,016 cDNA clones.

Reverse Northern Blot Analysis

A total of 928 plasmids were used for an initial differential screening (hybridization) using reverse Northern blot analysis. The cDNA inserts were equally loaded on two separate membranes. The 'tester' and 'driver' cDNA probes were hybridized separately and cDNA clones that differentially hybridized with the cold-induced and control probes were identified (Fig. 2). A total of 114 clones (12.3%) were confirmed to be differentially expressed in our screen. Clones that were not differentially expressed were eliminated. A set of 523 clones from the forward subtractive (cold-induced) library including clones differentially expressed or that showed no hybridization signal were sequenced. All sequences were deposited into GenBank with accession numbers ranging from EG354701 to EG354724 and from EL610007 to EL610516. The sequences were analyzed using DNASTAR software, and 238 unique sequences (84 repeat sequences and 154 singlets) were harvested. The EST sequences were then blasted online (http://www.ncbi.nlm.nih.gov/) using BlastX and BlastN and 74% of the sequences had homologous genes.

Tester (cold-treated) cDNA probe   Driver (control) cDNA probe

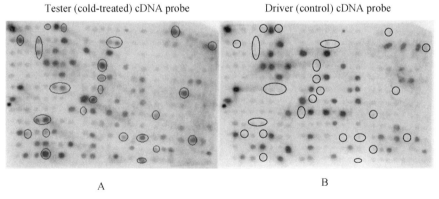

A   B

Fig. 2 Reverse Northern blotting of cDNA clones isolated from the cold-induced cDNA library. The cDNA inserts were spotted in duplicate onto Hybond N$^+$ membranes, and were hybridized with [α-^{32}P]dCTP-labeled 'tester' and 'driver' cDNA probes, respectively. *Circles* indicate clones showing strong differential hybridization signals between the two set of probes

Expression of the Selected cDNAs in Response to Cold Treatment

To confirm that gene expression corresponding to expressed sequence tags (ESTs) generated by SSH were differentially expressed in *M. falcata* under cold condition, eight representative clones were selected as probes for Northern blot analysis. These genes could be functionally classified as transcription factor (2D1), signal transduction (4C6), protein degradation (5F2), metabolism (22A8, 2C7, 23A12), stress responses (3D8), and unknown

Table 1 The identified cDNAs up-regulated by cold

Clone	Length (bp)	GenBank accession no.	Annotation	e-Value	Blast program
2D1	1,097	EG354707	EREBP-like protein	3.00E-22	Blast N
4C6	406	EG354715	Phosphoinositide phospholipase C	8.00E-39	Blast X
5F2	480	EG354717	FtsH protease	1.00E-64	Blast X
22A8	626	EG354701	Sucrose-phosphate synthase	6.00E-96	Blast X
2C7	573	EG354706	Sucrose synthase	3.00E-96	Blast X
23A12	610	EG354702	Myo-inositol-1-phosphate synthase	3.00E-104	Blast X
3D8	439	EG354712	Dehydrin-like protein	2.00E-83	Blast N
24G2	620	EH385063	Hypothetical protein	2.00E-47	Blast X

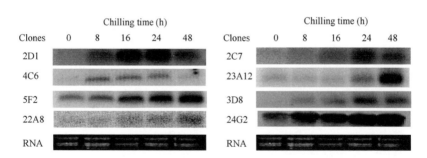

Fig. 3 Northern blot analysis of partial cDNA clones in response to cold. Total RNA (20 μg) from the leaves sampled at 0, 8, 16, 24 and 48 h after transfer to low temperatures was fractionated on a 1% agarose RNA gel and transferred to Hybond N$^+$ nylon membrane. DNA probes specific to the identified cDNA clones were radioactively labeled by [α-^{32}P]dCTP. Blots were hybridized and washed according to standard procedures

functions protein (24G2), from the cold-induced library Their annotation, accession numbers and length are summarized in Table 1. They were all induced in response to cold treatment. The expression pattern of these partial sequences is shown in Fig. 3. The clones showed a gradual increase in their expression with duration of cold exposure. Clones (2D1, 4C6, and 2C7) had an increased expression within 24 h, followed by a decreased expression at 48 h post cold treatment, while clone (23A12) had a significant expression only after 24 h of cold treatment (Fig. 3).

Discussion

Using SSH, we identified 238 unique gene transcripts (84 repeats and 154 singlets) with differential expression in *M. falcata* in response to cold stress. Eight representative transcripts were analyzed for the time course of their response to cold treatment. These include genes encoding proteins that are involved in signal transduction, protein degradation, central metabolism, general defence responses, and a protein with unknown functions.

The present work identified regulatory genes, including a transcription factor and a PLC gene. EREBP contains one AP2/EREBP domain, and is one of the largest groups of transcriptional factors (Riechmann et al. 2000). These genes can work downstream of abscisic acid and ethylene signaling pathways as well as environmental-stress responses such as cold, drought, and salt. PLC has important roles in the cellular signal transduction. It hydrolyses the minor lipid phosphatidylinositol 4,5-bisphosphate (PIP_2) into two second messengers: inositol 1,4,5-trisphosphate (IP_3) and diacylglycerol (DAG). The cold-induced expression of EREBP and PLC in *M. falcata* may regulate the expression of downstream genes in response to cold stress so that the plant can adapt to the stressed environmental conditions.

FtsH is a protease bound to the thylakoid membrane and is involved in the degradation of the photo damaged D1 protein of photosystem II (PSII) reaction center, in the context of repair from photoinhibition (Bailey et al. 2002; Lindahl et al. 2000), and in the degradation of unassembled thylakoid proteins (Ostersetzer and Adam 1997). The up-regulation of FtsH may protect PS II against to cold-induced damages in *M. falcata*.

The important role of sugars accumulation in the cold acclimation has long been recognized. Sucrose phosphate synthase (SPS) and sucrose synthase (SS) are two key enzymes of sucrose biosynthesis and metabolism pathway. Many investigations reveal that SPS and SS activity and sucrose content increased significantly under low temperature (Castonguay and Nadeau 1998). L-myo-inositol-1-phosphate synthase (MIPS) is the key enzyme for the synthesis of *myo*-inositol, an intermediate of raffinose and stachyose, which are two major sugars involved in cold acclimation of *Medicago* (Majumder et al. 1997; Cunningham et al. 2003). Our observation that the expression of SPS, SS, and MIPS genes was induced by cold demonstrate that sucrose and *myo*-inositol metabolism are also involved in the cold acclimation of *M. falcata*.

A dehydrin-like protein was identified in this work. Dehydrins have been proposed to have a protective function during abiotic stress (Campbell and Close 1997; Saavedra et al. 2006). They protect enzyme activity under cold or dehydration conditions (Lin and Thomashow 1992; Rinne et al. 1999; Sanchez-Ballesta et al. 2004); or act as radical scavengers (Hara et al. 2003) or as membrane stabilizers (Campbell and Close 1997; Koag et al. 2003). Recent studies have suggested a cold protective role for some members of this protein family (Hara et al. 2003). The cold-induced expression of a dehydrin-like protein might have a role in the tolerance of *M. falcata*.

In summary, our preliminary investigation indicates that SSH is an effective tool for identification of cold responsive genes, and suggests that *M. falcata* may rely on typical stress-inducible molecular mechanisms (e.g. signal transductions, transcriptional regulations, and stress defence genes) to acclimate to low temperatures. More sequencing and functional analysis of the sequences are under the way.

Acknowledgements

This work was supported by The National Basic Research Program of China (2007CB108905).

References

Bailey S, Thompson E, Nixon PJ, Horton P, Mullineaux CW, Robinson C, Mann NH (2002) A critical role for the Var2 FtsH homologue of *Arabidopsis thaliana* in the photosystem II repair cycle in vivo. J Biol Chem 277:2006–2011

Campbell SA, Close TJ (1997) Dehydrins: genes, proteins, and associations with phenotypic traits. New Phytol 137:61–74

Castonguay Y, Nadeau P (1998) Enzymatic control of soluble carbohydrate accumulation in cold-acclimated crowns of alfalfa. Crop Sci 38:1183–1189

Cunningham SM, Nadeau P, Castonguay Y, Laberge S, Volenee JJ (2003) Raffinose and stachyose accumulation, galactinol synthase expression, and winter injury of contrasting alfalfa germplasms. Crop Sci 43:562–570

Hara M, Terashima TF, Fukaya T, Kuboi T (2003) Enhancement of cold tolerance and inhibition of lipid peroxidation by citrus dehydrin in transgenic tobacco. Planta 217:290–298

Koag M-C, Fenton RD, Wilkens S, Close TJ (2003) The binding of maize DHN1 to lipid vesicles. Gain of structure and lipid specificity. Plant Physiol 131:309–316

Lin C, Thomashow MF (1992) A cold-regulated Arabidopsis gene encodes a polypeptide having potent cryoprotective activity. Biochem Biophys Res Commun 183:1103–1108

Lindahl M, Spetea C, Hundal T, Oppenheim AB, Adam Z, Andersson B (2000) The thylakoid FtsH protease plays a role in the light-induced turnover of the photosystem II D1 protein. Plant Cell 12:419–431

Majumder AL, Johnson MD, Henry SA (1997) L-myo-inositol-1-phosphate synthase. Biochim Biophys Acta 1348:245–256

Michaud R, Lehman WF, Rumbaugh MD (1988) World distribution and historical development. In: Hanson AA, Barnes DK, Hill RR (eds) Alfalfa and alfalfa improvement. American Society of Agronomy Inc., Madison, WI, USA, pp 25–91

Ostersetzer O, Adam Z (1997) Light-stimulated degradation of an unassembled Rieske FeS protein by a thylakoid-bound protease: the possible role of the FtsH protease. Plant Cell 9:957–965

Riechmann JL, Heard J, Martin G, Reuber L, Jiang C, Keddie J, Adam L, Pineda O, Ratcliffe OJ, Samaha RR, Creelman R, Pilgrim M, Broun P, Zhang JZ, Ghandehari D, Sherman BK, Yu G-L (2000) *Arabidopsis* transcription factors: genome-wide comparative analysis among eukaryotes. Science 290:2105–2110

Rinne PLH, Kaikuranta PLM, van der Plas LHW, van der Shoot C (1999) Dehydrins in cold acclimation apices of birch (*Betula pubescens* Ehrh.): production, localization and potential role in rescuing enzyme function during dehydration. Planta 209:377–388

Saavedra L, Svensson J, Carballo V, Izmendi D, Welin B, Vidal S (2006) A dehydrin gene in *Physcomitrella patens* is required for salt and osmotic stress tolerance. Plant J 45:237–249

Sanchez-Ballesta MT, Rodrigo MJ, Lafuente MT, Granell A, Zacarias L (2004) Dehydrin from citrus, which confers in vitro dehydration and freezing protection activity, is constitutive and highly expressed in the flavedo of fruit but responsive to cold and water stress in leaves. J Agric Food Chem 52:1950–1957

Volenec JJ, Cunningham SM, Haagenson DM, Berg WK, Joem BC, Wiersma DW (2002) Physiological genetics of alfalfa improvement past failures, future prospects. Field Crops Res 75:97–110

Genetic Diversity and Association Mapping of Three *O*-Methyltransferase Genes in Maize and Tropical Grasses

Baldomero Alarcon-Zuniga[1,3], Adriana Hernandez-Garcia[1], Elias Vega-Vicente[1], Cuauhtemoc Cervantes-Martinez[1], Marilyn Warburton[2] and Teresa Cervantes-Martinez[1]

[1]Universidad Autónoma Chapingo. Carr. México-Texcoco km. 38.5 Chapingo, Méx. 56230 Mexico
[2]Int. Applied Biotechnology Center, CIMMYT, El Batan, Texcoco 06600 Mexico
[3]Corresponding author, b_alarcon_zuniga@yahoo.com.mx

Abstract. Lignification and degradation of cell walls in tropical grasses have been considered complex metabolic processes. These involve a wide number of genes and interactions among structural carbohydrates per se, maturity, and monolignol precursor biosynthesis and content including siryngyl, guaiacyl, and p-coumaryl. Three major genes involved in lignin precursor biosynthesis are two caffeoyl-CoA 3-0-methyltransferase genes (*CCoAOMT1* and *CCoAOMT2*) and the aldehyde *O*-methyltransferase gene (*AldOMT*). The objective of this study was the identification of the number and frequencies of alleles and nucleotide diversity of the three *O*-methyltransferase genes in 79 maize S_9 inbred populations originating from Mexican biracial crosses and six tropical perennial grass species, and their association to cell wall degradability. The six tropical grasses genotypes were stargrass (*Cynodon plectostachyus*), bermudagrass (*C. dactylon*), palisadegrass (*Brachiaria brizantha*), kinggrass (*Pennisetum purpurem*), and guineagrass (*Panicum maximum*). Among the maize inbred populations, we identified two alleles in *CCoAOMT1* (polymorphic information content (PIC) = 0.23), eight in *CCoAOMT2* (PIC = 0.73), two in *AldOMT*-Exon1 (PIC = 0.36), three in *AldOMT*-Exon2 (PIC = 0.69) and five in *AldOMT*-Intron. All PCR primers amplified in the tropical grass genotypes except *AldOMT*-Exon1, and two (PIC = 0.31), four (PIC = 0.69), seven (PIC = 0.38), and eight alleles (PIC = 0.71) were found in the above mentioned genes, respectively. Phylogeny analysis clustered the maize populations into six groups strongly related to ecogeographical origin, and no association resulted within the tropical grasses (57% nucleotide identity of gene sequences). A negative genetic correlation between neutral detergent fiber, acid detergent fiber, crude protein and in vitro dry matter digestibility was found with *CCoAOMT2* and *AldOMT*-Exon2 ($r > 0.61$). We identified two candidate *CCoAOMT2* alleles and three *AldOMT* alleles that could contribute to the improvement of exotic maize germplasm for digestibility (and thus for use as a forage crop and production of bioethanol), but useless for the genotypes of tropical grasses studied herein.

Maize as Model Genome for Tropical Forage Crops

Maize is the premier organism for addressing fundamental biological questions in monocots, particularly the tropical cereal and forage crops, which contribute worldwide >40% of the calories in the human diet and ~35% in the animal diet as grain, silage and forage. A primary goal of the maize genomics community is to work towards an understanding the structure and function of the maize genome by developing and disseminating a comprehensive integrated physical and genetic mapping framework (Tomkins et al. 2002). CINVESTAV-IPN-Mexico showed the first maize genome coding sequence with 52,000 genes on June, 2007, of an ancient Mexican race Palomero Toluqueño (Herrera et al. 2007). However, a consolidated physical and genetic map of grain and forage maize for tropical environments in developing countries is still being developed. These maps will enable us to further investigations in tropical maize, including understand genetic backgrounds and similarities, heterotic patterns, gene discovery and function, and comparative genomic studies with other plant species, particularly both tropical and temperate grasses (Draye et al. 2001). In Mexico, corn farmers continue to cultivate local or criollo maize varieties in a traditional manner, and this contributes to the conservation and generation of in situ genetic diversity of maize (Cervantes et al. 1978; Reif et al. 2006). Most genetic diversity studies of Mexican farmer's varieties have considered geographical origins and morphological characteristics among gene pools or races, but there is still a lack of information about genetic diversity at the molecular level and gene alleles for genetic interactions.

CIMMYT has explored their maize genetic diversity pool using both morphological traits and molecular markers, constituted by racial complexes, and they concluded that multiracial gene populations have leaded them to inbreeding and loss of hybrid vigor (Reif et al. 2004, 2006). According with the previous results, it has been concluded that genetic diversity and gene interaction are highly depended on the genetic divergence and these are higher if the gene pool is conserved throughout racial heterotic groups per se, instead of multiracial populations. In Mexico, only seven races of maize (Conico, Chalqueño, Celaya, Oloton, Pepitilla, Tepecintle, Tuxpeño and Zapalote) out of 53 are commonly used for plant breeding purposes, "since most of the native populations present local adaptation, tall plants, poor root development, stalk quality, flowering asynchronicity, high frequency of plants without ears, and high susceptibility to inbreeding depression and ear rot" (Cervantes 1978). Currently, efforts are made to incorporate the genetic richness of the Mexican maize diversity into breeding programs. Alarcon-Zuniga et al. (2006) have improved maize tropical and temperate populations belonging to specific races per se, to perform well on desirable agronomic traits. Thus, we have hypothesized that the adapted and improved tropical exotic race populations per se are comparable to local improved material in terms of field performance, due to their high gene diversity richness combined with the genetic contribution of local material, resulting in a higher heterosis than expected from crosses within complexes racial pools (Carrera and Cervantes 2002).

Unified Tropical Grass Genome

Traditionally, genetic relationships between related species have been studied via cytological analysis of interspecific hybrids, and surveys of naturally occurring chromosomal markers (knobs) within populations (McClintock et al. 1981). Comparative analyses of DNA sequences add a powerful new technique for investigating the mode and tempo of chromosomal evolution and gene functional expression. The use of common sets of low-copy-number DNA markers, often coding sequences, in the mapping grass genomes has indicated that the gene content of different grass species does not vary greatly (Bennetzen and Freeling 1997). Studies among maize, sorghum, sugar cane, rye, wheat, and oat, have lead to the conclusion that grass species possess largely conserved gene content and a number of conserved non-coding genomic regions, and probably also exhibit extended regions of map colinearity within gene families, such as the lignin biosynthesis gene family. Thus, the observation of conserved gene content and order in the grass genomes gave rise the model that individual grass species could be viewed best as manifestations of a single grass lignin-genome. This may benefit a whole range of grasses into the tribe Andropogonae, such as *Brachiaria*, *Cynodon*, *Panicum*, *Pennisetum*, among others, considered as the most important tropical forage crop genera with limited digestibility and quality value.

Other potentially useful genetic traits in exotic grass species or ecotypes have been identified, including apomixis, which has been documented in 145 different species of exotic tropical grasses (Carman 1995). A first step toward learning more about these useful traits is to identify the gene diversity and richness (allele number, frequency and colinearity) in genomically comparable grass species, and determine whether different species appear to have orthologous locations and similar function. Standard sets of grass DNA markers could be used for this mapping. When comparative mapping experiments do not find the gene in the species where it was wanted, very wide crosses could be made to introgress important traits into genetically tractable exotic plants. This project seeks to implement genotypic and molecular approaches in several grass species to dissect genetic diversity, association mapping and gene tagging of agronomic and nutritional traits, using maize as model plant for tropical forage grasses.

Candidate Genes and Gene Approach for Forage Digestibility

Plant cell walls are a major energy source for ruminants and play an important role in forage utilization. Cell walls are composed of cellulose fibrils embedded within a matrix of lignin and hemicellulose; in addition, they contain inorganic solvents, phenolics and proteins. Digestion of intact cell walls is limited by the presence of lignin and phenolic acids within the cell-wall matrix. The most important candidate genes associated with the lignin pathway are: (a) caffeic acid *O*-methyltransferase (*COMt*, also called 5-hydroxyconiferilaldehide *O*-methyltransferase); (b) phenylalanine ammonia-lyase (*PAL*); (c) cinnamoyl *CoA* reductase (*CCR*); and (d) cinnamoyl alcohol dehydrogenase (*CAD*). Breeding for high digestibility in forage maize and grasses is an important goal because it would improve animal intake, growth rate, and milk production, but it is a long term process and breeders must screen many genotypes to find the desired phenotypes. There are also several biotech approaches for enhancing the digestibility of forage crops: (1) introgressing known mutants of the lignin pathway into highly productive germplasm; (2) downregulation of *COMt*, AldMOT, *PAL*, and *CAD* lignin genes, or other genes involved in the cellulose and hemicellulose pathways, via genetic engineering of selected genotypes; and (3) breeding for lower fiber and lignin concentrations with conventional and/or with marker-assisted selection.

Recent research with *brown midrib* (*bm*) mutants has demonstrated that lignin is a highly plastic compound, synthesized by a metabolic grid that results in a wide array of biochemical routes (Ralph 2006), and numerous potential sites at which genetic modification may occur (Bout and Vermerris 2003; Krakowsky et al. 2005). Most of the genetic variation that has been generated in novel lignin transgenic plants of the family grass Poaceae is parallel to that observed in spontaneous mutants, including variation in lignin concentration and composition (Casler et al. 2002). Many novel lignin transgenic events appear to mimic natural genetic polymorphism, including the *brown-midrib* mutants of maize, sorghum and other annual cereals (Casler et al. 2002; Bout and Vermerris 2003). Therefore, identifying allelic diversity, gene tagging and sequencing of natural genes involved with digestibility within maize populations may allow the development of simple PCR primers that may be screened among more populations; this may lead to the establishment if candidate genes in important metabolic pathways within the maize genome and related grass species, (such as forage tropical grasses). We present here initial results regarding gene diversity among maize populations for three lignin *O*-methyltransferase genes and their relationship to tropical perennial forage crops.

Molecular Breeding Approaches

The experimental plant material consisted of 79 S_9 maize inbred composite populations, originating from 125 bi-racial crosses of temperate and tropical Mexican races, and selected up to $F_{13\#}$ by parental crossing using mass selection, followed by selfing resulting in the inbred composites $F_{13\#}S_9$ (Carrera and Cervantes 2002). The following six tropical species were also included: stargrass (*Cynodon plectostachyus*), bermudagrass (*C. dactylon*), palisadegrass (*Brachiaria brizantha*), kinggrass (*Pennisetum purpurem*), and guineagrass (*Panicum maximum*). For the mapping association analysis, the five tropical species were considered as heterogeneous and heterozygous by origin. The 79 S_9 populations were planted in two locations at Chapingo and Tecamac, Mexico, during 2005 and 2006. The six tropical species were planted in Veracruz, Mexico, in 2006. Morphological and forage quality traits were scored in all plant materials. The molecular analysis consisted of PCR-amplification in three *O*-methyltransferase genes, (caffeoyl-CoA 3-0-methyltransferase genes *CCoAOMT1* and *CCoAOMT2*, and aldehyde *O*-methyltransferase gene *AldOMT*) related to the lignin metabolic pathway, as reported by Guillet-Claude et al. (2004). Primer pairs for the complete *CCoAOMT1* and *CCoAOMT2* coding sequences were used, and the *AldOMT* gene was partitioned into exon1, intron and exon2 amplicons (Guillet-Claude et al. 2004). Data analyses of agronomic and nutritional traits of both maize and tropical grasses were performed using the GLM, IML and MIXED procedures of SAS 9.1 for Windows (SAS Institute Inc. 1999). The modified Rogers's distance (MRD) between two inbred populations, gene diversity and phylogeny analysis were calculated according to Wright (1978), Goodman and Stuber (1983), and Weir (1996), respectively.

Preliminary analysis of morphological traits including plant height, number of ears, biomass components (leaf, stem and grain) and flowering time showed highly significant variation among all 79 inbred populations (Alarcon-Zuniga et al. 2005). Cluster analysis of morphological traits grouped the inbred populations into five important groups, which showed a high general combining ability (GCA) and specific combining ability (SCA), but not significant genotype-by-environment interaction, as reported by Alarcon-Zuniga et al. (2006). The five pairs of primers generated a total of 20 alleles in the 79 S_9 maize inbred populations. Among the maize inbred populations, we identified two alleles in *CCoAOMT1* (polymorphic information content (PIC = 0.23), eight in *CCoAOMT2* (PIC = 0.73), two in *AldOMT*-Exon1 (PIC = 0.36), three in *AldOMT*-Exon2 (PIC = 0.69) and five in *AldOMT*-Intron, similar as that reported by Reif et al. (2006). MRD phylogeny analysis in the maize inbreds, also clustered the populations into six groups strongly related to the ecogeographical origin, considering that allele diversity for *O*-methyltransferase genes might be associated to the maize center of origin and even to an "altitude and latitude" gradient (Fig. 1). All PCR primers amplified for the tropical grass genotypes with two (PIC = 0.31), four (PIC = 0.69), seven (PIC = 0.38), and eight alleles (PIC = 0.71), respectively, except *AldOMT*-Exon1 that failed PCR amplifications; confirming the broad genetic diversity among tropical genotypes. Nucleotide diversity for the complete sequence of all three *O*-methyltransferase genes was compared between the inbred populations in this study with those reported by Guillet-Claude et al. (2004). We identified 97% of nucleotide similarities in *CCoAOMT* genes to maize Polar Dent, F7025, MBS847 (GenBank Acc. 323272-77) and 96% similarity to lines EP1, B73 and B14, Rainbow Flint, among others (GenBank Acc. 323207/323299).

No association resulted between maize inbred populations with the six tropical grasses in the *O*-methyltransferase gene sequences (57% nucleotide identity of gene sequences), but 80% similar to *Sacharum officinarum COMT* genes (GenBank Acc. 231133.1 and 365419.1).

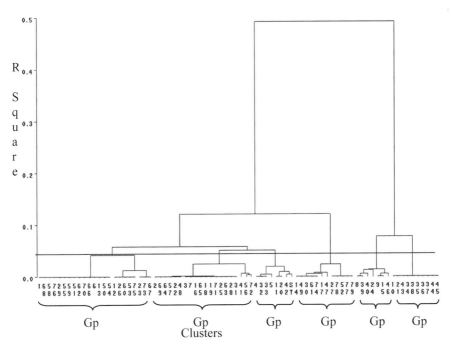

Fig. 1 Modified Roger's distance dendrogram of S7 inbred populations based on the allele frequencies of five primers of three *O*-methyltransfase genes. Each number corresponds to S9 inbred populations

The six clusters derived from the MDR phylogeny analysis using all 79 S_9 inbred populations were defined by the ecogeographical origin as follow: (1) northern tropics with Chapalote and Reventador as representatives; (2) northern and southern tropics, with Tuxpeno and Tepecintle; (3) south western tropics, with Vandeno and Comiteco; (4) southern tropics and temperate, with Olotillo and Zapalote as representatives; (5) southeast tropics represented by Nal-Tel and Harinoso de Ocho; and (6) southern and central temperate, with Celaya, Tabloncillo and Pepitilla. These results are similar as reported by Cervantes et al. (1978) considering morphological traits. To identify the association between genetic diversity of five primers originating of three *O*-methyltransferas genes (*CCoAOMT1*, *CCoAOMT2* and *AldOMT*) and the forage quality predictors (crude protein (CP), neutral detergent fiber (NDF), acid detergent fiber (ADF) and in vitro dry matter digestibility (IVDMD)), we randomly selected five S_9 inbred populations of each of the six groups (Table 1).

Table 1 Maize S9 inbred populations chosen to study the association between forage quality traits and three O-methyltransferase genes

Group	Number in dendrogram	Maize race identification	Group	Number in dendrogra	Maize race identification
I. Northern tropics	1	Tehua × Olotillo	IV. Southern tropics and temperate	14	Pepitilla × Harinoso 8
	68	Olotillo × Pepitilla		79	Zapalote × Tehua
	6	Blandito × Tuxpeño		17	Chapalote × Tablilla
	56	Chapalote × Tuxpeño		30	Reventador × Olotillo
	26	Reventador × Zapalote		27	Chapalote × Pepitilla
II. Northern and southern tropics	2	Tuxpeno × Comiteco	V. Southeast tropics	8	Pepitilla × Oloton
	73	Comiteco × Jala		47	Blandito × Pepitilla
	66	Reventador × Tuxpeño		24	Reventador × Tehua
	49	Tepecintle × Jala		41	Nal-Tel × Chapalote
	25	Reventador × Tehua		9	Blandito × Tablilla
III. South-western tropics	4	Reventador × Tuxpeno	VI. Southern and central temperate	10	Reventador × Oloton
	43	Vandeno × Oloton		46	Chapalote × Reventador
	34	Tepecintle × Celaya		39	Olotillo × Tabloncillo
	5	Nal-Tel × Comiteco		44	Celaya × Olotillo
	11	Chapalote × Bofo		38	Pepitilla × Celaya

The proximal analysis in whole plant of 30 maize inbred populations was estimated by NIRS (Near Infrared Spectrophotometry) analysis, considering a minimum value of 0.9 for the coefficient of determination and >95% of variances, for each quality estimator. No differences in forage quality estimators was found among groups, except for the neutral detergent fiber where group III (Southwest tropical inbred populations) had the lowest and group V (Southeast tropical inbred pops) the highest (Table 2).

Table 2 Whole plant proximal analysis of 30 maize S9 inbred populations, corresponding to five populations per group, based on five primers of three *O*-methyltransferase genes

Group	% CP[a]	% NDF[b]	% ADF[c]	IVDMD[d] 24 h	IVDMD 48 h	IVDMD 72 h						
I	6.61 A	76.13 AB		51.49 A	35.17 A	48.32 A	58.24 A
II	5.64 A	73.50 AB		51.45 A	38.24 A	50.33 A	58.27 A
III	5.72 A	71.84 B		48.90 A	38.10 A	50.42 A	58.67 A
IV	6.80 A	74.07 AB		51.42 A	36.04 A	47.01 A	55.97 A
V	5.39 A	79.00 A		52.32 A	35.57 A	47.98 A	59.66 A
VI	6.24 A	76.91 AB		50.58 A	34.43 A	48.01 A	57.55 A

Means with the same letter in the same column are statically equal with alpha = 0.05
[a]Crude protein
[b]Neutral detergent fiber
[c]Acid detergent fiber
[d]In vitro dry matter digestibility

Measurements taken from the forage quality predictors in all 30 S_9 inbred populations did not associate the estimators with any of the primers used to amplify the three *O*-methyltransferase genes (Table 3), except for

Table 3 R^2 (*above*) and probability values (*below*) of forage quality estimators in 30 maize S9 inbred populations and their association to three *O*-methyltransferase genes

Trait	CCoAOMT1	CCoAOMT2	AldOMT-E1	AldOMT-E2	AldOMT-I
CP	0.004 $P = 0.75$	0.047 $P = 0.25$	0.0009 $P = 0.87$	0.034 $P = 0.83$	0.012 $P = 0.55$
NDF	0.099 $P = 0.09$	0.16 $P = 0.23$	0.105 $P = 0.08$	0.081 $P = 0.17$	0.15 $P = 0.03$
ADF	0.0001 $P = 0.96$	0.002 $P = 0.85$	0.02 $P = 0.49$	0.004 $P = 0.74$	0.089 $P = 0.51$
IVDMD, 24h	0.012 $P = 0.56$	0.071 $P = 0.53$	0.042 $P = 0.27$	0.005 $P = 0.72$	0.15 $P = 0.04$
IVDMD, 48h	0.031 $P = 0.33$	0.072 $P = 0.64$	0.0004 $P = 0.91$	0.003 $P = 0.77$	0.095 $P = 0.51$
IVDMD, 72h	0.08 $P = 0.13$	0.0007 $P = 0.93$	0.009 $P = 0.61$	0.002 $P = 0.82$	0.083 $P = 0.38$

AldOMT-Intron with NDF ($P = 0.03$) and IVDMD at 24 h ($P = 0.04$), assuming that *AldOMT* can be used as a candidate gene for cell wall digestibility, as reported by Guillet-Claude et al. (2004), using European maize inbred lines.

The forage quality predictors in six tropical forage crops were consistent between harvest dates. However, differences among the six genotypes were showed for CP, NDF, and IVDMD at 24 and 72 h. As expected, kinggrass showed the highest quality, and bermuda grass the lowest. All PCR primers amplified in the tropical grass species except for *AldOMT*-Exon1 which did not amplify under any conditions. In the tropical grasses, *CCoAOMT1* amplified two alleles, (PIC = 0.31), *CCoAOMT2* amplified four (PIC = 0.69), *AldOMT*-Exon2 amplified seven (PIC = 0.38), and *AldOMT*-Intron amplified eight alleles (PIC = 0.71). Association mapping between the forage quality predictors of tropical grasses with the primers of the three maize *O*-methyltransferase genes showed that *CCoAOMT1* and *CCoAOMT2* were associated to IVDMD at 72 h ($R^2 = 0.23$, $P = 0.0069$; $R^2 = 0.17$, $P = 0.014$, respectively) and *CCoAOMT2* was associated to CP ($R^2 = 0.213$, $P = 0.01$). This association based on candidate genes in tropical grasses might be especially promising for marker assisted selection, although further validation is needed before reporting them to be robust functional markers, suggesting the construction of mapping populations in the tropical species to confirm the association between the primers and forage quality predictors.

Table 4 R^2 (*above*) and probability values (*below*) of forage quality estimators in six tropical grass genotypes and their association to three maize *O*-methyltransferase genes

Trait	*CCoAOMT1*	*CCoAOMT2*	*AldOMT*-Exon2	*AldOMT*-Intron
CP	0.07	0.213	0.03	0.0.39
	$P = 0.218$	$P = 0.01$	$P = 0.42$	$P = 0.61$
NDF	0.063	0.042	0.039	0.016
	$P = 0.179$	$P = 0.27$	$P = 0.74$	$P = 0.67$
ADF	0.023	0.091	0.039	0.031
	$P = 0.42$	$P = 0.22$	$P = 0.31$	$P = 0.53$
IVDMD, 24h	0.031	0.04	0.002	0.055
	$P = 0.35$	$P = 0.46$	$P = 0.83$	$P = 0.23$
IVDMD, 48h	0.072	0.017	0.001	0.053
	$P = 0.15$	$P = 0.51$	$P = 0.82$	$P = 0.18$
IVDMD, 72h	0.23	0.17	0.053	0.051
	$P = 0.0069$	$P = 0.014$	$P = 0.22$	$P = 0.24$

Conclusion

Breeding for forage quality in tropical grasses is laborious and expensive for large sets of individuals, particularly for apomictic and vegetatively propagated species. Novel biotech approaches can be used in different species, using information from the model genome species maize, which has highly conserved metabolic pathways, such as the lignin pathway. It is sufficiently similar to closely related genera of the tribe Andropogonae to be useful in providing functional markers for useful traits.

Acknowledgments

This study was completed with funds from the Mexican grating agency Fondo Sectorial SAGARPA-CONACYT, and the Maize Research Program at DGIP-Universidad Autonoma Chapingo; and thanks to Dr. Tarcicio Cervantes Santana for proving the maize germplasm. We also thanks to the excellent technical support of Dr. Maria M. Crosby-Guzman, Dr. Marcos Menesses-Mayo, and field and laboratory technical assistance of Mr. Andres Luna-Galindo and Ms. Lily X. Zelaya-Molina.

References

Alarcon-Zuniga B, Warburton M, Cervantes ST, Mendoza RM, Cervantes MT (2005) Phylogenetic relationship of tropical maize germplasm as revealed by SSR. ASA-CSSA-SSS International Annual Meeting, Salt Lake City, November 7–10, 2005 [CD-Rom computer file]

Alarcon-Zuniga B, Cervantes MT, Warburton M (2006) Heterosis and combining ability of tropical maize in Central Valley of Mexico: morphological and molecular characterization for silage. Proceeding of the International Plant Breeding Symposium, CIMMYT, August 20–25, Mexico City, p 24

Bennetzen JL, Freeling M (1997) The unified grass genome: synergy in synteny. Genome Res 7(4):301–306

Bout S, Vermerris W (2003) A candidate gene approach to clone the sorghum Brown midrid gene encoding caffeic acid *O*-methyltransferase. Mol Gen Genomics 269:205–214

Carman J (1995) Gametophitic angiosperm apomicts and the occurrence of polyspory and polyembrony among their relatives. Apomixis Newslett 8:39–53

Carrera VJA, Cervantes TS (2002) Tropical maize populations selected per se and crosses performance in high valleys. Agrociencia 36(6):693–701

Casler MD, Buxton DR, Vogel KP (2002) Genetic modification of lignin concentration affects fitness of perennial herbaceous plants. Theor Appl Genet 104: 127–131

Cervantes TS (1978) Recursos genéticos disponibles a Mexico. Sociedad Mexicana de Fitogenetica A.C. Chapingo, Mexico

Cervantes T, Goodman MM, Casas E, Rawlings OL (1978) Use of genetic effects and genotype by environmental interactions for the classification of Mexican races of maize. Genetics 90:339–348

Draye X, Lin YR, Qian X, Bowers JE, Burow GB, Morrell PL, Peterson DG, Presting GG, Ren S, Wing RA, Paterson AH (2001) Toward integration of comparative genetic, physical, diversity and cytomolecular maps fro grasses and grains, using the sorghum genome as a foundation. Plant Physiol 125:1325–1341

Goodman MM, Stuber CW (1983) Races of maize: VI. Isozyme variation among races of maize in Bolivia. Maydica 28:169–187

Guillet-Claude C, Birolleau-Touchard C, Manicacci D, Fourmann S, Barraud S, Carret V, Martinant JP, Barriere Y (2004) Genetic diversity associated with variation in silage corn digestibility for three O-methyltransferase genes involved in lignin biosynthesis. Theor Appl Genet 110:126–135

Herrera EL, Martinez O, Vielle CJP (2007) Descrifa IPN genoma del maíz. Crónica (in Spanish) 18(3):1–9

Krakowsky MD, Lee M, Coors JG (2005) Quantitative trait loci for cell-wall components in recombinant inbred lines of maize (Zea mays L.) I: Stalk tissue. Theor Appl Genet 11:337–346

McClintock B, Kato TAY, Blumenschein A (1981) Chromosome constitution of races of maize. Colegio de Potsgraduados Press, Mexico

Ralph J (2006). What makes a good monolignol substitute? In: T. Hayashi (editor) The Science and lore of the plant cell wall. Brown Walker Press, Boca Raton, FL, pp. 285–293

Reif JC, Xia XC, Melchinger AE, Warburton ML, Hoisington DA, Beck D, Bohn M, Frisch M (2004) Genetic diversity determined within and among CIMMYT maize populations of tropical, subtropical and temperate germplasm by SSR markers. Crop Sci 44:326–334

Reif JC, Warburton ML, Xia XC, Hoisington DA, Crossa J, Taba S, Muminovic J, Bohn M, Frisch M, Melchinger AE (2006) Grouping of accessions of Mexican races of maize revisited with SSR markers. Theor Appl Genet 113:177–185

SAS Institute Inc. (1999) SAS/STAT user's guide, Version 6, fourth edition. SAS Institute Inc., Cary, NC

Tomkins JP, Davis G, Main D, Yim Y, Duru N, Musket T, Goicoechea JL, Frisch DA, Coe EH Jr, Wing RA (2002) Construction and characterization of a deep-coverage bacterial artificial chromosome library for maize. Crop Sci 42:928–933

Weir BS (1996) Genetic data analysis II, second edition. Sinauer Associates, Sunderland, MA

Wright S (1978) Evolution and genetics of population. Vol. IV. University of Chicago Press, IL

QTL Analysis and Gene Expression Studies in White Clover

Michael T. Abberton[1,2], Athole Marshall[1], Rosemary P. Collins[1], Charlotte Jones[1] and Matthew Lowe[1]

[1]Legume Breeding and Genetics Team, Institute of Grassland and Environmental Research (IGER), Plas Gogerddan, Aberystwyth, SY23 3EB, UK
[2]Corresponding author, mla@aber.ac.uk

Abstract. In white clover (*Trifolium repens* L.) plant persistence, overwintering and grazing tolerance are to a significant extent determined by the presence of a dense network of horizontal stems or stolons. At IGER, we have developed new linkage maps of white clover specifically to facilitate the identification of quantitative trait loci (QTLs) for important components of stolon morphology. The parents of the F_1 mapping family were derived following two generations of divergent selection for stolon traits, in particular length and thickness. One parental genotype had thick, sparse stolons and one parent had thin profuse stolons. Two linkage maps were generated for each parental clone: one consisted of 16 linkage groups with 154 markers covering 710 cM and the other consisted 13 linkage groups with 86 markers covering 524 cM. There are 31 shared markers across the two maps. A total of nine QTLs were identified on the maps for the following traits: stolon width, internode length, petiole length, leaf length and width, plant spread and height. Potential QTLs associated with water use efficiency (WUE) have also been identified.

We have complemented this work by transcriptome analysis of gene expression changes in response to developing drought in both white clover and the model legume *Medicago truncatula* using the recently released *M. truncatula* Affymetrix Genechip[TM]. A number of genes which show up-regulation in response to drought in both species have been identified. These candidate genes are being confirmed by quantitative real time PCR and by mapping and assessing the extent of co-location with QTLs for WUE.

T. Yamada and G. Spangenberg (eds.), *Molecular Breeding of Forage and Turf*,
doi: 10.1007/978-0-387-79144-9_15, © Springer Science + Business Media, LLC 2009

Introduction

White clover (*Trifolium repens* L.) is the most important legume of temperate pastures. It is a grazing tolerant, perennial species grown with a companion grass on sheep, beef and dairy farms. White clover is an outbreeding, highly heterozygous allotetraploid ($2n = 4x = 32$) and these features inform both approaches to variety development and to genetic analysis.

A distinctive characteristic of this species is the horizontal stem or stolon which allows it to spread within a pasture. From the stolon nodes come adventitious roots, petioles and peduncles. A dense network of stolons, with many adventitious roots, is particularly important with respect to tolerance of grazing, especially by sheep. Thus the genetic control of stolon traits is an important target in terms of variety development in this species (Abberton and Marshall 2005). Previous quantitative trait loci (QTL) studies have focused on a range of morphological and agronomic traits including seed yield (Cogan et al. 2006; Barrett et al. 2005; Zhang et al. 2007) but have not considered stolon characteristics.

Stolons and their adventitious roots also have a major role in the ability of the plant to access water (Caradus 1981; Ennos 1985), and potentially in water use efficiency (WUE), a trait of growing importance in the face of predicted climate change impacts.

Mapping of Stolon Traits in White Clover

Parents of the mapping family used in this study were derived from two generations of divergent selection for stolon traits from the small leaf size variety AberCrest (Collins et al. 1997, 1998). This selection program classified individual genotypes according to number and thickness of stolons, producing four lines: thick profuse (TKPR), thick sparse (TKSP), thin profuse (TNPR) and thin sparse (TNSP). Initial crosses were made in all combinations and a F_1 family showing the greatest segregation for stolon traits was selected for mapping. This family was a cross between a plant from TKPR (the female) and a plant from TNSP (the male). Three hundred and fifty progenies were replicated three times and planted out in the field in 2004 for a detailed survey of the stolon characteristics, which were used for QTL analysis. Ten characteristics were measured which were divided into three categories:

(a) Stolon morphology: stolon length and width, internode and petiole length.

(b) Plant and leaf morphology: leaflet length and width, plant height and spread.

(c) Flowering date and height of peduncles.

Broad sense heritabilities (BSH) were calculated for these traits (see Table 1).

Table 1 Broad sense heritabilities of measured traits

Trait	BSH
Flowering date	0.39
Flower height	0.24
Plant height	0.32
Plant spread	0.30
Stolon width	0.41
Stolon number	0.38
Internode length	0.50
Leaf length	0.34
Leaf width	0.35
Petiole length	0.27

Ninety-six randomly selected F_1 progenies were also used in a preliminary experiment analyzing differences in WUE. This experiment involved a further five measurements of: mean total, root and shoot mass, root-shoot ratio and WUE of the plant.

Mapping was performed using amplified fragment length polymorphism (AFLP) and simple sequence repeat (SSR) markers. The tested SSR markers were derived from white clover genomic and EST SSR (Barrett et al. 2004; Kölliker et al. 2001; Jones et al. 2003), candidate gene sequences based on *M. truncatula* in the EU Grain Legume Integrated Project (GLIP) and ESTs of *M. truncatula*. Marker detection was performed by an ABI3700 genetic analyser. JoinMap® v3.0 was used to construct linkage maps for each parent and MapQTL® v4.0 was used to identify QTLs for the traits measured in the field and the WUE data on the linkage maps (Fig. 1).

Two maps were generated for each of the parental genotypes of the mapping family. The TKPR map (Map 1) has 154 markers over 710 cM on 16 linkage groups (LGs), on average 5 cM between markers. The TNSP map (Map 2) is shorter, with 86 markers on 13 LGs over 524 cM. A total of 12 QTLs were identified on 10 LGs of TKPR map (Map 3), while 11 QTLs were mapped on 8 LGs of TNSP map (Map 4).

Map 1 TKPR linkage map

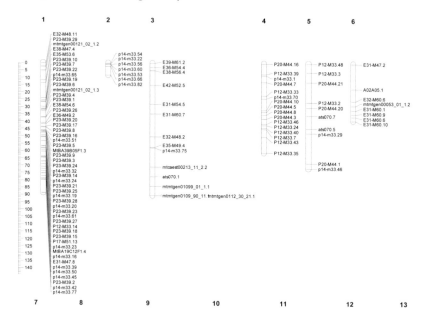

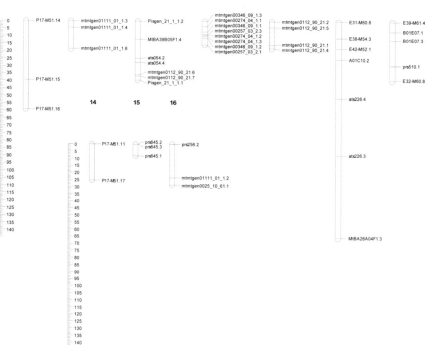

Map 2 TNSP linkage map

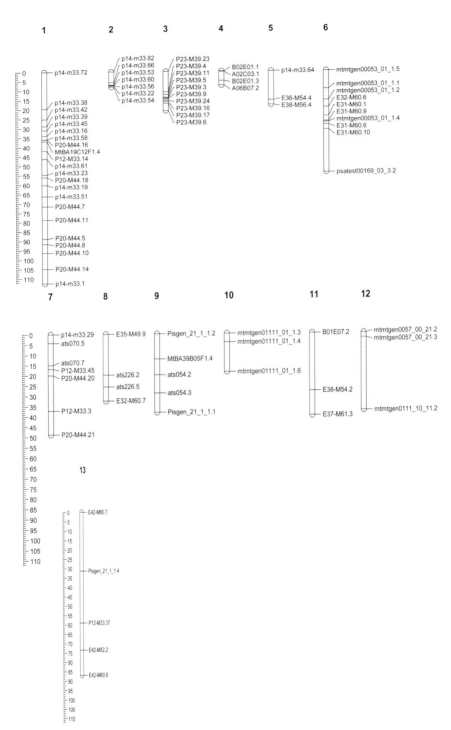

Map 3 TKPR QTL map

Map 4 TNSP QTL map

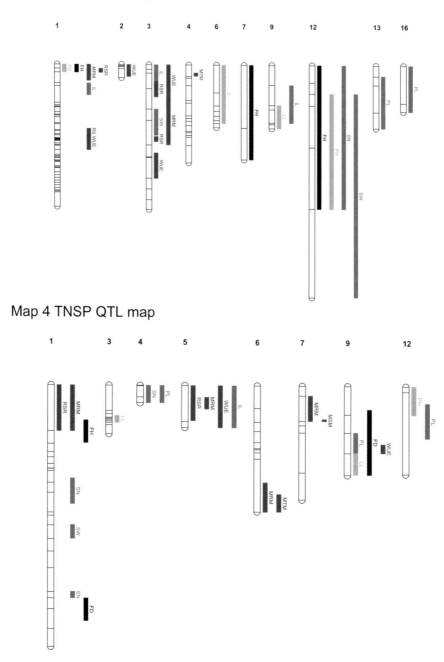

Fig. 1 Linkage maps (maps 1 and 2) linkage and QTL maps (maps 3 and 4) derived from the thin sparse (TNSP) and thick profuse (TKPR) parents

KEY
Stolon characteristics: *SW* stolon width; *SN* stolon number; *IL* internode length; *PL* Petiole length
Leaf/plant characteristics: *LL* leaf length; *LW* leaf width; *PH* plant height; *PS* plant spread
Water use efficiency characteristics: *WUE* water use efficiency; *MSM* mean shoot mass; *MRM* mean root mass; *MTM* mean total mass; *RSR* root:shoot ratio
Flowering characteristics: *FD* Flowering date; *FH* Flower height

Clearly this QTL analysis is preliminary and we are currently adding more markers to the map, carrying out further phenotyping and seeking to compare QTL positions with those obtained from other newly developed mapping families.

Gene Expression Changes in Response to Drought

It is now well recognized that, particularly for highly heterozygous outbreeders, QTL detection in mapping families derived from single pair crosses has considerable limitations in terms of developing molecular tools that can be effectively applied in breeding programmes, including the likelihood of recombination between marker and QTL. This has led to the development of a range of complementary approaches including linkage disequilibrium (LD) mapping and mapping of single nucleotide polymorphisms (SNPs) in candidate genes. Comparison of changes in gene expression profiles is one route to the identification of candidate genes.

We have carried out preliminary experiments with the *M. truncatula* Affymetrix Genechip[TM] to compare transcriptome changes between the two species in response to drought.

An experiment was carried out at IGER in which plants of white clover and *M. truncatula* were grown under well watered and developing drought. Conditions. Briefly, 18 plants each of white clover L. (W980108) and *M. truncatula* were selected and divided into two groups of nine plants (control and drought). The plants were either well watered (control) or subjected to drought conditions (drought) for 7 days. Control plants were maintained at field capacity (90–100%) by watering whilst the drought plants received a decreasing volume of water each day. The leaf relative water content (RWC) of each plant was measured daily and leaf samples collected at the same time for extraction of RNA. Two of the nine plants were selected for their relative drought tolerance and susceptibility from the RWC results. RNA was then extracted from the control and droughted samples of drought tolerant and drought susceptible plants of both species for hybridization to the *M. truncatula* Affymetrix Genechip™.

Hybridizations to the Affymetrix Genechip™ were carried out at the University of Bristol and the extent of hybridization is shown (Table 2).

Table 2 Number of probes hybridized and percentage hybridization from probes of *M. truncatula* Affymetrix chip with white clover and *M. truncatula*

Type of hybridization	Number of probes hybridized	Percentage hybridization
White clover/*M. truncatula*	7,598	12.4
White clover and *M. truncatula*	6,357	10.4
White clover only	1,241	2.0

Analysis was carried out using GeneSpring software. Genes were filtered to analyze present/marginal probes only and a tenfold change in expression level was taken as the cut off point. In *Medicago/Medicago* comparisons, 368 probes were down-regulated under drought and 264 up-regulated. For the *M. truncatula*/white clover hybridization 83 probes were down-regulated under drought and 309 up-regulated.

Selection of overlapping probes between the two sets of analyses gave a final set of 27 probes up-regulated in drought, and 31 probes down-regulated in drought in both white clover and *M. truncatula*. Homology searches were carried out and a number of genes of interest were identified e.g. Drought induced protein, Fiber protein Fb19, Homeobox-Leucine Zipper Protein, Low Affinity Glucose Transporter HXT4. Further evidence to support a function on drought tolerance for these and other genes is being sought through qRT-PCR analyis and mapping to assess the extent of co-location with QTL for drought tolerance and WUE.

Conclusions

Clearly both the studies presented here are preliminary and further work is needed before any firm conclusions can be drawn. However, they both show that the use of tools and resources derived from the model legume

M. truncatula has considerable potential and is likely to become important in studies underpinning precision breeding in white clover.

Acknowledgements

IGER is grant aided by the Biotechnology and Biological Sciences Research Council and this work was also financially supported by the Department for Environment, Food and Rural Affairs.

References

Abberton, M.T., Marshall, A.H. (2005) Progress in breeding perennial clovers for temperate agriculture. Centenary Review. Journal of Agricultural Science 143, 117–135

Barrett, B., Griffiths, A., Schreiber, M., Ellison, N., Mercer, C., Bouton, J., Ong, B., Forster, J., Sawbridge, T., Spangenberg, G., Bryan, G., Woodfield, D. (2004) A microsatellite map of white clover. Theoretical and Applied Genetics, 109, 596–608

Barrett, B.A., Baird, I.J., Woodfield, D.R. (2005) A QTL analysis of white clover seed production. Crop Science 45, 1844–1850

Caradus, J.R. (1981) Root growth of white clover (*Trifolium repens* L.) lines in glass-fronted containers. New Zealand Journal of Agricultural Research 24, 43–54

Cogan, N.O.I., Abberton, M.T., Smith, K.F., Kearney, G., Marshall, A.H., Williams, A., Michaelson-Yeates, T.P.T., Bowen, C., Jones, E.S., Vecchies, A.C., Forster, J.W. (2006) Individual and multi-environment combined analyses identify QTLs for morphogenetic and reproductive development traits in white clover (*Trifolium repens* L.). Theoretical and Applied Genetics 112, 1401–1415

Collins, R.P., Abberton, M.T., Michaelson-Yeates, T.P.T., Rhodes, I. (1997) Response to divergent selection for stolon characters in white clover (*Trifolium repens*). Journal of Agricultural Science 129, 279–285

Collins, R.P., Abberton, M.T., Michaelson-Yeates, T.P.T., Marshall, A.H., Rhodes, I. (1998) Effects of divergent selection on correlations between morphological traits in white clover (*Trifolium repens* L.). Euphytica 101, 301–305

Ennos, R.A. (1985) The significance of genetic variation for root growth within a natural population of white clover (*Trifolium repens* L.). Journal of Ecology 73, 615–624

Jones, E.S., Hughes, L.J., Drayton, M.C., Abberton, M.T., Michaelson-Yeates, T.P.T., Bowen, C., Forster, J.W. (2003) An SSR and molecular marker-based genetic map of white clover (*Trifolium repens* L.). Plant Science 165, 447–479

Kölliker, R., Jones, E.S., Drayton, M.C., Dupal, M.P. (2001) Development and characterization of simple sequence repeat (SSR) markers for white clover (*Trifolium repens* L.). Theoretical and Applied Genetics, 102, 416–424

Zhang, Y., Sledge, M.K., Bouton, J.H. (2007) QTL mapping of morphological traits and agronomic traits in white clover. Proceedings of International Plant and Animal Genomics XV. January 13–17, 2007, San Diego, CA, USA

Genome Mapping in Cool-Season Forage Grasses

Hongwei Cai[1,2,3], Maiko Inoue[1], Nana Yuyama[1] and Mariko Hirata[1]

[1]Forage Crop Research Institute, Japan Grassland Agriculture & Forage Seed Association, 388-5, Higashiakata, Nasushiobara, Tochigi 329-2742, Japan
[2]Department of Plant Genetics & Breeding, College of Agronomy and Biotechnology, China Agricultural University, 2 Yuanmingyuan West Road, Beijing 100094, China
[3]Corresponding author, hongwei_cai@yahoo.co.jp

Abstract. Molecular markers, specifically SSR markers, have found wide utility in genomic research. However, in forage grasses, the availability of such markers restricts and restrains the breeder/geneticist from taking full advantage of this frequently utilized technology. In this report, we review recent developments in our laboratory to generate or increase the number and availability of SSR markers for timothy (*Phleum pratense*), orchardgrass (*Dactylis glomerata*), Italian ryegrass (*Lolium multiflorum*) and zoysiagrass (*Zoysia japonica*). Until now, genomic research (including marker development, gene mapping, and quantitative trait loci (QTL) analysis in most forage grasses, with the exception of Lolium species, lagged behind that performed in major crops species such as rice and maize. To address this deficit, we have developed about 1,000 simple sequence repeat (SSR) markers for *Phleum pratense*, *Dactylis glomerata*, *Zoysia japonica* and *Lolium multiflorum* from SSR-enriched genomic libraries. Most of these SSR markers are polymorphic in a screening panel that included eight individuals of each species. From the polymorphic SSR makers we constructed a linkage map for hexaploid and diploid timothy. From the multi-alleles of hexaploid timothy SSR markers, we detected seven homologous linkage groups and found a homologous relationship between hexaploid and diploid timothy. We also developed SSR markers for Italian ryegrass of which 11 are specific to 12 Italian ryegrass cultivars developed in Japan. An orchardgrass SSR-linkage map is under construction.

T. Yamada and G. Spangenberg (eds.), *Molecular Breeding of Forage and Turf*,   173
doi: 10.1007/978-0-387-79144-9_16, © Springer Science + Business Media, LLC 2009

Introduction

Timothy (*Phleum pratense* L.), orchardgrass (*Dactylis glomerata* L.), and Italian ryegrass (*Lolium multiflorum* Lam.) are the most commonly cultivated forage grass species in Japan. Because of their large genome size and the autoploid genetic status of timothy and orchardgrass, little marker development or marker analysis has been forthcoming. On the other hand, marker development and genomic studies of two *Lolium* species, Italian ryegrass and perennial ryegrass, have an extensive publication history including marker methods such as restricted fragment length polymorphism (RFLP), amplified fragment length polymorphism (AFLP), simple sequence repeat (SSR), expressed sequence tag (EST), and single nucleotide polymorphism (SNPs). Several partial and complete linkage maps are available and quantitative trait loci (QTL) analysis is a relatively common undertaking in *Lolium* species. The genus *Zoysia* consists of approximately 11 species that are naturally distributed on coastal regions and grasslands in the Pacific (Kitamura 1989). Five species have been identified to range from southern Hokkaido to the southwest islands of Japan (Kitamura 1989). Zoysiagrass has a chromosome number of 40 and is a typically considered a cross-pollinated, allotetraploid species with a small genome size (421 Mbp for *Z. japonica*) (Forbes 1952; Arumuganathan et al. 1999). *Zoysia japonica* Steud., *Z. matrella* Merr., and *Z. tenuifolia* Willd. are utilized extensively as turfgrasses, and zoysiagrass (*Z. japonica*) has its primary use as a forage grass in Japan and other East Asian countries (Shoji 1983; Fukuoka 1989). In our laboratory, we focus on marker development, map construction and genome analysis of four important species, timothy, orchardgrass, Italian ryegrass and zoysiagrass.

Marker Development

Timothy

Until now, there have been no reports on molecular marker development in timothy. We have been developing SSR markers from a hexaploid timothy strain named SK, and we developed markers from four SSR-enriched genomic libraries to isolate 1,331 SSR-containing clones (Cai et al. 2003b; Table 1). From these clones, we developed 822 SSR markers by testing all 1,331 SSR primer pairs across a screening panel consisting of eight timothy clones to detect the level of polymorphism. We screened all 822 SSR markers for usefulness in diploid and tetraploid timothy. Approximately,

80% of the hexaploid timothy markers could be amplified in diploid or tetraploid timothy. In addition, we identified 664 SSR markers (80.8%) that could be amplified in all (2x, 4x, 6x) timothy. The remaining markers, 12 (1.5%) were amplified only in 2x and 6x timothy; 13 (1.6%) were amplified only in 4x and 6x timothy; and 133 (16.2%) were amplified only in 6x timothy. This result suggests that in the process of evolution from 2x to 4x to 6x forms, some minor genomic changes may have occurred.

Orchardgrass

We also developed SSR markers from four SSR-enriched genomic libraries of the tetraploid orchardgrass cultivar, Akimodori II (our unpublished data). From 969 unique SSR-containing clones, we designed 959 pairs of SSR primers. After screening with a panel consisting of eight tetraploid orchard-grass individuals, we found that 606 primers could amplify polymorphic products in the screening panel; and the other primers either generated monomorphic bands, multiple bands or, did not provide any amplification product (Table 1).

Table 1 Efficacy of working primers from four SSR libraries for respective species.

Library	Enrichment motif	No. of unique SSR clones	Primer evaluation			
			Polymorphic	Monomorphic	Multiple band	No amplification
Zoysiagrass						
A	CA/TG	140	125	7	1	7
B	GA/TC	803	718	11	6	68
C	AAG/TTC	125	82	20	18	5
D	AAT/TTA	95	78	3	10	4
Total		1163	1003	41	35	84
Italian ryegrass						
A	CA/TG	360	212	37	37	74
B	GA/TC	540	376	51	32	81
C	AAG/TTC	264	113	57	53	41
D	AAT/TTA	263	150	21	28	64
Total		1427	851	166	150	260
Orchardgrass						
A	CA/TG	205	94	17	21	73
B	GA/TC	539	398	19	16	106
C	AAG/TTC	121	65	23	14	19
D	TAGA/ATCT	104	49	11	5	39
Total		969	606	70	56	237
Timothy						
A	CA/TG	311	230	0	15	66
B	GA/TC	474	360	14	23	77
C	AAG/TTC	215	143	22	30	20
D	TAGA/ATCT*	331	207	14	19	91
Total		1331	940	50	87	254

* Most SSR motifs were GA/TC, AAT/TTA and TGA/ACT

Italian Ryegrass

Molecular markers have been reported in Italian ryegrass. These include RFLP (Inoue et al. 2004), sequence-tagged site (STS) (Inoue and Cai

2004), EST-cleaved amplified polymorphic sequence (CAPS) (Miura et al. 2007), SSR (Hirata et al. 2006), and resistance gene analog (RGA) (Ikeda 2005) markers. Hirata et al. (2006) developed 395 SSR markers from a CA repeat library and mapped some of them in a mapping population; however, several were duplicate sites and the total number of markers developed was limited. To increase the available number of SSR markers in Italian ryegrass, we used generated four SSR-enriched genomic libraries from which 851 polymorphic primers were identified following evaluation across a screening panel consisting of five individuals of Italian ryegrass, one of perennial ryegrass, one meadow fescue, and one tall fescue (Table 1). Following re-screening of the SSR markers from Hirata et al. (2006), some tall fescue EST-SSR primers (Saha et al. 2005), and other published perennial ryegrass SSR markers (Jones et al. 2001), we had a total of 1,172 working primers including above new developed SSR markers in our Italian ryegrass SSR marker database. Of these, 679, 581, and 682 primers successfully amplified SSR sites in perennial ryegrass, meadow fescue and tall fescue, respectively.

Zoysiagrass

Until now, a limited number of RFLP (Yaneshita et al. 1999) and SSR (Tsuruta et al. 2005) markers have been developed for zoysiagrass. To develop larger numbers of SSR markers for this genus, we used developed four SSR-enriched genomic libraries to isolate 1,163 unique SSR clones (Table 1). All library contained high percentages of perfect repetitive sequence clones, ranging from 67.1 to 96.0%; compound clones occurred with higher frequencies in the CA-enriched library (28.6%) and the AAT-enriched library (11.6%). From these clones, we developed 1,044 working SSR markers when we tested all 1,163 SSR primer pairs (Cai et al. 2005).

Database Construction

Filemaker 8.0 (FileMaker, Inc. Santa Clara, CA) was used to construct SSR marker databases for Italian ryegrass, timothy, orchardgrass, and zoysiagrass. The database included marker name, repeat motif, product size, primer sequences, and linkage group (if the marker was mapped). The sequence of the clone and the electrophoresis pattern (.TIF file) for the amplification were also included.

Linkage Map Construction

Timothy

A hexaploid timothy linkage map was constructed on the basis of AFLP and SSR markers using a pseudo-testcross F_1 population derived from two hexaploid timothy strains (#243 and #341) native to Japan. The female linkage map (map of 243) consists of 811 single dosage markers (SDMs) belonging to 64 linkage groups, and the male linkage map (map of 341) consists of 673 SDMs belonging to 48 linkage groups. On the basis of the SSR marker mapping, seven homologous linkage groups were detected (Cai et al. 2003a, data not shown). A diploid timothy linkage map was constructed from the SSR markers by using a pseudo-testcross F_1 population (PI302911 × PI319078). All 822 SSR markers developed from hexaploid timothy were used for screening of the parents' polymorphism, and a total of 363 (53.7%) markers were found to be segregating. The diploid timothy linkage maps, consisting of 226 SSR markers, belonged to seven linkage groups at the level of log of odds (LOD) = 5. The total length was 445 cM and the average distance between two flanking markers was about 3.0 cM (our unpublished data; Fig. 1). In addition, from the same SSR marker mapping, a comparative relationship between the seven diploid timothy linkage groups and the homologous linkage groups in hexaploid timothy was detected (Fig. 2a and b).

Orchardgrass

An SSR marker-based linkage map of orchardgrass is under construction. A pseudo-testcross F_1 population consisting of 88 individuals derived from a cross between two orchard grass cultivars, Akimidori II and Loki, are being utilized in the mapping population. Screening of 606 SSR markers between the two parents has revealed that 321 markers are polymorphic. The QTL responsible for heading date will also be analyzed in this population.

Italian Ryegrass

A high-density molecular linkage map of Italian ryegrass was constructed using a two-way pseudo-testcross F_1 population consisting of 82 individuals to analyze three types of markers: RFLP markers (genomic probes from

Italian ryegrass as well as heterologous anchor probes from other species belonging to the Poaceae family); AFLP markers (*Pst*I/*Mse*I primer combinations); and telomeric repeat associated sequence (TAS) markers. The 385 (mostly RFLP) markers that were selected from among the 1,226 original markers were grouped into seven linkage groups. The maps covered 1,244.4 cM, with an average of 3.7 cM between markers (Inoue et al. 2004). We are now placing SSR markers to this map using the same population.

Zoysiagrass

An AFLP-based molecular linkage map of zoysiagrass was constructed using an F$_2$ population consisting of 78 individuals and analyzing 471 AFLP markers derived from 126 *Pst*I/*Mse*I primer combinations. Of these markers, 364 were grouped into 26 linkage groups. The maps covered a total length of 932.5 cM, with an average spacing of 2.6 cM between markers (Cai et al. 2004). In addition, of 1,003 SSRs evaluated, 170 SSR markers that detected 172 loci segregated in the same mapping population. Of these 172 mapped loci, 150 were mapped to the 24 existing AFLP-based linkage groups, 11 were mapped to six new detected linkage groups (mostly small groups), and 11 could not be mapped to any linkage group. Five pairs of existing AFLP-based linkage groups were joined into five

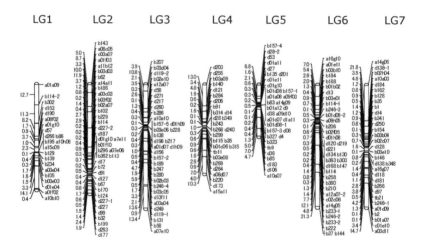

Fig. 1 SSR-based linkage map of diploid timothy population

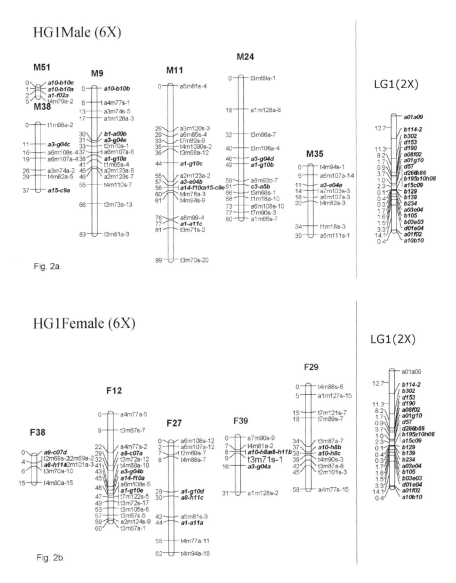

Fig. 2 Comparison of diploid timothy map and hexaploid timothy map. (**a**) HG1 male; (**b**) HG1 female. SSR markers were indicated as *italic* and **bold**

linkage groups on the basis of the SSR data. The SSR–AFLP-based linkage map consisted of 540 markers and covered a total map length of 1,187 cM, with an average spacing of 2.2 cM between markers (Cai et al. 2005).

Italian Ryegrass Cultivar Identification by Using SSR Markers and Mitochondrial and Chloroplast Genome SNP Markers

Italian ryegrass is the most widely cultivated annual forage grass in Japan. Many cultivars have been developed in Japan by public institutions and private companies. To identify specific markers for Italian ryegrass cultivars, we screened 1,172 SSR markers from *Lolium* and other species using eight bulks (ten individuals each) of 12 cultivars as template DNA and were able to detect 127 specific SSR primers for 12 cultivars using this approach. Following the analysis of 100 individuals for each cultivar, we selected 11 specific primers for 12 cultivars that had alleles specific for each cultivar. The frequencies of the specific alleles ranged from about 15–50% (unpublished data).

The organellar genome is usually well conserved among different plant species, and organellar markers are used as tools for studies of population genetics and evolution and for phylogenic analysis. To find cultivar-specific markers, we tested a total of 38 primer pairs from chloroplast genes of rice and other species and 30 primer pairs from mitochondrial genes. The analysis indicated that almost all the primers could provide good amplification products in Italian ryegrass; however, since no size polymorphism were identified among the 12 cultivars, an alternative approach was required. In an attempt to identify polymorphisms, amplification products were digested with *Alu*I, *Hae*III, *Rsa*I, *Hha*I, *Msp*I, and *Mse*I. Again, no polymorphisms were identified. As an additional avenue of investigation for polymorphism, we sequenced the PCR products to identify potential SNP markers. Sequencing of seven chloroplast markers and four mitochondrial markers successfully identified four SNP markers in the chloroplast genome and one SNP in the mitochondrial genome (Table 2). Comparison of the frequencies of these five SNP markers enabled us to clearly discriminate genetic differences across the 12 Italian ryegrass cultivars (our unpublished data).

We are now using EST markers to conduct a comparative genomic analysis of these four forage species against rice and wheat. We are conducting marker-assisted selection against such diseases as crown rust, blast resistance in Italian ryegrass; and purple spot and lodging resistance in timothy. We are also conducting QTL analysis of seed productivity and gene mapping of rust resistance in zoysiagrass.

Table 2 Italian ryegrass SNP markers detected from chloroplast and mitochondrial genomes

Name of genes	Genome	Forward primer	Reverse primer	Product size	Reference	SNP
rbcL	chloroplast	TGTCACCAAAAACAGAGACT	TTCCATACTTCACAAGCAGC	1387	Kishimoto et al. 2003	Yes
16S	chloroplast	ACGGGTGAGTAACGCGTAAG	CTTCCAGTACGGCTACCTTG	1375	Kishimoto et al. 2003	No
ORF100	chloroplast	AGTCCACTCAGCCATCTCTC	GGCCATCATTTTCTTCTTTAG	900	Kanno et al. 1993	Yes
PS-Id	chloroplast	AAAGATCTAGATTTCGTAAACA ACATAGAGGAAGAA	ATCTGCAGCATTTAAAAGGGTC TGAGGTTGAATCAT	550	Nakamura et al. 1998	No
trnS-trnI	chloroplast	CGAGGGTTCGAATCCCTCTC	AGAGCATCGCATTTGTAATG	1386	Demesure et al. 1995	Yes
trnT-trnF	chloroplast	CATTACAAATGCGATGCTCT	ATTTGAACTGGTGACACGAG	1754	Taberlet et al. 1991	Yes
trnL5'B-trnF	chloroplast	CGAAATCGGTAGACGCTACG	ATTTGAACTGGTGACACGAG	974	Taberlet et al. 1991	No
nad1(5)	mitochondria	GCCTATCCTAGATCTTCC	GAGCCATTGAAAGGTGAC	215	Robison and Wolyn 2002	No
cox1	mitochondria	CAGCTACCATGATCATAGCTG	GGATTTGACCTAAAGTTTCAGG	320	Robison and Wolyn 2002	No
18s-5s	mitochondria	GTGTTGCTGAGACCATGCGCC	ATATGGCGCAAGACGATTCC	1177	Al-Janabi et al. 1994	No
nad7(3)	mitochondria	GAACACGCTCATTCTTCAG	CGAGAAGCAAATTGTTGTG	320	Robison and Wolyn 2002	Yes

Acknowledgments

All the above research was supported by a grant from the Japan Racing Association. We also thank Mr. Fujimori M (Yamanashi Prefectural Dairy Experiment Station) and Dr. Tamaki H, and Mr. Yoshizawa A (Hokkaido Kitami Agricultural Experiment Station) for their cooperation. We also thank Prof. Yamada T (Hokkaido University, Japan) and Kindiger B (USDA, Agricultural Research Service, Grazing Lands Research Laboratory, USA) for their critical reading and valuable comments on the manuscript.

References

Al-Janabi SM, McClelland M, Petersen C, Sobral BWS (1994) Phylogenetic analysis of organellar DNA sequences in the Andropogoneae: Saccharinae. Theor Appl Genet 88: 933–944

Arumuganathan K, Tallury SP, Fraser ML, Bruneau AH, Qu R (1999) Nuclear DNA content of thirteen turfgrass species by flow cytometry. Crop Sci 39: 1518–1521

Cai HW, Yuyama N, Tamaki H, Yosizawa A (2003a) Development of SSR markers and detection of homologous linkage groups in hexaploid grass timothy (*Phleum pratense* L.). Plant, Animal & Microbe Genomes XI Conference, January, San Diego, CA

Cai HW, Yuyama N, Tamaki H, Yoshizawa A (2003b) Isolation and characterization of simple sequence repeat markers in the hexaploid forage grass timothy (*Phleum pratense* L.). Theor Appl Genet 107: 1337–1349

Cai HW, Inoue M, Yuyama N, Nakayama S (2004) An AFLP-based linkage map of Zoysiagrass (*Zoysia japonica* Steud.). Plant Breed 123: 543–548

Cai HW, Inoue M, Yuyama N, Takahashi W, Hirata M, Sasaki T (2005) Isolation, characterization and mapping of simple sequence repeat markers in Zoysiagrass (*Zoysia* spp.). Theor Appl Genet 112: 158–166

Demesure B, Sodzi N, Petit RJ (1995) A set of universal primers for amplification of polymorphic non-coding regions of mitochondrial and chloroplast DNA in plants. Mol Ecol 4: 129–131

Forbes I Jr. (1952) Chromosome numbers and hybrids in Zoysia. Agron J 44: 147–151

Fukuoka H (1989) Breeding of *Zoysia* spp. J Jpn Soc Turfgrass Sci 17: 183–190 (in Japanese)

Hirata M, Cai HW, Inoue M, Yuyama N, Miura Y, Komatsu K, Takamizo T, Fujimori M (2006) Development of simple sequence repeat (SSR) markers and construction of an SSR-based linkage map in Italian ryegrass (*Lolium multiflorum* Lam.). Theor Appl Genet 113: 270–279

Ikeda S (2005) Isolation of disease resistance gene analogs from Italian ryegrass (*Lolium multiflorum* Lam.). Grassl Sci 51: 63–70

Inoue M, Cai HW (2004) Sequence analysis and conversion of genomic RFLP markers to STS and SSR markers in Italian Ryegrass (*Lolium multiflorum* Lam.). Breed Sci 54: 245–251

Inoue M, Gao ZS, Hirata M, Fujimori M, Cai HW (2004) Construction of a high-density linkage map of Italian ryegrass (*Lolium multiflorum* Lam.) using restriction fragment length polymorphism, amplified fragment length polymorphism, and telomeric repeat associated sequence markers. Genome 47: 57–65

Jones ES, Dupal MP, Kölliker R, Drayton MC, Forster JW (2001) Development and characterization of simple sequence repeat (SSR) markers for perennial ryegrass (*Lolium perenne* L.). Theor Appl Genet 102: 405–415

Kanno A, Watanabe N, Nakamura I, Hirai A (1993) Variations in chloroplast DNA from rice (*Oryza sativa*): differences between deletions mediated by short direct-repeat sequences within a single species. Theor Appl Genet 86: 579–584

Kishimoto S, Aida R, Shibata M (2003) Identification of chloroplast DNA variations by PCR-RFLP analysis in *Dendranthema*. J Jpn Soc Hort Sci 72: 197–204

Kitamura F (1989) The climate of Japan and its surrounding areas and the distribution and classification of zoysiagrasses. Int Turfgrass Soc Res J 6: 17–21

Miura Y, Hirata M, Fujimori M (2007) Mapping of EST-derived CAPS markers in Italian ryegrass (*Lolium multiflorum* Lam.). Plant Breed 126: 353–360

Nakamura I, Kameya N, Kato Y, Yamanaka S, Jomori H, Sato YI (1998) A proposal for identifying the short ID sequence which addresses the plastid subtype of higher plants. Breed Sci 47: 385–388

Robison M, Wolyn D (2002) Complex organization of the mitochondrial genome of petaloid CMS carrot. Mol Genet Genomics 268: 232–239

Saha MC, Mian R, Zwonitzer JC, Chekhovskiy K, Hopkins AA (2005) An SSR- and AFLP-based genetic linkage map of tall fescue (*Festuca arundinacea* Schreb.). Theor Appl Genet 110: 323–336

Shoji S (1983) Species ecology of *Zoysia* grass. J Jpn Soc Turfgrass Sci 12: 105–110 (in Japanese)

Taberlet P, Gielly L, Pautou G, Bouvet J (1991) Universal primers for amplification of three non-coding regions of chloroplast DNA. Plant Mol Biol 17: 1105–1109

Tsuruta SI, Hashiguchi M, Ebina M, Matsuo T, Yamamoto T, Kobayashi M, Takahara M, Nakagawa H, Akashi R (2005) Development and characterization of simple sequence repeat markers in *Zoysia japonica* Steud. Grassl Sci 51: 249–257

Yaneshita M, Kaneko S, Sasakuma T (1999) Allotetraploidy of *Zoysia* species with 2n=40 based on a RFLP genetic map. Theor Appl Genet 98: 751–756

Understanding the Genetic Basis of Flowering and Fertility in the Ryegrasses (*Lolium* spp.)

Ian Armstead[1,4], Bicheng Yang[1,2,3], Susanne Barth[2], Lesley Turner[1], Leif Skøt[1], Athole Marshall[1], Mervyn Humphreys[1], Ian King[1] and Danny Thorogood[1]

[1]Institute of Grassland and Environmental Research, Aberystwyth, UK
[2]Teagasc, Crops Research Centre, Oak Park, Carlow, Ireland
[3]School of Biosciences, University of Birmingham, UK
[4]Coresponding author, ipa@aber.ac.uk

Abstract. Ryegrasses and fescues of the *Lolium/Festuca* complex form the basis of many temperate grassland agricultural and turf systems and the ability to manipulate flowering and fertility within these grasses is of considerable importance. Genetic studies can identify multiple quantitative trait loci affecting these traits which can be used in the development of marker-assisted selection protocols for combining favourable alleles. By applying a knowledge of plant comparative genetics and genomics, it is possible to increase the information content of these studies by 'cross-referencing' between the *Lolium/Festuca* grasses and other crop and model species (e.g. rice, Triticeae cereals, *Arabidopsis*). Syntenic genomic regions, new genetic markers and candidate genes can, thus, be identified.

Introduction

Ryegrasses are predominately cultivated as forage and amenity grasses, where the vegetative phenotype is the primary performance measure for the crop. As a consequence, most breeding effort has been focused on improving the vegetative characteristics related to the use of grasses for forage and turf. In this context, the transition from vegetative to reproductive growth in grasses involves profound physiological changes which can have considerable impact on field performance, influencing traits such as

T. Yamada and G. Spangenberg (eds.), *Molecular Breeding of Forage and Turf,*
doi: 10.1007/978-0-387-79144-9_17, © Springer Science + Business Media, LLC 2009

biomass production, digestibility and palatability. Under normal environmental conditions in a perennial crop, most ryegrasses will flower on an annual basis. Thus, in the light of the inherent genetic variation that exists for flowering phenology within the ryegrasses and, increasingly, in the likelihood of accelerating climate change, there is a requirement to be able to control and predict flowering variation within ryegrass populations and varieties in order to promote economic and environmental sustainability.

In addition to its effect on agronomic performance, consistent and predictable flowering is also of considerable significance in terms of seed production, as commercial propagation of ryegrasses is almost exclusively by seed. Consequently, seed yield *per se* is an important trait and seed producers are reluctant to take the risk of growing poor seed-yielding cultivars even if they show superior performance for other characteristics. In spite of this, direct selection for fertility-associated traits has received little attention, possibly because seed production that is dependent on preferable allocation of resources from vegetative to reproductive organs is likely to be detrimental to agronomic traits such as good vegetative yield and quality. In that context, improved fertility will be negatively correlated with agronomic performance. However, seed-set (i.e. the proportion of florets that produce a seed, *sensu strictu* caryopsis) and seed retention are two aspects of reproduction that are independent of vegetative growth performance traits and these offer opportunities for the breeder to improve seed-yield without compromising other quality aspects, particularly as considerable variation for seed-set has been observed in perennial ryegrass and is common within most forage grass species.

Another factor in determining fertility within the ryegrasses is the presence of a two loci genetic self-incompatibility system (Thorogood et al. 2002). The effect of incompatibility on fertility is not likely to be of major significance in *Lolium/Festuca*, as out-crossing and frequency dependent selection effectively maintains multiallelism at both incompatibility loci at a population level. However, where plant improvement schemes include crosses and selections within a narrower genetic base (or even a degree of inbreeding) self-incompatibility systems may be a barrier to full exploitation. Thus, evaluations of fertility within these grasses should be made in the context of the possible effects of these incompatibility systems.

Comparative Genetics of Flowering and Fertility

Because of the key significance of flowering in determining yield in most crop plants and because of the amenability of this trait to experimental evaluation, the genetic control of flowering has been studied in detail in both crop and model systems. Comparisons of the genetic control

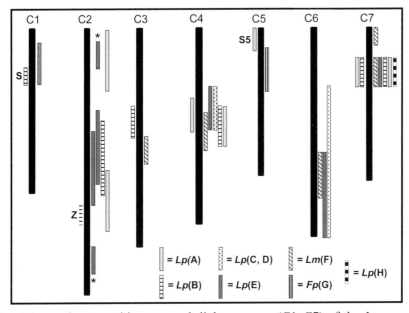

Fig. 1 Approximate positions on each linkage group (C1–C7) of the *L. perenne* genetic map of regions associated with fertility (*left*) and flowering (*right*). S, Z and S5 identify regions associated directly or through synteny with gametophytic incompatibility loci. The relative positions of identified regions were derived from the following studies: (**A**) Armstead et al. (2004, 2005); Thorogood et al. (2005, unpublished); (**B**) Thorogood et al. (2002, unpublished); (**C**) Yamada et al. (2004); (**D**) Shinozuka et al. (2005); (**E**) Jensen et al. (2005); (**F**) Inoue et al. (2004); (**G**) Ergon et al. (2006); (**H**) Skøt et al. (2005, 2007). Relative linkage group lengths were derived from study A. *Lp* = *L. perenne*, *Lm* = *L. multiflorum*, *Fp* = *F. pratensis*. *The orientation of C2 in study E is not established relative to the other studies

pathways from these various studies have shown similarities both within the monocots and between monocots and dicots (Cockram et al. 2007). Similar detailed genetic studies within *Lolium/Festuca* are still being developed, but it has been established that many of the key genes identified in other species are also present within *Lolium/Festuca*. Therefore, established models for flowering control in species such as wheat, barley, rice and *Arabidopsis* can form the basis for the determination of flowering control in *Lolium/Festuca*.

Comparisons of quantitative trait loci (QTL) mapping and linkage disequilibrium (LD) studies which have evaluated flowering within *Lolium/Festuca* indicate considerable variability between different populations in the control of flowering (Fig. 1). While it is apparent that genomic regions from all of the linkage groups can have significant effects, some genomic

Table 1 *Lolium/Festuca* genetic studies on flowering (1.1) and fertility (1.2)

Study[a]	Species	Population name	Origin	QTL C4	Candidate gene[b]	QTL C7	Candidate gene[b]
1.1							
A	*Lp*	WSC/F2	UK	Yes	?	Yes	*Hd3a/Hd1*
B	*Lp*	ILGI	UK	Yes	?	Yes	*Hd3a/Hd1*
C	*Lp*	ILGI	Japan	Yes	*Vrn-1*	No	–
D	*Lp*	ILGI	Japan	Yes	*Ck2α*	–	–
E	*Lp*	VrnA F2	Denmark	Yes	*Vrn-1*	Yes	*Hd3a/Hd1*
F	*Lm*	NN	Japan	Yes	?	Yes	*Hd3a/Hd1*
G	*Fp*	BF14/16·HF2/7	Norway	No	–	No	–
H	*Lp*	LD	UK	–	–	Yes	*Hd3a/Hd1*
1.2							
A	*Lp*	WSC/F2	UK	Yes	*S33(t)*	Yes	*Sn⁵*
B	*Lp*	ILGI	UK	No	–	Yes	*Sn⁵*

Lp L. Perenne; *Lm* L. Multiflorum; *Fp* Festuca pratensis
[a]Study references are as described in Fig. 1
[b]Candidate gene identities are either derived from the individual studies or suggested by cross-comparisons of the different mapping populations

regions, particularly on chromosomes 4 and 7, are associated with flowering QTL across a number of the studies, indicating cross-environmental significance. These same regions are also associated with fertility QTL (defined as seed-set/flowering head) in the two studies that have evaluated this trait (Fig. 1). There are a number of reasons why variation in flowering time might directly lead to a variation in fertility, such as seasonal environmental fluctuation in temperature, humidity and pollen availability. However, it is also possible that closely linked but distinct genetic determinants may also be involved. By using more detailed comparative genetics to analyze these regions on C4 and C7, clarification of the underlying genetic control and subsequent identification of candidate genes is possible (Table 1).

Flowering and Fertility QTL on *Lolium* C4

Using common markers to align the genetic maps from the different *Lolium/Festuca* populations (Fig. 1 and Table 1), the five studies that identified flowering QTL on C4 seemed to identify two distinct regions. One region, directly associated with a vernalization response, co-segregates with the position of the *L. perenne* orthologue of the Triticeae gene *VRN-1*, a gene known to have a role in the vernalization response (Jensen et al. 2005). A second possible candidate gene mapping to this region is *CK2α*, the gene which determines the heading-date QTL HD6 in rice and which is also known to be a determinant of the flowering response in *Arabidopsis* (Shinozuka et al. 2005). There is, so far, no obvious candidate gene underlying the region associated with the second flowering QTL on *Lolium* C4, though comparative genetics suggests that there are putatively equivalent rice QTL associated with this syntenic region on rice C3.

A significant fertility QTL was only reported in one *L. perenne* study on C4 (study A, see Fig.1 and Table 1) and, while the heading-date and fertility QTL were co-incident to a certain extent, the regions of major effect were separated by c. 10 cM, suggesting distinct genetic control. This fertility QTL co-located with a pollen viability QTL in the same population and as fertility was evaluated under open-pollinated conditions, pollen viability was unlikely to have had a direct effect on fertility. Thus, if a single effective gene underlying this QTL influences both seed-set and pollen viability, it may be a gene that has an overall effect on reproductive development. Comparative analysis of the syntenic region of rice C3 identifies a number of QTL for fertility-associated traits (e.g. spikelet-fertility and grain yield) and, interestingly, it also contains a fine-mapped locus, *S33(t)*, which determines pollen sterility (Jing et al. 2007). When the underlying gene at *S33(t)* is identified, this will represent a candidate gene for pollen viability and, possibly, overall fertility on *Lolium* C4.

Flowering and Fertility QTL on *Lolium* C7

Four of the *Lolium/Festuca* QTL studies identified QTL for flowering in a similar region on C7 (Fig. 1 and Table 1). It has been established that this region of *Lolium* C7 contains two candidate genes, *Hd3a/FT* and *Hd1/CO*, which can have significant effects on flowering phenotype and the *L. perenne* LD study identified significant LD associated with both these genes in relation to flowering time (Armstead et al. 2004, 2005; Skøt et al. 2007). The syntenic region of rice C6 also contains these candidate genes and associated flowering QTL, heading-date 3 (*Hd3*) and 1 (*Hd1*). While direct experimental evidence that these genes determine these QTL in *Lolium/Festuca* is still lacking, on-going characterization of allelic variation and selective crossing should help to clarify this situation.

Fertility QTL were reported for this same region of *Lolium* C7 for both the populations that evaluated this trait (Fig. 1 and Table 1). In contrast to C4, there was no clear separation between the flowering and fertility main effects in terms of genetic position and it is possible that variation in fertility was directly determined by variation in flowering time in both studies. However, comparative mapping between *L. perenne* and rice indicates that a rice locus, $S5^n$ (Ji et al. 2005), alleles of which can influence fertility, particularly in *indica* × *japonica* rice hybrids, may also be present in this region of *L. perenne* and the potential effect of such a gene cannot be discounted in *Lolium/Festuca*. As with *S33(t)* on rice C3, fine mapping and positional cloning approaches should identify the gene underlying $S5^n$ on rice C6, thus allowing for direct evaluation in grass populations.

Conclusion

Flowering and fertility in ryegrasses and fescues are important traits due to their effects on quality aspects of forage, turf and seed production and, thus, economic and environmental sustainability. QTL studies based on different *Lolium/Festuca* populations indicate that there are a number of genetic components to the control of flowering and fertility, though particular genomic regions seem to be important across populations and environments. Common models for the genetic control of flowering in other crop and model species suggests similar mechanisms may be active in *Lolium/Festuca* and comparative genetics can be usefully applied in identifying candidate genes for *Lolium/Festuca* flowering QTL. While the characterization of the genetic control of flowering and fertility is in its early stages in *Lolium/Festuca*, similar experimental approaches to those used in other crop and model species, based upon and understanding of comparative genetics and genomics, are likely to prove useful.

Acknowledgements

This work was supported by the BBSRC, UK.

References

Armstead IP, Turner LB, Farrell M, Skøt L, Gomez P, Montoya T, Donnison IS, King IP, Humphreys MO (2004). Synteny between a major heading-date QTL in perennial ryegrass (*Lolium perenne* L.) and the *Hd3* heading-date locus in rice. Theor Appl Genet 108:822–828

Armstead IP, Skøt L, Turner LB, Skøt K, Donnison IS, Humphreys MO, King IP (2005) Identification of perennial ryegrass (*Lolium perenne* L.) and meadow fescue (*Festuca pratensis* Huds.) candidate orthologous sequences to the rice *Hd1(Se1)* and barley *HvCO1* CONSTANS-like genes through comparative mapping and microsynteny. New Phytol 167:239–247

Cockram J, Jones H, Leigh FJ, O'Sullivan D, Powell W, Laurie DA, Greenland AJ (2007) Control of flowering time in temperate cereals: genes, domestication, and sustainable productivity. J Exp Bot 58:1231–1244

Ergon Å, Fang C, Jørgensen Ø, Aamlid TS, Rognli OA (2006) Quantitative trait loci controlling vernalisation requirement, heading time and number of panicles in meadow fescue (*Festuca pratensis* Huds). Theor Appl Genet 112:232–242

Inoue M, Gao ZS, Hirata M, Fujimori, M, Cai HW (2004) Construction of a high-density linkage map of Italian ryegrass (*Lolium multiflorum* Lam.) using restriction fragment length polymorphism, amplified fragment length polymorphism, and telomeric repeat associated sequence markers. Genome 47:57–65

Jensen LB, Andersen JR, Frei U, Xing YZ, Taylor C, Holm PB, Lübberstedt TL (2005) QTL mapping of vernalization response in perennial ryegrass (*Lolium perenne* L.) reveals co-location with an orthologue of wheat *VRN1*. Theor Appl Genet 110:527–536

Ji Q, Lu JF, Chao Q, Gu MH, Xu ML (2005) Delimiting a rice wide-compatibility gene *S-5^n* to a 50 kb region. Theor Appl Genet 111:1495–1503

Jing W, Zhang W, Jiang L, Chen L, Zhai H, Wan J (2007) Two novel loci for pollen sterility in hybrids between the weedy strain Ludao and the *Japonica* variety Akihikari of rice (*Oryza sativa* L.). Theor Appl Genet 114:915–925

Shinozuka H, Hisano H, Ponting RC, Cogan NOI, Jones ES, Forster JW, Yamada T (2005) Molecular cloning and genetic mapping of perennial ryegrass casein protein kinase 2 alpha-subunit genes. Theor Appl Genet 112:167–177

Skøt L, Humphreys MO, Armstead I, Heywood S, Skøt KP, Sanderson R, Thomas ID, Chorlton KH, Hamilton NRS (2005) An association mapping approach to identify flowering time genes in natural populations of *Lolium perenne* (L.). Mol Breed 15:233–245

Skøt L, Humphreys J, Humphreys MO, Thorogood D, Gallagher J, Sanderson R, Armstead I, Thomas I (2007) Association of candidate genes with flowering time and water soluble carbohydrate content in *Lolium perenne*. Genetics 177:535–547

Thorogood D, Kaiser WJ, Jones JG, Armstead I (2002) Self-incompatibility in ryegrass 12. Genotyping and mapping the *S* and *Z* loci of *Lolium perenne* L. Heredity 88:385–390

Thorogood D, Armstead IP, Turner LB, Humphreys MO, Hayward MD (2005) Identification and mode of action of self-compatibility loci in *Lolium perenne* L. Heredity 94:356–363

Yamada T, Jones ES, Cogan NOI, Vecchies AC, Nomura T, Hisano H, Shimamoto Y, Smith KF, Hayward MD, Forster JW (2004) QTL analysis of morphological, developmental, and winter hardiness-associated traits in perennial ryegrass. Crop Sci 44:925–935

Improving Selection in Forage, Turf, and Biomass Crops Using Molecular Markers

E. Charles Brummer[1,3] and Michael D. Casler[2]

[1]Center for Applied Genetic Technologies, Crop and Soil Science Department, University of Georgia, 111 Riverbend Rd., Athens, GA 30602, USA
[2]USDA-ARS, U.S. Dairy Forage Research Center, Madison, WI 53706-1108, USA
[3]Corresponding author, brummer@uga.edu

Abstract. Selection of improved forage, turf, and bioenergy crops is optimized if measuring the phenotype of interest is rapid, inexpensive, and repeatable. Phenotyping remains the most difficult issue to resolve for many important traits, including biomass yield, abiotic stress tolerance, and long-term persistence. The identification of molecular markers may augment phenotypic selection if markers are identified that are closely linked to or at genes controlling the traits of interest. Simply inherited traits can be easily manipulated with marker assisted selection (MAS), but using markers in more complex situations requires additional thought. In this paper, we put the use of molecular markers into the context of typical perennial forage and turf breeding programs. Identifying markers based on bi-parental mapping populations is likely not the best way to implement a MAS program, although this approach is useful to introgress alleles from wild germplasm. Instead, a more practical approach may be the use of association mapping, measuring both phenotypes and markers directly on the plants in the breeding nursery. Complications of this method include the limited amount of information on linkage disequilibrium that is available for breeding populations, but the increasing availability of gene identification methods and the use of single nucleotide polymorphism (SNP) markers may enable the use of association mapping in many cases. Applying the information to breeding may be done to assist selection, to prescreen plants to determine those on which field-based phenotypic data will later be collected, and to make rapid off-season selections. The practical applications of markers to the breeding programs are discussed.

T. Yamada and G. Spangenberg (eds.), *Molecular Breeding of Forage and Turf,*
doi: 10.1007/978-0-387-79144-9_18, © Springer Science + Business Media, LLC 2009

"There have been many attempts at indirect selection for DMY [dry matter yield], but none have been really successful."
Wilkins and Humphreys (2003)

Introduction

In the words of the eminent forage and turfgrass breeder Glenn Burton, "Increased yield is the 'bottom line' in most plant breeding programs" (Burton 1982). Nevertheless, forage yield in most species has risen slowly, if at all, (e.g., Wilkins and Humphreys 2003) and the gains that have been seen in some crops may have more to do with improvement in secondary traits, such as disease resistance, than in yield per se (e.g., Lamb et al. 2006). Some reasons for this lack of progress have been discussed previously (Brummer 2005). The purpose of this paper is to discuss breeding methods that may lead to an improvement in yield or other quantitatively inherited traits and ways that both selection methodology and molecular marker-assisted selection (MAS) can be fine-tuned to make the breeding program more effective and efficient.

When thinking about MAS in forage, turf, or biomass crops, we need to differentiate between two possible uses. First, markers may be used to manipulate traits controlled by single genes or by transgenes. We will not discuss this use here because MAS is straightforward in these situations, and except for the complexity introduced by polyploidy in many forage crops, the improvement of these traits using markers to facilitate introgression or accelerate backcrossing is unquestioned.

Instead, we will address the second use of markers, namely the use of markers within a recurrent selection program to improve the population for yield or other complex traits. The structure of the paper will be as follows: first, we will address aspects of the selection methods used to improve yield and attempt to highlight the best approach for this goal. Second, we will discuss the complexities surrounding the identification of molecular marker – trait associations that will be useful in the context of a breeding program. And finally, we will try to merge the first two aspects into a consideration of how markers could be employed in a recurrent selection breeding program.

Recurrent Selection

The main goal in a typical forage breeding recurrent selection program is to *increase the frequency* of desirable alleles (and allele combinations) within a population to effect trait improvement. In other words, the goal of recurrent selection is genetic gain, which simply put, is the improvement of a population for a given trait or suite of traits. Genetic gain is influenced by two main components (1) the fraction of the population that is selected and (2) the ability of individuals in the population to transmit their phenotype to their progeny. Thus,

$$\Delta G = h^2 S \tag{1}$$

where ΔG is genetic gain (i.e., response to selection), h^2 is the narrow sense heritability, and S is the selection differential (Falconer and Mackay 1996). Narrow sense heritability, the ratio of additive genetic variance (σ_A^2) to phenotypic variance (σ_P^2), describes the amount of heritable variance that is present in the population. A trait with high narrow sense heritability is expected to respond readily to selection. The additive genetic variance is a property of the population, and the phenotypic variance includes genetic (both additive and non-additive), environmental, and genotype × environment components. The latter two components can be decreased by increasing the number of replications and environments in which the breeding material is tested, and by employing experimental designs or spatial analyses that reduce error variance. Thus, the narrow sense heritability is not a static feature for a given trait, but can vary depending on the population being evaluated and the characteristics of the evaluation nursery. The selection differential S is the product of the intensity of selection k and the phenotypic standard deviation of the population (Falconer and Mackay 1996).

A third factor must also be considered when comparing breeding methods for genetic gain: the parental control c. Parental control indicates whether the actual plants to be recombined include alleles only from the selected fraction of the population or if they also include alleles from non-selected plants as well. For example, in a half-sib family selection program, the best performing families are identified. If the parents of the selected half-sib lines are recombined, then all progeny of the recombination block will receive alleles only from selected plants. In contrast, plants from within the best half-sib families could be recombined instead of the parents of the family. These within-family plants will only have half of their alleles from the selected plants (those from their maternal parent), because their paternal parent could have been any of the original plants under evaluation. Thus, only the maternal parent is controlled in the recombination block, so the progeny will only receive ½ of their alleles from the selected plants. Therefore, $c = ½$ if individual plants from within families are recombined, but $c = 1$ if original parental plants are recombined. Based on the above considerations, (1) can be rewritten generically as follows:

$$\Delta G = \frac{kc\sigma_A^2}{\sigma_P} \qquad (2)$$

In general, two main types of recurrent selection methods can be implemented (1) phenotypic recurrent selection [PRS] based on individual plant evaluation (e.g., mass selection) or (2) genotypic recurrent selection based on progeny testing (e.g., half-sib family selection [HSFS]) (Fehr 1987). Family selection methods assess a genotype's genetic merit by testing that genotype's progeny in replicated, multi-location trials. This testing results in a higher heritability for the trait of interest than is possible with PRS, and typically leads to better genetic gain.

Despite the importance of forage yield, few experiments have actually measured the gain from selection for yield per se, and virtually none has compared multiple methods (Table 1). Selection for yield has generally resulted in yield improvement, but not always. In particular, individual plant selection using some form of PRS, including Burton's (1982) recurrent restricted phenotypic selection (RRPS), often results in no gain. Thus, empirical results suggest that family selection is the preferred method to increase forage yield, but the paucity of actual selection experiments does not clearly indicate the best type of family selection to be conducted.

Half-sib family recurrent selection is probably the most commonly used family-based selection method for a number of reasons. First, production of half-sib family seed is straightforward using a polycross nursery. The desired plants are intercrossed in isolation, typically in a replicated design which promotes random mating among genotypes. Seed from each maternal genotype represents a half-sib family because all the progeny have the same maternal parent, but the paternal contribution comes from the members of the entire population. In a polycross of replicated parental clones, considerable quantities of seed can be produced, facilitating the evaluation of seeded rows or even swards, more closely simulating actual production conditions than individual spaced plants. And finally, half-sib families are preferred over other family types because they are easier to produce than a

Table 1 Progress from selection for forage yield in various grass and legume species using different selection methods

Species[a]	Method[b]	Cycles	Gain	References
Orchardgrass	AWFS	1	Yes	Casler et al. (2002)
Orchardgrass	RRPS	3	No	Shateryan et al. (1995)
Pensacola bahiagrass	RRPS	22	Yes	Burton and Mullinex (1998)
Perennial ryegrass	PRS	3	Yes/no[c]	Hayward (1983); Hayward and Vivero (1984)
Perennial ryegrass	PRS+HSFS	4	Yes	Wilkins and Humphreys (2003)
Rye	RRPS[d]	4	Yes	Bruckner et al. (1991)
Smooth bromegrass	PRS	1	No	Carpenter and Casler (1990)
Switchgrass	PRS	1	Yes	Missaoui et al. (2005)
Switchgrass	HSPT	1	Yes	Rose et al. (2007)
Timothy	RRPS	3	No	Shateryan et al. (1995)
Wheat	PRS[d,e]	1	Yes	Uddin et al. (1993)
Alfalfa	FSFS[f,g]	2	Yes	Katepa-Mupondwa et al. (2002)
Alfalfa	PRS[f]	1	Yes	Salter et al. (1984)
Cicer milkvetch	HSPT[d]	1	Yes	Townsend (1981)

[a]Orchardgrass = *Dactylis glomerata* L.; Pensacola bahiagrass = *Paspalum notatum* Flugge var. *saure* Parodi; Perennial ryegrass = *Lolium perenne* L.; Rye = *Secale cereale* L.; Smooth bromegrass = *Bromus inermis* L.; Switchgrass = *Panicum virgatum* L.; Timothy = *Phleum pratense* L.; Wheat = *Triticum aestivum* L.; Alfalfa = *Medicago sativa* L.; Cicer milkvetch = *Astragalus cicer* L.

[b]PRS=phenotypic recurrent selection; RRPS=recurrent restricted phenotypic selection (Burton 1982); AWFS=among-and-within-half-sib family selection; HSPT=half-sib progeny test selection; HSFS=half-sib family selection, recombination unit unspecified; FSFS=full-sib family selection

[c]Progress under spaced-planted but not sward conditions, where yield actually decreased after selection

[d]Based on visual selection

[e]Selection among lines in F_4 populations

[f]Selection not based on field evaluation

[g]Recombination based on remnant seed of selected full-sib families

series of paired crosses for full-sib progeny, and they avoid the well-known inbreeding depression and self incompatibility problems of many forage species that prevent or severely limit the production of selfed progeny families.

In half-sib family selection, families are evaluated in plots of spaced plants, seeded rows, or swards. Variations in HSFS are largely based on which plants are recombined. The two major alternatives are (1) the half-sib progeny test (HSPT), in which the parents of the selected families are recombined, and (2) among and within family selection (AWFS), in which selected plants from within the progeny plots are recombined (Vogel and Pedersen 1993; Fig. 1). Selection within families may be based on the same trait as that measured on the entire row (e.g., each plant is measured for yield independently, and the sum of yields per plot is the family value for that replication) or on a trait that is (thought to be) correlated with the trait measured among families (e.g., yield could be measured on families, but visual vigor ratings used as a proxy for yield within families). This latter situation may be common if measuring a trait is particularly difficult, time-consuming, or onerous on individual plants and a related trait can be more easily evaluated with the resources available. If this correlation is strong – that is, visual selection accurately predicts yield – then within family selection based on that criterion will be useful. However, this will usually be unlikely because if it were highly correlated, then selection for the more easily measured trait would be done without measuring yield at all, and this is generally not the case. A much fuller description of selection within families for correlated traits and the conditions that need to be met to make this a viable selection method are discussed in Casler and Brummer (2008).

In the HSPT, families are measured for the trait of interest and selection is based on family performance. However, only ¼ of the additive genetic variation in the population is partitioned among families, and therefore,

$$\Delta G = \frac{k(1)\frac{1}{4}\sigma_A^2}{\sigma_P} \tag{3}$$

Because ¾ of the additive genetic variation in the population resides within families, selecting within families has intuitive merit. Thus,

$$\Delta G = \frac{k_a(0.5)\frac{1}{4}\sigma_A^2}{\sigma_{P_a}} + \frac{k_w(0.5)\frac{3}{4}\sigma_A^2}{\sigma_{P_w}} \tag{4}$$

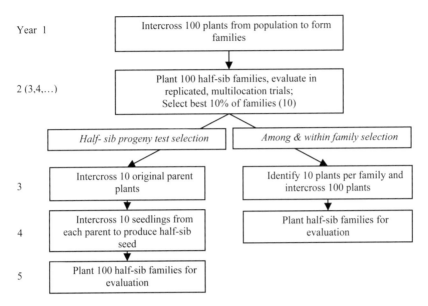

Fig. 1 Schematic diagram of half-sib progeny test selection and among and within family selection

where the subscripts a and w denote among and within families, respectively. Note that polysomic inheritance slightly shifts the balance in favor of the latter, since more variation is partitioned within and less among families. This is due to the presence of nonadditive sources of variance in these equations in polysomic polyploids (Gallais 2003, p. 200).

Vogel and Pedersen (1993) concluded that AWFS was about 3.6 times as efficient as HSPT. To reach their conclusion, they made two assumptions that are unlikely to be met in practice (1) that the selection intensities are the same for among and within family selection and (2) that the phenotypic variance among families equals that within families. In reality, the selection intensity among families may be greater than that used within families (although this need not be the case), and the phenotypic variance among (unreplicated) plants within families will almost certainly be considerably larger than that among (replicated) families.

In order for AWFS to be better than HSPT, the second term in (4) would need to be larger than the gain added by controlling both parents in the HSPT ($c = 1$ in (3) vs. $c = \frac{1}{2}$ in (4)). Let's assume that the trait is yield, that the among-family selection intensity is 10%, and that we set the additive genetic variance equal to 10. The ratio of gain using the HSPT to gain using AWFS is shown in Table 2 for a series of heritabilities both within and among families for the trait under consideration, and under two different selection intensities within families. AWFS improves relative to HSPT as the selection intensity within families increases and as the within-family heritability rises relative to the among-family heritability.

Selection intensity may not be strong within families, particularly if spaced-plants are used. Assume we evaluate 100 families per cycle; if we select the ten best families (10%), we will need to identify ten plants per family to recombine in order to result in 100 families for the next cycle. Thus, if the selection intensity within families is also 10%, we would need to evaluate a total of 100 plants per family (across replications and locations), or, in other words, have a nursery size of 10,000 plants. The question facing the breeder is whether resources are better spent evaluating one population with 100 plants per family, two populations with 50 plants per family, or some other arrangement. These considerations will need to be answered on a case-by-case basis, but the result is that different methods may be more effective in some situations than in others.

Table 2 Ratio of gain from half-sib progeny test (HSPT) selection to gain from among and within family selection (AWFS) based on different heritabilities for the trait of interest within and among families when selecting either (a) 50% or (b) 20% of plants within families. For convenience, additive variance was assigned a value of 10

(a) 50%		Heritability within families						
		0.01	0.05	0.10	0.20	0.30	0.40	0.50
Heritability among families	0.50	1.67	1.40	1.24	1.07	0.97	0.90	0.84
	0.60	1.70	1.43	1.28	1.12	1.02	0.94	0.89
	0.70	1.72	1.46	1.32	1.15	1.05	0.98	0.93
	0.80	1.73	1.49	1.35	1.19	1.09	1.02	0.96
	0.90	1.75	1.51	1.37	1.21	1.12	1.04	0.99
	1.00	1.76	1.53	1.40	1.24	1.14	1.07	1.02

(b) 20%		0.01	0.05	0.10	0.20	0.30	0.40	0.50
Heritability among families	0.50	1.49	1.14	0.96	0.79	0.70	0.64	0.59
	0.60	1.53	1.18	1.01	0.84	0.74	0.68	0.63
	0.70	1.55	1.22	1.05	0.88	0.78	0.71	0.66
	0.80	1.58	1.25	1.08	0.91	0.81	0.74	0.69
	0.90	1.60	1.28	1.11	0.94	0.84	0.77	0.72
	1.00	1.61	1.30	1.14	0.96	0.86	0.79	0.74

Note that an additional one to two seasons will be required for HSPT to produce families for the next cycle of selection after the parental plants are recombined. In some species, recombination could be done in the greenhouse so that the additional season would not affect the gain per year, but in other situations, particularly where plants need to be vernalized, this may not be possible, and the gain *per year* may be improved for AWFS relative to HSPT in many cases.

In summary, AWFS ranges from somewhat better to somewhat worse than HSPT on a per-cycle basis depending on the particulars of the breeding program. This concept is explored in more detail elsewhere under a larger range of conditions and for full-sib selection as well as half-sib selection (Casler and Brummer 2008). Thus, the large advantage of AWFS compared to HSPT as shown in Vogel and Pedersen (1993) is not true under any realistic scenarios. However, although HSPT may be as good as or better than AWFS in some cases, it requires saving the parental plants, which many breeding programs do not or cannot do. And finally, for the purposes of this paper, the pendulum may swing decidedly toward AWFS with the incorporation of molecular markers into the breeding program, as we discuss below.

Identifying Marker-Trait Associations

Marker-trait associations have been typically identified by developing a segregating population, such as the progeny of a biparental cross, which is evaluated for agronomic traits in replicated field trials and scored for molecular marker genotypes. These types of populations have been successfully used to genetically map quantitative trait loci (QTL) for a variety of traits in numerous forage crops, such as yield, plant height, and regrowth in alfalfa (Robins et al. 2007a,b).

Although this strategy has worked well to begin the process of localizing particular QTL, it suffers from two potential problems when placed in the context of improving a population's mean performance. First, the parents chosen for QTL mapping typically are *not* selected randomly from a single population, but rather, they are selected based on their differences for particular traits of interest. Thus, the first question is whether QTL identified in the biparental population are even segregating in the breeding population undergoing recurrent selection. Second, and perhaps more importantly, the population under selection is composed of a heterogeneous mixture of individuals each of which is heterozygous at many loci, and collectively the population may have many alleles at any given locus. Therefore, the few QTL identified in any single biparental mapping population are unlikely to represent all the major QTL within a given breeding population. Thus, while QTL identified in a given population *may* be useful as markers in a marker-assisted recurrent selection program, it is likely that they will not be sufficient for the job.

The alternative approach is to use association mapping (Hirschhorn and Daly 2005; Forster et al. 2007), whereby markers are integrated directly into the breeding program, with genotypes being determined on the parents of the families being evaluated for phenotypic traits. This enables the direct association of markers with phenotypes in the breeding population.

The use of association mapping faces several hurdles that may limit its utility. First, and most importantly, it revolves around the extent of linkage disequilibrium (LD) in the population. LD occurs when alleles at two loci appear together more commonly (or more rarely) than would be expected based on random association (Hartl and Clark 2007). Loci that are more closely located on a chromosome (or that have a low level of recombination between them) will be more likely to be in LD than loci located further apart. The genetic distance over which LD is present along a chromosome depends on the population history: populations that have undergone many rounds of recombination will show less LD than populations that have had little recombination.

In experiments examining large cross-sections of germplasm, LD typically decays over the length of a gene (e.g., Remington et al. 2001). This means that to identify markers close enough to the genes controlling the trait and with sufficient LD to show the association, the marker density needed to identify a significant number of the loci controlling a complex trait in the population would need to be extremely high (e.g., The Wellcome Trust Case Control Consortium 2007). In most forage crops, this level of marker saturation is simply not feasible at this time. Not only are insufficient numbers of markers available, but the cost of evaluating tens or hundreds of thousands of markers in a population is prohibitive. However, association mapping successfully identified flowering time genes in natural populations of perennial ryegrass (Skøt et al. 2005), suggesting that the utility of the approach will vary depending on species and plant materials used.

In general, breeding populations are not broad cross-sections of a crop's germplasm base, but rather represent a highly selected subset of germplasm.

These populations likely have undergone recent bottlenecks, which potentially create an initially high level of LD, and subsequently, the population has probably had relatively few recombination events. Furthermore, selection itself creates LD by favoring chromosome blocks that possess larger numbers of favorable alleles (Comstock 1996). Therefore, in breeding populations, LD may extend over significantly large regions, perhaps as much as several centimorgans. If this is the case, then the use of a reasonable number of markers, even simple sequence repeat (SSR) markers, would enable the detection of marker-trait associations. Support for this idea has recently been shown in sugarcane, where amplified fragment length polymorphism (AFLP) and SSR markers associated with disease resistance have been identified in a set of breeding clones (Wei et al. 2006).

Particular genes, so-called *candidate genes* (Thornsberry et al. 2001; Forster et al. 2007), may be expected to be associated with the trait of interest. Candidate genes can be targeted for marker development, especially based on single nucleotide polymorphisms (SNP), because they may be expected to underlie QTL controlling the trait being studied. These markers would be expected to be in LD with any functional polymorphism in the candidate gene, and ideally are the causative polymorphism itself (Andersen and Lübberstedt 2003). Candidate gene markers are useful for association mapping in any population, but would be particularly valuable in populations that don't have extensive LD and for crops that have few other markers available. Mapping candidate genes doesn't guarantee identification of important QTL, of course, but it does raise the likelihood of finding useful associations if a sufficient density of whole-genome markers cannot be realized.

We are currently conducting an association mapping experiment in tetraploid alfalfa. The population consists of approximately 200 individuals that were clonally propagated and planted into replicated field trials at four locations and were measured for yield. Concurrently, the individuals are being genotyped using SSR markers selected to cover the entire alfalfa genome. Based on very preliminary results of data from only a single location and fewer than 40 SSR markers, we were able to identify several markers associated with yield and stem cell wall composition (Brummer unpublished results). The alleles amplified by these markers had both positive and negative effects on yield. Importantly, several of these markers had been previously associated with yield in a bi-parental mapping population (Robins et al. 2007a,b). While these results are quite preliminary, they suggest that LD extends for sufficiently long distances in this breeding population to identify at least some marker-trait associations. If this is the case in many breeding populations, then a relatively small number of markers will be useful in detecting QTL for yield. However, because the amount of total variation for any phenotypic trait that can be accounted for by the markers is relatively small, breeders will need to pyramid multiple QTL for the trait of interest within a single population.

Using Markers in Recurrent Selection

Given that we can identify markers associated with traits of interest, either in biparental mapping populations or by association mapping, how do we effectively integrate them into a recurrent selection scheme to improve the rate of genetic gain? The problem is not trivial. Because we are attempting to improve mean performance of an outcrossing population, rather than developing an improved inbred line, MAS is necessarily more complicated than in many of the major row crops.

Before we get too far into the discussion, we need to note that alleles are most valuable when they are low in frequency. As allele frequency increases in the population, the value of the allele for improving the population diminishes until the allele is fixed, *when it has no value at all!* Breeding value represents the ability of a particular individual to pass its phenotype (and hence its alleles) to the progeny. The breeding value of an individual will be lower when its alleles are common than when they are rare in the population (Falconer and Mackay 1996). As we concentrate desirable QTL alleles through MAS, we will need to remap the population to identify new sets of marker alleles that explain the remaining variation in yield (or other traits). The result is a sequential ratcheting of yield by identifying desirable alleles, moving them toward fixation, and then identifying a new set of alleles.

How do we integrate markers into the recurrent selection scheme? Assume we have developed our families for phenotypic evaluation in the field. We can assay markers on the parental plants and then conduct an association analysis as described above. The success of association mapping depends not only on having sufficient marker density, but also in having robust phenotypic data. For traits like yield, these data must come from replicated field trials across locations, in order to have sufficient heritability to detect marker-trait associations.

Because we have good phenotypic data and because markers are only going to explain a portion of the phenotypic variation, selection of the best families should be done based on the phenotypic data alone. That is, the marker-trait associations are *of no use* in choosing families to recombine, because we can select directly on phenotypes and make better choices. However, the marker data will be useful in a different way, and one that we typically don't consider in forage breeding. The marker data provide information on the relatedness of the various parental clones, and thus, we could use these relationships to generate a better estimate of the phenotypic value of individual parents, much in the same way that animal breeders or plant breeders who maintain pedigree information have done. The methods for doing this are beyond the scope of this paper, but further work on this aspect of markers in forage programs is warranted.

In the HSPT, markers could be used to select the plants to be used in forming families for the next cycle of selection. After a polycross, a number of individuals from each maternal plant need to be intercrossed to generate families for evaluation. These are typically selected randomly, but with markers, the requisite number could be selected as seedlings from a larger population based on the markers identified in the previous cycle. This would help to concentrate marker alleles at this point, and for marker alleles that are not present in particular parental genotypes, increase the chances that they will be present in that parent's progeny. Using our example here, perhaps 100 individuals from each of the ten parents could be genotyped with five markers, for a total of 1,000 plants and 5,000 marker assays, a reasonable number for a moderately sized breeding program.

Markers could be put to more constructive use in AWFS by helping to select plants that have desirable alleles within the families that have the most desirable phenotypes. Heritability on a single plant basis is typically low for traits like yield, so using markers would likely help make better choices than phenotype alone. By using markers within families, we can select individuals that have particular marker alleles that are not present in that family's maternal parent. By choosing plants with particular markers in these families, we can potentially add more desirable alleles to the next generation than would be possible with HSPT, because all plants in the population could then contribute desirable marker alleles and not just the selected parents. If we select ten plants from desirable families out of a nursery of 50 plants per family, then we only have 500 plants to genotype. We only need to genotype these 500 plants with the small number of markers that were significantly associated with the trait, which is a number that will be feasible for most breeding programs in the very near future, if not already.

With any method, an off-season selection cycle could be performed based only on markers to further concentrate desirable alleles before another field season. Depending on the species, this may be quite possible, and would essentially add a cycle of selection without increasing the conventional cycle time used in traditional selection. The advantage of this extra cycle of selection and recombination depends on the ability of the markers alone to select for the phenotype desired, which may not always be the case (Moreau et al. 2004).

Some cautionary notes are warranted. First, we want to concentrate desirable marker alleles, but our initial selection of families is based on phenotypic, not genotypic, data. Thus, the parental plants of the selected families may not carry the desirable marker alleles at all loci, and in the worst case, may not have any desirable alleles at all. Part of the problem is the marker allele frequency in the population. For example, a marker may be significantly associated with yield, but it may also only be present in 5% of the parental clones. If we select 10%, clearly at least half will not have the allele. Another aspect of the problem is that markers only explain a fraction of the variation in the trait. If more markers could be added, thereby explaining a larger percentage of the variation, then selected parents would be more likely to have at least some of the desirable alleles.

Nevertheless, we face a very real numbers problem in normal breeding programs. For instance, consider two unlinked genes, each with two alleles at equal frequency in the population. Only 6.25% of the population will be homozygous for both loci and only 56% will have at least one allele at each locus. The problem becomes exponentially worse as we add more marker loci on which to select at each stage in the breeding program. This leads back to the somewhat discouraging conclusion that many parental clones will not have all the alleles that are being selected, and thus, moving these alleles to fixation will take time. This suggests also that the best strategy would be to not focus on rare alleles but rather to concentrate on moving a few (e.g., 2–5) alleles that are already relatively common in the population toward fixation at a time, and picking up subsequent alleles in later cycles of selection.

Conclusions

First, recurrent selection programs need to be designed so that the traits of interest – e.g., yield – can be improved. This will require the adoption of family-based selection methods, in general, in order to make genetic gain. However, the choice of selection method and recombination unit need to be developed in light of anticipated genetic gains in biological characteristics of the species, and physical restrictions on the breeding program and its facilities.

Second, genetic markers may be applied to the breeding program at several stages, but the utility of the markers may not be obvious. The use of markers to generate a relationship matrix among the parental clones will lead to better estimation of phenotypic value, and hence to better gain. Markers can be used to help select plants within families, making AWFS more effective, and can be used to conduct marker-only selection in off-season nurseries, potentially accelerating genetic gain. Finally, using markers in recurrent selection will require mapping, selection, and remapping as the first alleles undergoing selection approach fixation.

The open question at this time is whether adding markers into the system will speed genetic gain, and if it does, whether it is worth the cost.

References

Andersen JR, Lübberstedt T (2003) Functional markers in plants. Trends Plant Sci 8:554–560

Bruckner PL, Raymer PL, Burton GW (1991) Recurrent phenotypic selection for forage yield in rye. Euphytica 54:11–17

Brummer EC (2005) Thoughts on breeding for increased forage yield. In: O'Mara FP, Wilkins RJ, 't Mannetje L, Lovett DK, Rogers PAM, Boland, TM (eds) XX International Grassland Congress: Offered Papers. Wageningen Academic Publishers, Wageningen, the Netherlands, p. 63

Burton GW (1982) Improved recurrent restricted phenotypic selection increases bahiagrass forage yields. Crop Sci 22:1058–1061

Burton GW, Mullinex BG (1998) Yield distributions of spaced plants within Pensacola bahiagrass populations developed by recurrent restricted phenotypic selection. Crop Sci 38:333–336

Carpenter JA, Casler MD (1990) Divergent phenotypic selection response in smooth bromegrass for forage yield and nutritive value. Crop Sci 30:17–22

Casler MD, Brummer EC (2008) Expected genetic gains for among-and-within-family selection methods in perennial forage crops. Crop Sci 48:890–902

Casler MD, Fales SL, McElroy AR, Hall MH, Hoffman LD, Undersander DJ, Leath KT (2002) Half-sib family selection for forage yield in orchardgrass. Plant Breed 121:43–48

Comstock RE (1996) Quantitative genetics with special reference to plant and animal breeding. Iowa State University Press, Ames, IA

Falconer DS, Mackay TFC (1996) Introduction to quantitative genetics, 4th edition. Longman, Harlow, England

Fehr W (1987) Principles of cultivar development. Vol. 1 Theory and technique. Macmillian, New York

Forster JW, Cogan NOI, Dobrowolski MP, Francki MG, Spangenberg GC, Smith KF (2007) Functionally-associated molecular genetic markers for temperate pasture plant improvement. In: Henry RJ (ed) Advances in plant genotyping. CABI Press, Wallingford, Oxford, UK

Gallais A (2003) Quantitative genetics and breeding methods in autopolyploid plants. INRA, Paris

Hartl DL, Clark AG (2007) Principles of population genetics. 4th edition, Sinauer Associates, Sunderland, MA

Hayward MD (1983) Selection for yield in *Lolium perenne*. I. Selection and performance under spaced plant conditions. Euphytica 32:85–95

Hayward MD, Vivero JL (1984) Selection for yield in *Lolium perenne*. II. Performance of spaced plant selections under competitive conditions. Euphytica 33:787–800

Hirschhorn JH, Daly MJ (2005) Genome-wide association studies for common diseases and complex traits. Nat Rev Genet 6:95–108

Katepa-Mupondwa FM, Christie BR, Michaels TE (2002) An improved breeding strategy for autotetraploid alfalfa (*Medicago sativa* L.). Euphytica 123:139–146

Lamb JFS, Sheaffer CC, Rhodes LH, Sulc M, Undersander DJ, Brummer EC (2006) Forage yield and quality of alfalfa cultivars released from the 1940s through the 1990s. Crop Sci 46:902–909

Missaoui AM, Fasoula VA, Bouton JH (2005) The effect of low plant density on response to selection for biomass production in switchgrass. Euphytica 142:1–12

Moreau LA, Charcosset A, Gallais A (2004) Experimental evaluation of several cycles of marker assisted selection in maize. Euphytica 137:111–118

Remington DL, Thornsberry JM, Masuoka Y, Wilson LM, Whitt SR, Doebley J, Kresovich S, Goodman MM, Buckler ES (2001) Structure of linkage disequilibrium and phenotypic associations in the maize genome. Nat Genet 98:11479–11484

Robins JG, Bauchan GR, Brummer EC (2007a) Genetic mapping forage yield, plant height, and regrowth at multiple harvests in tetraploid alfalfa (*Medicago sativa* L.). Crop Sci 47:11–16

Robins JG, Luth D, Campbell TA, Bauchan GR, He C, Viands DR, Hansen JL, Brummer EC (2007b) Mapping biomass production in tetraploid alfalfa (*Medicago sativa* L.). Crop Sci 47:1–10

Rose LW, Das MK, Fuentes RG, Taliaferro CM (2007) Effects of high- vs. low-yield environments on selection for increased biomass yield in switchgrass. Euphytica 156:407–415

Salter R, Melton B, Wilson M, Currier C (1984) Selection in alfalfa for forage yield with three moisture levels in drought boxes. Crop Sci 24:345–349

Shateryan D, Coulman BE, Mather DE (1995) Recurrent restricted phenotypic selection for forage yield in timothy and orchardgrass. Can J Plant Sci 75:871–875

Skøt L, Humphreys MO, Armstead I, Heywood S, Skøt KP, Sanderson R, Thomas ID, Chorlton KH, Sackville Hamilton NR (2005) An association mapping approach to identify flowering time genes in natural populations of *Lolium perenne* (L.). Mol Breed 15:233–245

The Wellcome Trust Case Control Consortium (2007) Genome-wide association study of 14,000 cases of seven common diseases and 3,000 shared controls. Nature 44:661–678

Thornsberry JM, Goodman MM, Doebley J, Kresovich S, Nielsen D, Buckler ES (2001) *Dwarf8* polymorphisms associate with variation in flowering time. Nat Genet 28:286–289

Townsend CE (1981) Breeding cicer milkvetch for improved forage yield. Crop Sci 21:363–366

Uddin N, Carver BF, Krenzer EG (1993) Visual selection for forage yield in winter-wheat. Crop Sci 33:41–45

Vogel KP, Pedersen JF (1993) Breeding systems for cross-pollinated perennial grasses. Plant Breed Rev 11:251–274

Wei X, Jackson PA, McIntyre CL, Aitken KS, Croft B (2006) Associations between DNA markers and resistance to diseases in sugarcane and effects of population substructure. Theor Appl Genet 114:155–164

Wilkins PW, Humphreys MO (2003) Progress in breeding perennial forage grasses for temperate agriculture. J Agric Sci 140:129–150

Interpretation of SNP Haplotype Complexity in White Clover (*Trifolium repens* L.), an Outbreeding Allotetraploid Species

Kahlil A. Lawless[1,3], Michelle C. Drayton[1,3], Melanie C. Hand[1,3], Rebecca C. Ponting[1,3], Noel O.I. Cogan[1,3], Timothy I. Sawbridge[1,3], Kevin F. Smith[2,3] Germán C. Spangenberg[1,3], John W. Forster[1,3,4]

[1]Department of Primary Industries, Biosciences Research Division, Victorian AgriBiosciences Centre, La Trobe University Research and Development Park, Bundoora, VIC 3083, Australia
[2]Department of Primary Industries, Biosciences Research Division, Hamilton Centre, Private Bag 105, Hamilton, VIC 3300, Australia
[3]Molecular Plant Breeding Cooperative Research Centre, Australia
[4]Corresponding author: e-mail: john.forster@dpi.vic.gov.au

Abstract. Single nucleotide polymorphisms (SNPs) within genic sequences provide the basis for functionally-associated genetic marker development. Gene-associated SNP discovery in white clover has been based on cloning and sequencing of PCR amplicons from parents and progeny of two-way pseudo-testcross mapping families. Target genes were selected from functional categories including phyto-hormone metabolism, nodulation, cell wall biosynthesis, metal binding, flavonoid biosynthesis and organic acid biosynthesis. Sequence alignments revealed haplotypic complexity that may be attributable to both paralogous gene structure and homoeologous sequence variation between sub-genomes. A high proportion of predicted allelic SNPs failed to verify in a Mendelian transmission test, confirming the prevalence of non-homologous variation. Incidence and frequency of haplotypes within and between genotypes was determined and interpreted in terms of models of sequence evolution and isolation. Methods for enhanced recovery of genome- and gene-specific sequences from white clover based on computational analysis, exploitation of large-insert DNA libraries and comparison with progenitor sequences are proposed and discussed.

T. Yamada and G. Spangenberg (eds.), *Molecular Breeding of Forage and Turf*,
doi: 10.1007/978-0-387-79144-9_19, © Springer Science + Business Media, LLC 2009

Introduction

Identification of sequence variation within coding and regulatory regions of genes permits the development of single nucleotide polymorphism (SNP) genetic markers. Gene-associated SNP loci can be evaluated for co-location on genetic maps with quantitative trait loci (QTLs) for putatively related agronomic traits (Forster et al. 2004; Spangenberg et al. 2005), as well as correlation with phenotypic diversity in association mapping strategies (Dobrowolski and Forster 2006). Verified haplotype–phenotype associations are then suitable for implementation in direct selection of superior allele content (Sorrells and Wilson 1997) in germplasm improvement programs. These strategies are likely to be of high value for species with complex varietal development systems based on multiple parent polycrosses, such as outbreeding forages (Dobrowolski and Forster 2006).

In vitro SNP discovery for perennial ryegrass (*Lolium perenne* L.) has been based on amplicon cloning and sequencing from the heterozygous parental genotypes of a genetic mapping population, followed by sequence alignment (Cogan et al. 2006). A minority (c. 25%) of predicted SNPs from this diploid species failed to show allelic segregation among F_1 progeny, and presumably arose from paralogous rather than homologous sequence alignments. The incidence of such effects is likely to increase for allopolyploid species, in which homoeologous sequence variation between sub-genomes is additional to paralogous sequence variation within gene families (Cronn and Wendel 1998; Somers et al. 2003). In vitro SNP discovery from allopolyploids has so far been confined to inbreeding plant species such as wheat (Bryan et al. 1999; Caldwell et al. 2004). Multiple sequence haplotypes obtained from single homozygous genotypes may be confidently attributed to non-homologous genes, and use of aneuploids such as wheat nullisomic-tetrasomic (NT) substitution lines can assign different haplotypes to specific chromosomes (Caldwell et al. 2004). A similar strategy has been used for in silico-predicted wheat SNPs (Somers et al. 2003), permitting identification of putative homoeologous sequence variants (HSVs).

White clover is an allotetraploid forage legume (2n = 4x = 32) thought to be derived from progenitors similar to the closely related contemporary species *T. occidentale* D.E. Coombe (Western clover) and *T. pallescens* Schreber (Ellison et al. 2006). High levels of intrapopulation genetic diversity (George et al. 2006) and intragenotype heterozygosity (Kölliker et al. 2001; Jones et al. 2003) have been observed in white clover. In addition, aneuploid lines for one-step assignment to chromosomes are not available. The combination of variability between paralogous, homoeologous and homologous sequences is likely to complicate in vitro SNP discovery and identification of HSVs.

The incidence of haplotype complexity in PCR-generated amplicons has been evaluated for a sample of white clover genes represented by full-length cDNAs. The selected genes correspond to a range of functional categories including phytohormone metabolism, nodulation, cell wall biosynthesis and metal binding, and with a special emphasis on flavonoid and organic acid biosynthesis. The results have been interpreted in terms of models for gene family structure and used to inform strategies for refined SNP discovery.

Methods and Results

In Vitro SNP Discovery Process

A total of 43 genes from a range of functional categories were selected for SNP discovery, including: acyanogenic β-glucosidase, auxin response factor, nodulin, chlorophyll a/b binding protein, peroxidase, urocanase, metallothionein, flavonoid biosynthesis genes (chalcone isomerise, chalcone synthase, phenylalanine lyase, etc.), and organic acid biosynthesis genes (citrate synthase, phosphoenolpyruvate carboxylase, and malate dehydrogenase). Locus amplification primer (LAP) pair sets were designed for each gene and used to obtain PCR products from the parents of the $F_1(Haifa_2 \times LCL_2)$ and $F_1(S184_6 \times LCL_6)$ genetic mapping families (Cogan et al. 2007). Amplicons were obtained from a total of 35 genes. Twelve clones were sequenced for each amplicon from each of the target genotypes. A total of 29.4 kb of consensus resequenced genomic DNA was obtained, at an average of 840 bp per template gene. The average length of the sequenced amplicons was 718 bp, with a range from 182 to 1,765 bp. A total of 4.2 kb was contributed by 38 predicted introns, occurring at an average frequency of one per 774.7 bp. Insertion-deletion (indel) events as large as 288 bp were observed, generally located within intronic sequences. A larger number of smaller predicted indels were also observed. High levels of haplotypic complexity were identified, in excess of the number expected for locus-specific amplification.

Putative SNPs within and between parental genotypes were identified for each predicted contig, and a total of 129 single nucleotide primer extension (SNuPe) interrogation primers from 34 contigs (an average of 3.8) were designed for verification. Polymorphism was assessed using the parental genotypes and selected F_1 progeny of the relevant mapping population. A total of 106 SNP assays failed to reveal clear allelic segregation, generally due to monomorphism. A total of 23 SNPs in 20 contigs (due to multiple SNPs in the TrACO2, TrCHId and TrPALb genes) revealed putative allelic variation. A total of 18% of the predicted SNPs consequently achieved primary verification.

Analysis of Haplotype Complexity

Due to the high level of attrition (c. 80%) observed during in vitro SNP discovery, further intensive analysis was performed on haplotype complexity in selected genes. The presence of two, four, or more haplotypes per genotype is compatible with amplification from both alleles of a single homologous locus, each of single copy homoeoloci in the sub-genomes of the allotetraploid, and paralogous gene structure within and between genomes, respectively. Chimeric products may also be generated by PCR-mediated recombination and can potentially confound the analysis. A schematic format was developed to represent haplotype-diagnostic base variants and respective frequencies, numbers of variants defining haplotypes, and presence within parental and progeny genotypes. Data for a representative gene is shown in Fig. 1.

Haplotype	Base position 1272	1275	1284	1299	1305	1308	1320	1356	1398	1449	1464	1476	1506	1515	1518	1551	1560	1608	1620	1650	1653
1	C	A	A	G	C	C	C	C	T	T	T	A	A	T	C	T	T	G	T	T	T
2	C	A	A	G	C	C	C	C	T	A	C	G	G	C	T	G	C	C	C	T	T
3	C	A	A	G	C	C	C	C	T	A	C	G	G	C	T	G	C	C	C	A	T
4	C	A	A	G	C	C	C	C	T	A	C	G	G	C	T	G	C	C	C	T	T
5	T	G	G	G	C	C	C	C	T	T	C	G	G	C	T	G	C	C	C	A	T
6	T	G	G	T	T	T	T	T	C	T	C	G	A	T	C	T	T	C	C	A	T
7	T	G	G	T	T	T	T	T	C	T	C	G	A	T	C	T	T	G	T	T	T
8	T	G	G	T	T	T	T	T	C	T	T	A	A	T	C	T	T	G	T	T	T
9	T	G	G	T	T	T	T	T	T	T	T	G	A	T	C	T	T	G	T	T	T

Fig. 1 Representation of nucleotide variation detected in the TrPALf (phenylalanine lyase) gene in the form of distinct haplotypes recovered from the single plant genotype Haifa$_2$. Minimal substitutional base changes that distinguish between all haplotypes are indicated in *grey shading*

Discussion

Effect of Non-homologous Gene Structure on SNP Verification

Although in vitro SNP discovery based on cloning and sequencing of PCR amplicons from divergent genotypes followed by sequence alignment is labour-intensive, the process allows direct determination of SNP haplotype structure within amplicons, and is, in principle, capable of discriminating between paralogous sequences. In the equivalent process, a relatively small set of predicted perennial ryegrass SNPs were identified as base variants between non-homologous gene sequences. By contrast, assessment of a similar number of predicted white clover SNPs (129 compared to 238) eliminated a much higher proportion (82%). These comparisons suggest that in vitro discovery is more vulnerable to non-homologous sequence alignment in white clover than perennial ryegrass. The two species belong to different angiosperm families (Fabaceae and Poaceae, respectively) and may differ in terms of degree of intragenomic gene duplication. Apart from specific gene duplication, which may occur in response to selective pressures, whole-genome duplication events have occurred frequently during angiosperm evolution. At least one such event apparently predates the divergence of monocots and dicots, while a second event predates the divergence of the *Arabidopsis thaliana* lineage from that of other dicots (Bowers et al. 2003). A major palaeoduplication event is estimated to have taken place c. 70 million years ago prior to the radiation of the Poaceae (Paterson et al. 2004). A similar event has been inferred within the Papilionoideae sub-family of the Fabaceae, prior to divergence of the Galegoid and Phaseoloid clades (Doyle and Luckow 2003). Although the incidence and frequency of such events may contribute to differing paralogy effects, the most obvious contributory factor is the allotetraploid nature of white clover.

Factors Influencing Haplotype Recovery

Amplicon cloning and sequencing is required to reliably discriminate between heterozygous allelic variants and multiple haplotypes from para-logous sequences (Edwards et al. 2006). However, sufficient independent clones are required to recover a reasonable sample of target sequences, and this number depends on expectations of allelic proportions, as well as the potential biasing effects of allele-specific PCR competition and paralogous gene structure. The simplest model is that of two different alleles (or haplotypes) at a single locus represented in the amplification product: sequencing of eight independent clones predicts a 99.5% probability of recovery of at least one clone from each allele (Cogan et al. 2006). For homoeologous loci with distinct heterozygous haplotypes at each locus, the model is equivalent to that developed for recovery of each allele from a simplex (ABCD) configuration in an autotetraploid such as potato. A sample of 12 clones, as used in this study, would obtain an 87.5%

probability of recovery of at least one clone from each allele (Simko 2004), while homozygosity at one or both duplicated loci would increase the probability to over 90%. The levels of haplotype complexity per genotype seen for representative genes in this study demonstrate that the homologous or homoeologous sequence models are often inappropriate. Models based on three or more heterozygous loci would predict lower confidences of full haplotype recovery from sequencing of 12 clones. Predictability is further reduced by the likelihood of biased recovery due either to PCR competition arising from priming site polymorphism, and differential cloning efficiency. The process of PCR-mediated recombination (Judo et al. 1998) which has been shown to occur in allotetraploid cotton (*Gossypium hirsutum* L.: Cronn et al. 2002), provides a further complicating factor. These considerations suggest that a larger clone sample would have permitted more effective haplotype recovery. However, such benefits must be compared to the time constraints and cost associated with DNA sequencing, especially for multiple target amplicons. Advances in massively parallel DNA sequencing technology will permit much deeper sampling of haplotype complexity, given the ability to address individual genotypes.

Diversity of Gene Family Structure in White Clover

The observed complexity of sequence haplotype structure in white clover is compatible with both paralogous and homoeologous gene amplification. Multicopy families within genomes are anticipated for many of the gene classes analyzed in this study. PCR amplification of low-copy genes from allopolyploid species usually recovers sequences from each homoeolocus (Cronn and Wendel 1998). However, the strategies used to define HSVs in inbreeding species are not appropriate for white clover. The combination of high levels of sequence heterozygosity within and between putative homoeoloci confounds simple visual assessment of haplotype origin. In addition, the close phylogenetic relationship between the putative progenitor genomes of white clover (Ellison et al. 2006) suggests that the sub-genomes may be highly similar at the DNA sequence level, further complicating identification of allelic variation. Such similarities appear to extend beyond protein coding components of genes, based on the relatively similar rates of verification for exonic and non-exonic (primarily intron-located) SNPs in this study. The process of predicted SNP validation by segregation analysis consequently provides the most direct method for discrimination of homologous and non-homologous sequences.

Strategies for Enhanced In Vitro SNP Discovery in White Clover

Refined methods are clearly required to improve the efficiency of in vitro SNP discovery and will require discrimination of genome- and gene-specific sequences. Gene-specific sequences may be identified from bacterial artificial chromosome (BAC) libraries rather than full-length cDNA or expressed sequence tag (EST) collections. A BAC library with c. sixfold genome coverage (Spangenberg et al. 2005) is available for this purpose. Comparison of sequences from independent BACs selected with template genes will allow directed primer design to locus-specific features.

The identification of putative progenitor genomes (Ellison et al. 2006) provides a more robust approach to homoeologous haplotype discrimination, based on generation and alignment of amplicons from contemporary *T. occidentale* and *T. pallescens* genotypes and subtraction of related haplotypes. The method has so far been successfully applied to nine genes involved in response to abiotic stresses, in concert with deeper sampling of haplotypes (24 clones per target genotype). Allelic nucleotide variants, HSVs and paralogous sequence variants (PSVs) have been discriminated by this process. Amplicons derived from *T. occidentale* show a close affinity to those from *T. repens*, defining an O sub-genome, while amplicons derived from *T. pallescens* show weaker affinities, defining a P′ sub-genome. A high level of coincidence (>60%) was observed between predicted SNPs and HSVs, providing an explanation for the high level of attrition during validation observed in this study.

Conclusions

The results of this study demonstrate the problems that can arise during in vitro resequencing-based SNP discovery in a highly heterozygous, allopolyploid species of equivocal origin lacking well-developed genetic analysis tools. Nonetheless, successful strategies for enhanced discovery have been on the basis of the current study and will be applicable to other species of this nature such as tall fescue (*Festuca arundinacea* Schreb.), strawberry (*Fragaria* × *ananassa* Duch.) and kiwifruit (*Actinia deliciosa* L.). White clover is a valuable and economically significant component of temperate pastoral grazing systems (Frame and Newbould 1986), and gene-associated SNP discovery will be critical for development of diagnostic markers for breeding improvement, justifying the required investment in technology development.

References

Bowers JE, Chapman BA, Rong J, Paterson AH (2003) Unravelling angiosperm genome evolution by phylogenetic analysis of chromosomal duplication events. Nature 422:433–438

Bryan GJ, Stephenson P, Collins A, Kirby J, Smith JB, Gale MD (1999) Low levels of DNA sequence variation among adapted genotypes of hexaploid wheat. Theor Appl Genet 99:192–198

Caldwell KS, Dvorak J, Lagudah ES, Akhunov E, Luo M-C, Wolters P, Powell W (2004) Sequence polymorphism in polyploid wheat and their D-genome diploid ancestor. Genetics 167:941–947

Cogan NOI, Ponting RC, Vecchies AC, Drayton MC, George J, Dobrowolski MP, Sawbridge TI, Spangenberg GC, Smith KF, Forster JW (2006) Gene-associated single nucleotide polymorphism (SNP) discovery in perennial ryegrass (*Lolium perenne* L.). Mol Genet Genomics 276:101–112

Cogan NOI, Drayton MC, Ponting RC, Vecchies AC, Bannan NR, Sawbridge TI, Smith KF, Spangenberg GC, Forster JW (2007) Validation of *in silico*-predicted genic single nucleotide polymorphism in white clover (*Trifolium repens* L.). Mol Genet Genomics 277:413–425

Cronn RC, Wendel JF (1998) Simple methods for isolating homoeologous loci from allopolyploid genomes. Genome 41:756–762

Cronn R, Cedroni M, Haselkorn T, Gorver C, Wendel JF (2002) PCR-mediated recombination in amplification products derived from polyploid cotton. Theor Appl Genet 104:482–489

Dobrowolski MP, Forster JW (2006) Chapter 9: Linkage disequilibrium-based association mapping in forage species. In: Oraguzie NC, Rikkerink E, Gardiner SE, De Silva NH (eds) Association Mapping in Plants, Springer, New York, pp. 197–209

Doyle JJ, Luckow MA (2003) The rest of the iceberg. Legume diversity and evolution in a phylogenetic context. Plant Physiol 131:900–910

Edwards D, Forster JW, Cogan NOI, Batley J, Chagné D (2006) Chapter 4: Single nucleotide polymorphism discovery in plants. In: Oraguzie NC, Rikkerink E, Gardiner SE, De Silva NH (eds) Association Mapping in Plants, Springer, New York, pp. 53–76

Ellison NW, Liston A, Steiner JJ, Williams WM, Taylor NL (2006) Molecular phylogenetics of the clover genus (*Trifolium* – Leguminosae). Mol Phylogenet Evol 39:688–705

Forster JW, Jones ES, Batley J, Smith KF (2004) Molecular marker-based genetic analysis of pasture and turf grasses. In: Hopkins A, Wang Z-Y, Sledge M, Barker RE (eds) Molecular breeding of forage and turf, Kluwer, Dordrecht, the Netherlands, pp. 197–239

Frame J, Newbould P (1986) Agronomy of white clover. Adv Agron 40:1–88

George J, Dobrowolski MP, van Zijll de Jong E, Cogan NOI, Smith KF, Forster JW (2006) Assessment of genetic diversity in cultivars of white clover

(*Trifolium repens* L.) detected by simple sequence repeat polymorphism. Genome 49:919–930

Jones ES, Hughes LJ, Drayton MC, Abberton MT, Michaelson-Yeates TPT, Forster JW (2003) An SSR and AFLP molecular marker-based genetic map of white clover (*Trifolium repens* L.). Plant Sci 165:531–539

Judo MSB, Wedel AB, Wilson C (1998) Stimulation and suppression of PCR-mediated recombination. Nucleic Acids Res 26:1819–1825

Kölliker R, Jones ES, Drayton MC, Dupal MP, Forster JW (2001) Development and characterisation of simple sequence repeat (SSR) markers for white clover (*Trifolium repens* L.). Theor Appl Genet 102:416–424

Paterson AH, Bowers JE, Chapman BA (2004) Ancient polyploidization predating divergence of the cereals, and its consequences for comparative genomics. Proc Natl Acad Sci USA 101:9903–9908

Simko I (2004) One potato, two potato: haplotype association mapping in autotetraploids. Trends Plant Sci 9:441–448

Somers DJ, Kirkpatrick R, Moniwa M, Walsh A (2003) Mining single-nucleotide polymorphisms from hexaploid wheat ESTs. Genome 49:431–437

Sorrells ME, Wilson WA (1997) Direct classification and selection of superior alleles for crop improvement. Crop Sci 37:691–697

Spangenberg GS, Forster JW, Edwards D, John U, Mouradov A, Emmerling M, Batley J, Felitti S, Cogan NOI, Smith KF, Dobrowolski MP (2005) Future directions in the molecular breeding of forage and turf. In: Humphreys MO (ed) Molecular breeding for the genetic improvement of forage crops and turf, Wageningen Academic Publishers, the Netherlands, pp. 83–97

Development of Marker-Assisted Selection for the Improvement of Freezing Tolerance in Alfalfa

Yves Castonguay[1,2], Jean Cloutier[1], Réal Michaud[1], Annick Bertrand[1] and Serge Laberge[1]

[1]Soils and Crops Research and Development Centre, Agriculture and Agri-Food Canada, 2560 Hochelaga Boulevard, QC, Canada G1V 2J3
[2]Corresponding author, castonguayy@agr.gc.ca

Abstract. Marker-assisted selection (MAS) accelerates conventional breeding approaches in the improvement of multigenic traits. We used a bulk segregant analysis (BSA) approach to identify genetic polymorphisms closely associated to cold adaptation among populations of alfalfa (*Medicago sativa* L.) recurrently selected for increased tolerance to freezing (TF). Using bulk DNA samples from cultivar Apica (A-TF0) and populations (A-TF2 and A-TF5) derived from that initial background, we observed both the intensification and the disappearance of several DNA fragments in response to selection pressure. Subsequent assessment of freezing tolerance of individual genotypes confirmed a close relationship between some of these polymorphisms and freezing tolerance. Our results illustrate that the combination of BSA and populations recurrently selected for the improvement of polygenic traits are effective tools to develop MAS applications in alfalfa.

Introduction

Lack of winter hardiness of alfalfa (*Medicago sativa* L.) is largely attributable to insufficient freezing tolerance and greatly reduces the reliability of this forage legume in northern climates. The development of winter hardy cultivars has been historically based on selection within field nurseries of genotypes that survived particularly severe winters. Due to spatial and temporal variability of environmental conditions allowing adequate screening, plantations must often be maintained several years and replicated at multiple locations.

Methods allowing for a more rapid and accurate identification of genotypes with superior adaptation to subfreezing temperatures would significantly assist plant breeding programs in their efforts to develop cultivars better adapted to harsh winter conditions. We recently applied a recurrent selection protocol entirely performed under environmentally-controlled conditions to develop alfalfa populations selectively improved for their tolerance to freezing (TF). Several cycles of recurrent phenotypic selection have been performed in various populations and new synthetic populations have been produced using elite genotypes (Castonguay et al. 2006). Significant increases in freezing tolerance in advanced cycles of selection translated into superior field survival in the spring. Although selection for freezing tolerance under environmentally-controlled conditions is more predictable than field tests, the process remains relatively lengthy since each recurrent cycle takes nearly a year to complete and that numerous cycles are sometimes required. Further acceleration of the selection process could be achieved by complementation of our freezing stress selection protocol with marker-assisted selection (MAS) approaches. A key outcome of the analysis of genomes is undoubtedly the identification of functional DNA variants responsible for the genetic component of phenotypic variation (Morgante 2006). DNA markers meet several of the desirable properties for the analysis of genetic diversity within populations including abundance, even distribution within the genome, reproducibility and more importantly, their stability regardless of tissue, timing and environmental conditions at sampling (Weising et al. 2005).

DNA Polymorphisms Between TF Populations

Recent analysis of DNA sequence variations among 20 accessions of *Arabidopsis thaliana* revealed substantial differences in their genic content with "major-effect changes" affecting nearly 10% of protein-coding genes (Clark et al. 2007). Interestingly, genes mediating interaction with the environment showed exceptionally high rate of polymorphisms. In that perspective, TF populations recurrently improved for their freezing tolerance constitute unique resources to probe the genetic bases of superior freezing tolerance in alfalfa. Considering that the development of TF populations was solely targeted towards the improvement of tolerance to subfreezing temperatures, there is a high likelihood that DNA polymorphisms that vary in intensity in response to selection pressure are linked to loci that control this trait. We thus used a bulk segregant analysis (BSA) approach as described by Michelmore et al. (1991) as a method to identify DNA markers linked to freezing tolerance variations among TF populations. Pooled DNA samples from segregating genotypes within each TF population were screened for DNA polymorphisms using restriction fragment length polymorphisms (RFLP) analysis of candidate genes putatively involved in the cold acclimation process and the sequence-related amplified polymorphism (SRAP) technique. BSA analysis of RFLP profiles of candidate genes has been previously used to identify differences in alleles frequency between contrasted phenotypes derived from a given genetic background (Quarrie et al. 1999). The SRAP technique is a PCR-based marker system that preferentially targets coding sequences at random (Li and Quiros 2001). Forward and reverse primers that respectively allow preferential amplification of exonic and intronic regions uncover polymorphic sequences resulting from variations in the length of introns, promoters and spacers among genotypes or populations. SRAP which has been shown to be effective in gene tagging in several species, is highly reproducible and comparatively less expensive to develop than other types of markers (Cravero et al. 2007).

Clear polymorphisms that intensified with the number of selection cycles were uncovered using RFLP analysis of a number of candidate genes including homologs of galactinol synthase (Castonguay et al. 2006) and other genes involved in central metabolism or known to be responsive to environmental changes. Both positive and negative polymorphisms were observed in response to selection pressure (data not shown). This indicates that improvement of freezing tolerance relies on both the increase in the frequency of favorable alleles and the elimination of unfavorable ones within populations of alfalfa. Alternatively, several anonymous DNA fragments that vary in intensity in response to recurrent selection have been uncovered with 42 combinations of SRAP primer pairs. As observed with RFLP analysis of candidate genes, polymorphisms either positively or negatively related to selection pressure were also identified with the SRAP technique. An illustration of a positive polymorphic variation in genic composition is illustrated in Fig. 1. Amplification profiles obtained with bulk DNA samples from closely related TF populations were highly similar confirming the high degree of reproducibility of the technique. However, we can see that a DNA fragment (identified by an arrow) that was initially undetectable in pooled samples from the initial background (A-TF0) and early cycles of selection (A-TF2) was markedly amplified in A-TF5. Further screening of this polymorphic amplicon in individual genotypes within each TF population revealed that the differential amplification observed with bulk DNA samples reflects differences in the frequency of its occurrence between the populations (Fig. 1). Nearly 30% of A-TF5 genotypes showed a positive amplification of the fragment as opposed to about 5% in A-TF0 and A-TF2. We repeatedly observed a similar relationship between changes in the intensity of DNA polymorphisms between bulk samples and variations in the frequency of these polymorphisms among individual genotypes that constitute these pools. Although this remains to be validated, this observation suggests that SRAP analysis of bulk samples provide a quantitative assessment of allele frequency within heterogeneous populations.

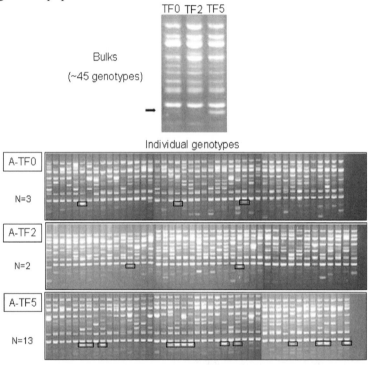

Fig. 1 SRAPs of bulk samples (~45 genotypes/bulk) and individual genotypes from alfalfa populations A-TF2 and A-TF5 recurrently selected for superior freezing tolerance within the cultivar 'Apica' (A-TF0). The number of genotypes (N) that are positive for the polymorphic fragment is indicated. Genomic DNA was quantitated by visual assessment. SRAPs were amplified with the F13:R15 (forward:reverse) primer pair as described in Vandemark et al. (2006). Amplicons were separated on a 2% (w/v) agarose gel in 1× Tris Borate EDTA running buffer pH 8.0 (70 V for 3 h). Ethidium bromide (1.0 µg/ml) was added to the gel before electrophoresis. Gel images were captured on a gel BioDoc-IT System (UVP, Upland, CA)

Identification of Markers Linked to Freezing Tolerance

In order to identify DNA polymorphisms potentially useful as markers, we established the relationship between their presence and freezing tolerance. For that purpose, we determined the freezing tolerance for each of ~45 genotypes within each population using clonal propagules acclimated to natural winter conditions in an unheated greenhouse. A typical quantitative response with freezing tolerance ranging from almost complete sensitivity to almost complete tolerance to freezing at −12°C was observed within each TF population (Fig. 2). However, genotypic assessment confirmed the expected increase in the frequency of tolerant genotypes and the decrease in the frequency of sensitive genotypes in A-TF5 as compared to A-TF0 (data not shown).

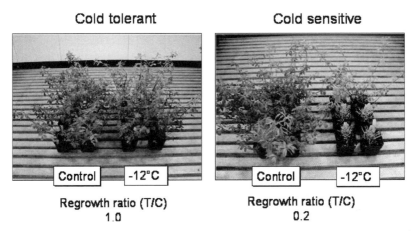

Fig. 2 Assessment of genotypic variability for freezing tolerance within alfalfa populations. Clonal propagules were cold acclimated under natural winter conditions in an unheated greenhouse. Freezing tolerance was estimated as the ratio of regrowth of six clonal propagules exposed to a single test temperature of −12°C (T) over that of six unstressed controls (C) immediately transferred to growth conditions. A ratio value near 0 reflects almost complete sensitivity to the freezing stress whereas a ratio of ~1 indicates full tolerance

To increase the probability of finding markers closely associated to freezing tolerance, we selected polymorphic amplicons that not only markedly varied in frequency between TF populations but that also showed contrasted amplification responses between the nine most freezing sensitive genotypes within A-TF0 and the nine most freezing tolerant genotypes within A-TF5. Using that approach we have selected four primer pair combinations that yielded divergent polymorphisms between genotypes of contrasted cold adaptation. We subsequently investigated the relationship between the accumulation of these four polymorphisms in the plant genome and freezing tolerance. Grouping of genotypes from A-TF0, A-TF2 and A-TF5 on the basis of the number of the four polymorphisms that they cumulate revealed a gradual increase in freezing tolerance with the number of polymorphisms (Table 1). Such response could be indicative of an additive effect of cumulative polymorphisms on the plants capacity to withstand subfreezing temperatures.

Table 1 Relationship between the number of polymorphisms and freezing tolerance within the alfalfa cultivar 'Apica'. Genotypes from populations A-TF0, A-TF2 and A-TF5, were grouped on the basis of the number of polymorphisms that they cumulate from none up to a maximum of 4. The number of genotypes and the average ratio of regrowth (an index of freezing tolerance described in Fig. 2) are indicated for each group of polymorphisms. A regrowth ratio value near 0 reflects almost complete sensitivity to a $-12°C$ freezing stress whereas a ratio of ~1 indicates superior tolerance

Polymorphisms	Genotypes	Regrowth ratio
>2	12	0.67
2	20	0.61
1	49	0.55
0	47	0.50
Total	128	0.55

Conclusions and Future Directions

BSA screening of heterogeneous populations recurrently selected for improved freezing tolerance in combination with the SRAP technique allowed us to uncover unique variations in the DNA composition of these genetic pools. Subsequent screening of several polymorphisms on the basis of their relationship with freezing tolerance led to the identification of four DNA fragments closely associated with cold adaptation. Whether these differences in genic content between TF populations are causally related to cold adaptation and to what extent they contribute to improve this trait are questions that need to be addressed. For that purpose, we have recently generated progenies combining the four selected polymorphisms at frequencies that are close and that should exceed those observed in A-TF5. Evaluation of freezing tolerance of these MAS-derived progenies and comparisons with A-TF0 and A-TF5 will allow us to assess the contribution of these four genomic regions to the gains in freezing tolerance achieved with A-TF5.

Acknowledgements

The technical assistance of Josée Bourassa, Lucette Chouinard, Pierre Lechasseur, Josée Michaud, Marie-Claude Pépin and Jean Auger is greatly appreciated.

References

Castonguay Y, Cloutier J, Laberge S, Bertrand A, Michaud R (2006) A bulk segregant approach to identify genetic polymorphisms associated with cold tolerance in alfalfa. In: Chen THH, Uemura M, Fujikawa S (eds). Cold hardiness in plants: molecular genetics, cell biology and physiology. CAB International, Wallingford, UK, pp. 88–102

Clark RM, Schweikert G, Toomajian C, Ossowski S, Zeeler G, Shinn P, Warthmann N, Hu TT, Fu G, Hinds DA, Chen H, Frazer KA, Huson DH, Schölkopf B, Nordborg M, Rätsch G, Ecker JR, Weigel D (2007) Common sequence polymorphisms shaping genetic diversity in *Arabidopsis thaliana*. Science 317: 338–342

Cravero V, Martín E, Cointry E (2007) Genetic diversity in *Cynara cardunculus* determined by sequence-related amplified polymorphism markers. J Am Soc Hort Sci 132: 208–212

Li G, Quiros CF (2001) Sequence-related amplified polymorphism (SRAP), a new marker system based on a simple PCR reaction: its application to mapping and gene tagging in *Brassica*. Theor Appl Genet 103: 455–461

Michelmore RW, Paran I, Kesseli RV (1991) Identification of markers linked to disease-resistance genes by bulked segregant analysis: a rapid method to detect markers in specific genomic regions by using segregating populations. Proc Natl Acad Sci USA 88: 9828–9832

Morgante M (2006) Plant genome organisation and diversity: the year of the junk! Curr Opin Biotechnol 17: 168–173

Quarrie SA, Vesna LJ, Kovacevic D, Steed A, Pekic S (1999) Bulk segregant analysis with molecular markers and its use in improving drought resistance in maize. J Exp Bot 50: 1299–1306

Vandemark GJ, Ariss JJ, Bauchan GA, Larsen RC, Hughes TJ (2006) Estimating genetic relationship among historical sources of alfalfa germplasm and selected cultivars with sequence related amplified polymorphisms. Euphytica 152: 9–16

Weising K, Nybom H, Wolff K, Kahl G (2005) DNA fingerprinting in plants. Principles, methods, and applications, second edition, CRC Press, Boca Raton, New York

Comparative Analysis of Disease Resistance Between Ryegrass and Cereal Crops

Geunhwa Jung[1,6], Young-Ki Jo[2], Sim Sung-Chur[3], Reed Barker[4], William Pfender[4] and Scott Warnke[5]

[1]Department of Plant, Soil and Insect Sciences, University of Massachusetts, Amherst, MA 01003, USA
[2]Department of Plant Pathology and Microbiology, Texas A&M University, College Station, TX 77843, USA
[3]Department of Horticulture and Crop Science, Ohio State University, Wooster, OH 44691, USA
[4]USDA-ARS, Oregon State University, Corvallis, OR 97331, USA
[5]USDA-ARS, Floral and Nursery Plants Research Unit, Beltsville, MD 20705, USA
[6]Corresponding author, jung@psis.umass.edu

Abstract. Perennial ryegrass (*Lolium perenne* L.) is one of the important forage and turf grasses in temperate zones in the world. Gray leaf spot caused by the fungus *Pyricularia oryzae* has recently become a serious problem on perennial ryegrass for golf course fairways. The causal agent also causes rice blast disease on rice, as well as foliar diseases on wheat and barley. Crown and stem rust caused by *Puccinia* spp. are also important for forage- and turf-type perennial ryegrass and seed production. In addition, foliar diseases caused by *Bipolaris* species, are common and widespread on graminaceous plants. Despite a recent advancement of molecular markers for forage and turf grasses, effective utilization of genetic information available in cereal crops will significantly lead to better understanding of the genetic architecture of disease resistance in ryegrass. Quantitative trait loci (QTL) analysis based on a three-generation interspecific ryegrass population detected a total of 16 QTLs for resistance to the four pathogens. Those QTL were compared with 45 resistance loci for the same or related pathogens previously identified in cereal crops, based on comparative genome analysis using a ryegrass genetic map and a rice physical map. Some pathogen-specific QTLs identified in ryegrass were conserved at corresponding genome regions in cereals but coincidence of QTLs for disease resistance in ryegrass and cereals was not statistically significant at the genome-wide comparison. In conclusion, the conserved synteny of disease resistance loci will facilitate transferring genetic resources for disease resistance between ryegrass

T. Yamada and G. Spangenberg (eds.), *Molecular Breeding of Forage and Turf,*
doi: 10.1007/978-0-387-79144-9_21, © Springer Science + Business Media, LLC 2009

and cereals to accommodate breeding needs for developing multiple disease resistance cultivars in ryegrass.

Introduction

Perennial ryegrass (*Lolium perenne* L.) is one of the most important forage and turf grasses in temperate climate zones in the world and diploid (2n = 2x = 14), outcrossing and self-incompatible species. As a member of the Festuceae tribe of the Pooideae subfamily of the Poaceae family (Yaneshita et al. 1993), it is taxonomically related to oat (*Avena sativa* L.), barley (*Hordeum vulgare* L.), wheat (*Triticum aestivum* L. em Thell), and rice (*Oryza sativa* L.) (Kellogg 2000). Its positive attributes include excellent forage quality that makes them the most important pasture grass species in temperate regions (Jones et al. 2002). In addition, its fast establishment and versatility as a turfgrass (Hannaway et al. 1999) as well as its bright green color, upright growth habit, and tolerance of low mowing height, contribute to its wide and popular use on golf course fairways, home lawns, and athletic fields.

Rust diseases caused by *Puccinia* spp. are important for cereal crops, particularly oat (*Avena sativa* L.) (Portyanko et al. 2005), wheat (*Triticum aestivum* L.) (Spielmeyer et al. 2003), and barley (*Hordeum vulgare* L.) (Brueggeman et al. 2002) as well as for forage-type perennial ryegrass and seed production. Molecular markers associated with resistance loci to crown rust (*P. coronata* Corda) have been identified in ryegrass (Dumsday et al. 2003; Muylle et al. 2005; Sim et al. 2007).

Gray leaf spot has become a serious problem on perennial ryegrass (Viji et al. 2001) since the time it was first reported on golf course fairways in Pennsylvania in 1992 (Landschoot and Hoyland 1992). The causal agent, the ascomycete fungus *Magnaporthe grisea*, also causes rice blast disease on rice (*Oryza sativa* L.), as well as foliar diseases on many grasses, such as blast on wheat (Viji et al. 2001), barley (Sato et al. 2001), and gray leaf spot on other turf and forage grasses such as tall fescue (*Festuca arundinacea* Schreb.), St. Augustinegrass [*Stenotaphrum secundatum* (Walt.) Kuntze] and Italian ryegrass (Viji et al. 2001). A valuable tool to study the genetics of both complete and partial disease resistance is the genetic linkage map, which has been widely used to study genetics of resistance to *M. grisea* in rice (Chen et al. 2003; Fukuoka and Okuno 2001; Tabien et al. 2002; Wang et al. 1994), barley (Sato et al.

2001), perennial ryegrass (Curley et al. 2005), and Italian ryegrass (*L. multiflorum*) (Miura et al. 2005).

Foliar, crown and root diseases caused by *Bipolaris*, *Drechslera* and *Exserohilum* species that were once referred to as the same genus, *Helminthosporium*, due to the similar epidemiology and symptoms, are common and widespread on graminaceous plants. These fungi have a broad host range, from turfgrass to most cereal crops including barley, wheat, oat, maize (*Zea mays* L.) and rice (Sivanesan 1987). Leaf spot disease caused by *Bipolaris sorokiniana* (Sacc.) Shoemaker progresses on perennial ryegrass under extended favorable environmental conditions. Lesions spread along the entire leaf blade and in severe cases, the fungi damage crowns and result in killing of numerous tillers in a process known as melting-out. Several qualitative and quantitative resistance genes to *Bipolaris* and *Drechslera* spp. have been identified in barley (Kutcher et al. 1996; Richter et al. 1998) and wheat (Sharma et al. 2004).

Breeding multiple disease resistant cultivars is one of the best disease management strategies in cereal crops, although this may take considerable time and require understanding of the genetics of host resistance. In rice, many quantitative and qualitative resistance genes have been identified (Wang et al. 1994; Fukuoka and Okuno 2001; Tabien et al. 2002). Important gene loci conferring broad-spectrum resistance have also been found in rice (Wisser et al. 2005). Development of molecular markers tightly linked to disease resistance provides a means to pyramid multiple resistant genes in elite cultivars. Combining both quantitative and qualitative resistance is ideal for managing different races or multiple diseases by precluding rapid breakdown of resistance by pathogens.

Recently more genetic linkage and traits mapping have been studied in forage and turf grasses, which provide the basis of marker-assisted selection (Hayward et al. 1998; Bert et al. 1999; Foster et al. 2001; Jones et al. 2002; Warnke et al. 2004; Curley et al. 2005; Chakraborty et al. 2005; Sim et al. 2007), and allows comparative genome analysis with model cereals (Jones et al. 2002; Sim et al. 2005). The ryegrass genome has conserved syntenic relationships with genomes of rice and oat, and most highly with the wheat genome (Sim et al. 2005). The conserved synteny and collinearity observed among genomes of grass species (Devos and Gale 2000) make it possible to transfer valuable genetic information from well-studied cereal crops to ryegrass. Effective utilization of important genetic information available in cereal crops will also lead to better understanding of the genetic architecture of disease resistance that are important targets for genetic manipulation and eventually crop improvement in ryegrass.

In the recent studies, quantitative trait loci (QTLs) for resistance to gray leaf spot (Curley et al. 2005), crown rust (Sim et al. 2007), and leaf spot and stem rust (Jung et al. 2006) have been detected in a three-generation interspecific (*L. perenne* × *L. multiflorum*) ryegrass mapping population (MFA × MFB). Comparative analyses of resistance to four fungal pathogens identified in this ryegrass population were conducted with those previously reported in cereal crops to test whether pathogen-specific resistance loci are conserved between ryegrass and cereal crops.

Materials and Methods

Phenotypic Evaluation

One hundred and sixty-nine progeny of three-generation interspecific MFA × MFB ryegrass mapping population (Warnke et al. 2004) were evaluated for susceptibilities to gray leaf spot in the growth chamber (Curley et al. 2005) and to crown rust in the field (Sim et al. 2007). In addition, leaf spot susceptibility was evaluated both in a controlled environment chamber and in the field and stem rust in the growth chamber (Jung et al. 2006). All the experiments were conducted twice with a randomized complete block design.

QTL Analysis of Multiple Disease Resistance

The linkage map of the MFA × MFB population used for QTL analysis was updated from the map previously developed by Sim et al. (2005). A total of 152 markers (43 RAPD and 109 RFLP) were mapped on the MFA genetic map and 135 markers (28 RAPD and 107 RFLP) were mapped on the MFB genetic map. The sources of heterologous cDNA probes for RFLP markers were oat (CDO), barley (BCD), rice (RZ) and creeping bentgrass (*Agrostis stolonifera* L.)(Ast).

Three QTL analyses as described in Curley et al. (2005), including Kruskal-Wallis, interval mapping, and multiple QTL mapping (MQM), were performed based on phenotypic data of disease susceptibility using the software program, MapQTL5 (Kyazma, Wageningen, Netherlands).

Comparative QTL Analysis of Multiple Disease Resistance Between Ryegrass and Cereal Crops

The locations of QTLs for resistance to four diseases detected in the MFA × MFB ryegrass population were compared with those to the same or closely related pathogen species previously identified in cereal crops. Forty-five loci for resistance to net blotch [*Drechslera teres* (Sacc.) Shoemaker related with *B. sorokiniana*], stem rust (*P. graminis*), rice blast and crown rust previously detected in rice, oat, wheat, rye, ryegrass or barley were located on the rice physical map. Comparative QTL analysis between ryegrass and cereal crops was based on expressed sequence tags (ESTs)-RFLP markers common in both the ryegrass genetic map and the rice physical map. To increase the number of common RFLP markers closely linked to QTLs for resistance in cereal crops, appropriate bridge maps were searched at web-based cereal databases: Gramene (http://www.gramene.org) and GrainGenes (http://wheat.pw.usda.gov/GG2/index.sthml).

Results

Assessment of Susceptibilities to Leaf Spot, Stem Rust, Gray Leaf Spot and Crown Rust

Significant genotypic effects in the MFA × MFB population were detected in the susceptibilities to these four diseases. Phenotype distribution indicated that susceptibilities to the diseases were quantitative and transgressive. The Spearman's rank correlation coefficient was used to measure monotone association among susceptibilities to leaf spot, stem rust, gray leaf spot and crown rust. Significant negative correlations between leaf spot (inoculation assays and field trials) and crown rust (field trials) were detected. However, no significant correlation between leaf spot and either stem rust (inoculation assays) or gray leaf spot (inoculation assays) was found. The significant phenotypic correlation remained consistent between different datasets of the same disease.

Loci for Multiple Disease Resistance

A total of 16 QTLs for resistance to leaf spot (seven), stem rust (one), gray leaf spot (four) and crown rust (four) were distributed on six LGs. The total phenotypic variation explained by the individual QTL ranged from 7 to 18%. Chi-square goodness-of-fit tests failed at the $P = 0.05$ to reject the null hypothesis that those QTLs overlap on the ryegrass linkage map by chance. However, two genomic regions on LGs 4 and 6 contained QTLs for resistance to multiple pathogens, stem rust, leaf spot, crown rust, and gray leaf spot (Fig. 1).

Due to a highly syntenic and collinear relationship between the rice and ryegrass genomes, most EST-RFLP markers mapped in ryegrass are in their syntenic, homeologous chromosomes of rice. The rice physical map could serve as a framework upon which disease resistance loci identified

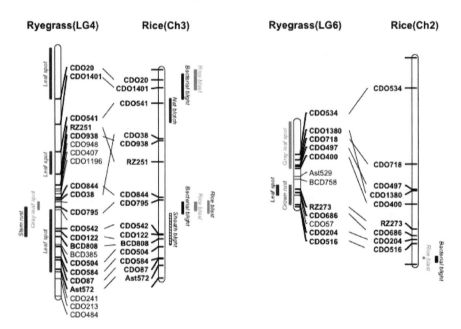

Fig. 1 Integrated disease resistance map of ryegrass and rice using common RFLP markers (*bold*). The sources of cDNA probes for RFLP markers are oat (CDO), barley (BCD), rice (RZ) and creeping bentgrass (Ast). Disease resistance loci are located on six linkage groups of the ryegrass genetic map and nine corresponding homeologous chromosome segments of rice. However, only ryegrass LGs 4 and 6 are shown in the figure. QTLs for disease resistance are located along a side of the ryegrass or rice map, and are labeled with common names of the following diseases, net blotch (*Drechslera teres*) on barley; leaf spot (*Bipolaris sorokiniana*), stem rust (*Puccinia graminis* subsp. *graminicola*), and crown rust (*Puccinia coronata* f.sp. *lolli*) on ryegrass; bacteria blight (*Xanthomonas oryzae* pv. *oryzae*), sheath blight (*Rhizoctonia solani*), and rice blast (*Magnaporthe grisea*) on rice

from various cereal crops could be readily located using sequences of RFLP/SSR markers tightly linked to disease resistance loci. A total of 45 loci conferring resistance to rice blast, crown rust, stem rust and net blotch identified in various Poaceae species were located throughout nine chromosomes of rice in Fig. 1.

Coincidence of QTLs for disease resistance in ryegrass with loci for disease resistance in cereals was tested for statistical significance using Cohen's kappa (Cohen 1960). Except syntenic regions ($P = 0.054$) for QTL for leaf spot resistance in ryegrass and blotch resistance in barley, QTLs for resistance to the other diseases in ryegrass were not significantly associated with loci for the corresponding disease resistance in cereals at the genome-wide comparison.

Discussion

Our comparative QTL analysis of resistance to multiple pathogens between ryegrass and cereal crops was greatly facilitated by the increased availability of publicly accessible genetic databases of cereal crops. Given the high level of syntenic and collinear relationship in Poaceae, molecular markers with known DNA sequences from different grass species can be readily located on the rice physical map as long as their homologous sequences are present in the rice genome.

A genome-wide comparison of QTLs on ryegrass map indicated no statistically support ($P = 0.05$) for colocalization of QTLs for resistance to biotrophic (stem rust and crown rust) and necrotrophic (gray leaf spot and leaf spot) diseases, which have fundamental differences in parasitism. Most QTLs are pathogen-specific but two genomic regions on LGs 4 and 6 might be associated potentially with multiple disease resistance or clusters of resistance loci for biotrophs or necrotrophs.

However, disease resistance genes between ryegrass and cereals seem to be conserved. The LG4 containing three QTLs for resistance to leaf spot, gray leaf spot and stem rust has a syntenic relationship with a segment of rice chromosome 3, which contains QTLs for resistance to multiple pathogens, including rice blast, sheath blight caused by *Rhizoctonia solani* Kühn and bacteria leaf blight caused by *Xanthomonas oryzae* pv. *oryzae* (Ishiyama) (Wisser et al. 2005). Curley et al. (2005) previously reported three QTLs on LGs 2, 3 and 4 for gray leaf spot resistance in ryegrass were located in the syntenic regions of QTLs for rice blast resistance in rice. Similarly, Sim et al. (2007) reported that two QTLs for crown rust field resistance were located in the syntenic regions of LGs 2 and 7 in ryegrass where loci for crown rust resistance in two different perennial ryegrass populations (Dumsday et al. 2003; Muylle et al. 2005) or oat (Wight et al. 2004) have also been detected.

The current comparative QTL analysiss of ryegrass with cereals was possible based on heterologous RFLP probes from rice, oat, barley and creeping bentgrass. However, applicability of RFLP is limited by the requirement of conserved sequences and the level of genetic polymorphism detected by restriction enzymes, since not all RFLP probes give similar transferability when applied to different Poaceae species (Sim et al. 2005). Also, macro-colinearity based on RFLP does not always predict micro-colinearity, which requires extensive sequence information (Sorrells et al. 2003). Recently, highly transferable PCR markers such as EST-derived simple sequence repeats (SSRs) have become available and will supple-ment RFLP-based comparative genome studies in Poaceae (Saha et al. 2004; Yu et al. 2004). Functionally associated PCR-based markers or molecular markers tightly linked with QTL for disease resistance will facilitate marker-assisted selection in ryegrass by pyramiding pathogen-specific or multiple disease resistance genes.

Acknowledgements

We thank Dr. Mike Casler, Dr. Laurel Cooper and Larry Kramer for their inputs on this manuscript and Eva Goldwater for statistical consultation. We gratefully acknowledge the financial support from the United States Golf Association.

References

Bert PE, Charmet G, Sourdille P, Hayward MD, Balfourier F (1999) A high-density molecular map for ryegrass (*Lolium perenne* L.) using AFLP markers. Theor Appl Genet 99:445–452
Brueggeman R, Rostoks N, Kudrna D, Kilian A, Han F, Chen J, Druka A, Steffenson B, Kleinhofs A (2002) The barley stem rust-resistance gene *Rpg1* is a novel disease-resistance gene with homology to receptor kinases. Proc Natl Acad Sci USA 99:9328–9333

Chakraborty N, Bae J, Warnke S, Chang T, Jung G (2005) Linkage map construction in allotetraploid creeping bentgrass (*Agrostis stolonifera* L.). Theor Appl Genet 111:795–803

Chen H, Wang S, Xing Y, Xu C, Hayes PM, Zhang Q (2003) Comparative analyses of genomic locations and race specificities of loci for quantitative resistance to *Pyricularia grisea* in rice and barley. Proc Natl Acad Sci USA 100:2544–2549

Cohen J (1960) A coefficient of agreement for nominal scales. Educ Psychol Meas 20:37–46

Curley J, Sim SC, Warnke S, Leong S, Barker R, Jung G (2005) QTL mapping of resistance to gray leaf spot in ryegrass. Theor Appl Genet 111:1107–1117

Devos KM, Gale MD (2000) Genome relationships: the grass model in current research. Plant Cell 12:637–646

Dumsday JL, Smith KF, Forster JW, Jones ES (2003) SSR-based genetic linkage analysis of resistance to crown rust (*Puccinia coronata* f. sp. *lolii*) in perennial ryegrass (*Lolium perenne*). Plant Pathol 52:628–637

Foster JW, Jone ES, Kölliker MC, Drayton JL, Dumsday MP, Dupal KM, Guthridge KM, Mahoney NL, van Zijll de Jong E, Smith KF (2001) Development and implementation of molecular markers for forage crop improvement. In: Spangenberg G (ed) Molecular breeding of forage crops. Kluwer, Dordrecht, pp 101–133

Fukuoka S, Okuno K (2001) QTL analysis and mapping of *pi21*, a recessive gene for field resistance to rice blast in Japanese upland rice. Theor Appl Genet 103:185–190

Hannaway D, Fransen S, Cropper J, Teel M, Chaney M, Griggs T, Halse R, Hart J, Cheeke P, Hansen D, Klinger R, Lane W (1999) Perennial ryegrass (*Lolium perenne* L.). Oregon State University Extension Publication PNW503

Hayward MD, Forster JW, Jones JG, Dolstra O, Evans C, McAdam NJ, Hossain KG, Stammers M, Will JAK, Humphreys MO, Evans GM (1998) Genetic analysis of *Lolium*. I. Identification of linkage groups and the establishment of a genetic map. Plant Breed 117:451–455

Jones ES, Mahoney NL, Hayward MD, Armstead HI, Jones JG, Humphreys MO, King IP, Kishida T, Yamada T, Balfourier F, Charmet G, Forster JW (2002) An enhanced molecular marker based genetic map of perennial ryegrass (*Lolium perenne*) reveals comparative relationships with other Poaceae genomes. Genome 45:282–295

Jung G, Jo Y, Barker R, Warnke S (2006) Comparative analysis of disease resistance in turfgrass and cereals. ASA-CSSA-SSSA 70th Annual Meeting (Abstract)

Kellogg E (2000) The grasses: a case study in macroevolution. Annu Rev Ecol Syst 31:217–238

Kutcher HR, Bailey KL, Rossnagel BG, Legge WG (1996) Identification of RAPD markers for common root rot and spot blotch (*Cochliobolus sativus*) resistance in barley. Genome 39:206–215

Landschoot P, Hoyland B (1992) Gray leaf spot of perennial ryegrass turf in Pennsylvania. Plant Dis 76:1280–1282

Miura Y, Ding C, Ozaki R, Hirata M, Fujimori M, Takahashi W, Cai HW, Mizuno K (2005) Development of EST-derived CAPS and AFLP markers linked to a gene for resistance to ryegrass blast (*Pyricularia* sp.) in Italian ryegrass (*Lolium multiflorum* Lam.)

Muylle H, Baert J, Van Bockstaele E, Pertijs J, Rolda'n-Ruiz I (2005) Four QTLs determine crown rust (*Puccinia coronata* f. sp. *lolii*) resistance in a perennial ryegrass (*Lolium perenne*) population. Heredity 95:348–357

Portyanko VA, Chen G, Rines HW, Phillips RL, Leonard KJ, Ochocki GE, Stuthman DD (2005) Quantitative trait loci for partial resistance to crown rust, *Puccinia coronata*, in cultivated oat, *Avena sativa* L. Theor Appl Genet 111:313–324

Richter K, Schondelmaier J, Jung C (1998) Mapping of quantitative trait loci affecting *Drechslera teres* resistance in barley with molecular markers. Theor Appl Genet 97:1225–1234

Saha MC, Mian MA, Eujayl I, Zwonitzer JC, Wang L, May GD (2004) Tall fescue EST-SSR markers with transferability across several grass species. Theor Appl Genet 109:783–791

Sato K, Inukai T, Hayes PM (2001) QTL analysis of resistance to the rice blast pathogen in barley (*Hordeum vulgare*). Theor Appl Genet 102:916–920

Sharma RC, Sah SN, Duveiller E (2004) Combining ability analysis of resistance to Helminthosporium leaf blight in spring wheat. Euphytica 136:341–348

Sim S, Chang T, Curley J, Warnke SE, Barker RE, Jung G (2005) Chromosomal rearrangements differentiating the ryegrass genome from the Triticeae, oat, and rice genomes using common heterologous RFLP probes. Theor Appl Genet 110:1011–1019

Sim S, Diesburg K, Casler M, Jung G (2007) Mapping and comparative analysis of QTL for crown rust resistance in an Italian × perennial ryegrass population. Phytopathology 97:767–776

Sivanesan A (1987) Graminicolous species of bipolaris, curvularia, drechslera, exserohilum and their teleomorphs. C.A.B International, Wallingford, Oxon

Sorrells ME, La Rota M, Bermudez-Kandianis CE, Greene RA, Kantety R, Munkvold JD, Miftahudin, Mahmoud A, Ma X, Gustafson PJ, Qi LL, Echalier B, Gill BS, Matthews DE, Lazo GR, Chao S, Anderson OD, Edwards H, Linkiewicz AM, Dubcovsky J, Akhunov ED, Dvorak J, Zhang D, Nguyen HT, Peng J, Lapitan NL, Gonzalez-Hernandez JL, Anderson JA, Hossain K, Kalavacharla V, Kianian SF, Choi DW, Close TJ, Dilbirligi M, Gill KS, Steber C, Walker-Simmons MK, McGuire PE, Qualset CO (2003) Comparative DNA sequence analysis of wheat and rice genomes. Genome Res 13:1818–1827

Spielmeyer W, Sharp PJ, Lagudah ES (2003) Identification and validation of markers linked to broad-spectrum stem rust resistance gene *Sr2* in wheat (*Triticum aestivum* L.). Crop Sci 43:333–336

Tabien E, Li Z, Paterson H, Marchetti A, Stansel W, Pinson M (2002) Mapping QTLs for field resistance to the rice blast pathogen and evaluating their individual and combined utility in improved varieties. Theor Appl Genet 105:313–324

Viji G, Wu B, Kang S, Uddin W, Huff DR (2001) *Pyricularia grisea* causing gray leaf spot of perennial ryegrass turf: population structure and host specificity. Plant Dis 85:817–826

Wang GL, Mackill DJ, Bonman JM, McCouch SR, Champoux MC, Nelson RJ (1994) RFLP mapping of genes conferring complete and partial resistance to blast in a durably resistant rice cultivar. Genetics 136:1421–1434

Warnke SE, Barker RE, Jung G, Sim SC, Rouf Mian MA, Saha MC, Brilman LA, Dupal MP, Forster JW (2004) Genetic linkage mapping of an annual × perennial ryegrass population. Theor Appl Genet 109:294–304

Wight CP, O'Donoughue LS, Chong J, Tinker NA, Molnar SJ (2004) Discovery, localization, and sequence characterization of molecular markers for the crown rust resistance genes *Pc38*, *Pc39* and *Pc48* in cultivated oat (*Avena sativa* L). Plant Breed 14:349–361

Wisser RJ, Sun Q, Hulbert SH, Kresovich S, Nelson RJ (2005) Identification and characterization of regions of the rice genome associated with broad-spectrum, quantitative disease resistance. Genetics 169:2277–2293

Yaneshita M, Ohmura T, Sasakua T, Ogihara Y (1993) Phylogenetic relationships of turfgrasses as revealed by restriction fragment analysis of chloroplast DNA. Theor Appl Genet 87:129–135

Yu JK, La Rota M, Kantety RV, Sorrells ME (2004) EST derived SSR markers for comparative mapping in wheat and rice. Mol Genet Genomics 271:742–751

White Clover Seed Yield: A Case Study in Marker-Assisted Selection

Brent Barrett[1,3], Ivan Baird[2] and Derek Woodfield[1]

[1]AgResearch Ltd, Grasslands Research Centre, Private Bag 11008, Palmerston North, New Zealand
[2]AgResearch Ltd, Lincoln Research Centre, Private Bag 4749, Christchurch, New Zealand
[3]Corresponding author, brent.barrett@agresearch.co.nz

Abstract. Genetic gain from phenotypic selection in open-pollinated forage species is constrained by the inability to accurately use phenotype to estimate genotype, prior to parent selection for polycrossing. The use of marker-assisted selection (MAS) offers the potential to accelerate genetic gain by partially overcoming this constraint. White clover (*Trifolium repens* L., 2n = 4x = 32) is an open-pollinated, high-quality, perennial forage legume with complex inheritance of traits underpinning pasture persistence, seed production, animal productivity, and animal health. Our legume improvement programme has utilised seed production in white clover as a case study in the application of MAS in outbred forage species, using microsatellite markers linked to quantitative trait loci (QTL) of moderate resolution. The QTL SY03-D2 on the distal end of group D2, was used to explore marker:trait associations in 12 breeding pools, leading to opportunities to conduct reselection experiments, and to monitor response to genotypic selection criteria in experimental polycrosses. Each breeding pool was sampled with 90 or more individuals grown out in an unreplicated field trial to assess seed yield traits, as per standard practice in our cultivar development programme. DNA samples were tested with up to three microsatellite markers associated with the QTL. Significant ($p < 0.01$) marker:trait associations were observed in 8 of the 12 breeding pools, with the most informative polymorphisms accounting for differences of 30–69% in the mean seed yield values within breeding pools. These data suggest that value can be realised from the current investment in genomics for MAS in white clover, given QTL of moderate resolution, and widely used marker platforms such as microsatellites.

T. Yamada and G. Spangenberg (eds.), *Molecular Breeding of Forage and Turf*,
doi: 10.1007/978-0-387-79144-9_22, © Springer Science + Business Media, LLC 2009

241

Introduction

Forages underpin pastoral productivity around the world, feeding the majority of an annual production of 61 M tonnes of beef, 8.6 M tonnes of sheep meat, and 549 M tonnes of cow's milk (www.fao.org). Most temperate pastures are mixed swards of perennial ryegrass (*Lolium perenne* L.) and white clover (*Trifolium repens* L.). There are claims of significant progress in breeding temperate forages for improved herbage yield, persistence, stress tolerance, and forage quality (Easton et al. 2001; Woodfield 1999; Woodfield and Easton 2004). However, on farm trials have reported less substantial changes in forage performance over time, as measured by animal productivity (Chapman et al. 1993; Crush et al. 2006).

This largely unrealised opportunity to make substantial improvements in livestock productivity through forage plant breeding may in part be due to general challenges in estimating and selecting genotype on the basis of phenotype. There is a further disconnection between phenotype and genotype in highly heterozygous species where deleterious recessive alleles often go undetected during parent selection for polycrossing, unless extensive and laborious progeny tests are undertaken prior to parent selection. Marker-assisted selection (MAS) may help realise this opportunity.

DNA marker technologies have proven valuable in a range of applications, including MAS. While livestock (Powell and Norman 2006) and some crops (Francia et al. 2005) already benefit from accelerated genetic improvement thanks to MAS, there are only a few published reports exploring MAS in open pollinated forage grasses and legumes. The few reports in forages and analogous systems indicate that there are several ways in which DNA markers may add value to forage breeding programmes, including screening out deleterious recessives (Stendal et al. 2006), and optimising the level of genetic diversity within or among parent genotypes (Hayes et al. 2006; Kölliker et al. 2005).

The use of DNA technologies for development of MAS in forages was mooted over two decades ago (Helentjaris et al. 1985), and there has been a steady uptake of enhanced marker technology platforms by forage researchers, including transitions through the range of marker platforms including RFLPs, RAPDs, AFLPs, microsatellites, and now into SNPs with an aim toward developing functional markers (Cogan et al. 2006b, 2007). While there has been a great deal of activity in the development of genomics resources in forages (Sawbridge et al. 2003), deployment of advanced marker platforms, construction of enhanced genetic linkage maps (Barrett et al. 2004; Faville et al. 2004; Jones et al. 2002; Zhang et al. 2007) and ongoing discovery of major genes and QTL (Barrett et al. 2005a,b; Cogan et al. 2006a; Xiong et al. 2006; Yamada et al. 2004) in forage crops, there are few indications in the literature that applications are being extended through to MAS.

In part, MAS uptake in forages may be hindered by the complex genetics of open pollinated species. While this complexity may deter some, the open pollinated breeding habits are in many ways an asset in terms of realising value from MAS, in that most breeding pools are highly variable (10- to 100-fold variation among individuals for key traits), and harbour a number of deleterious recessive alleles within each genotype. This observation is strengthened by mapping data from pair-cross populations, in which parents commonly harbour multiple recessive QTL alleles of substantial effect (accounting for >15% of variation for a trait) (Barrett et al. 2005b; Cogan et al. 2006a). These observations suggest that by cleaning up these gene pools, the return from MAS may be substantial. Models in analogous systems suggest that the rate of genetic gain may be doubled with MAS relative to current practice of phenotypic selection (Sonesson 2007). A complement to the development of advanced technologies for MAS is that forage breeding systems have also evolved to include simple crossing plans which are amenable to MAS (Tamaki et al. 2007).

Seed yield in white clover is a trait of high heritability ($h^2 > 0.70$) with significant and stable QTL known (Barrett et al. 2005b; Woodfield et al. 2004), which is of value to the seeds industry and is essential for a superior forage to be successfully delivered to the pastoral industry (Widdup et al. 2004). Seed yield values realised on farm are often below the genetic potential that can be achieved before significantly hindering performance under grazing (Woodfield et al. 2004).

Seed yield is an ideal case study in MAS in open pollinated forage species. Herein we describe work to determine the frequency and magnitude of seed yield marker:trait associations in white clover breeding pools, as a step toward full utilisation of MAS in the species.

Methods

Plant Materials

Practices routinely used in our commercial breeding population evaluations were employed in this study. Briefly, representative samples (n > 90) of plants from 12 multi-parent complex breeding pools were evaluated for seed yield in unreplicated field trials on a research farm in Lincoln, New Zealand. Each breeding pool sample was grown for a single year, with each plant maintained in a 0.20 m^2 ring to ensure comparable estimates of seed yield per unit area. Two breeding pools were sampled in 2003–2004, seven breeding pools in 2004–2005, and three breeding pools in 2005–2006. A DNA sample from each plant was archived to FTA cards (Whatman, USA) for subsequent purification and microsatellite marker genotyping in our laboratory in Palmerston North.

Phenotypic Evaluations

Seed yield was considered to have two component traits, yield per inflorescence, and numbers of inflorescences per plant in a fixed area (Woodfield et al. 2004). At maturity, the number of inflorescences per plant was recorded. Plants were then harvested, dried, and seed rubbed out and cleaned prior to weighing. Data recorded were seed yield (SY = g/plant) and inflorescence density (ID = inflorescences/plant). A derived measure of yield per inflorescence (YI = g seed per inflorescence) was obtained by mathematical division.

Genotypic Evaluations

Up to three microsatellite markers associated with the QTL SY03-D2 (Barrett et al. 2005b) were used to genotype each individual for which we had a recorded phenotype. Each microsatellite locus was not duplicated within homoeologous genome, but was duplicated across homoeologues. As locus and homoeologue assignments were not known for most amplicons, all PCR amplicon polymorphism data were collected and tested under the assumption that they are from the locus and homoeologue of interest.

Statistical Analysis

Polymorphism at each PCR amplicon size category was tested for effect on trait values using a method implemented in GenStat v9.0.(Lawes Agricultural Trust) Briefly, each amplicon size category was used to classify individuals within pools into two classes: those with the amplicon present and those with the amplicon absent. Individual plants with missing genotype data for a specific amplicon category were excluded. The polymorphism effect on trait values were tested using the Student's T test, with a significance threshold of $p < 0.01$ to reduce the number of false positives accumulated under multiple testing. Furthermore, for a marker:trait association to be declared significant, it had to effect both YI and SY, with polymorphism effects conserved across traits (i.e. amplicon state had to be associated with the same directional effect for both YI and SY). Analysis was restricted to those amplicons with a presence:absence ratio between 0.15 and 0.85. Only the marker:trait association with the greatest magnitude of effect on SY in each breeding pool was reported.

Results and Discussion

Twelve white clover breeding pools were tested for marker:trait associations using microsatellite markers associated with a seed yield QTL of moderate resolution, which had been discovered in an independent full-sib mapping population of $n = 182$ individuals genotyped at a mean density of ~5 cM (Barrett et al. 2005b).

Extensive variation among and within breeding pools was observed for SY, YI, and ID, with 10-fold and even 100-fold differences observed between the minimum and maximum values for both SY and ID traits within populations (Table 1). Frequency distributions of the populations indicate a range in mean seed yield potential, with some populations (e.g. 1, 10, 12) skewed toward low performance while other populations show a generally normal distribution (Fig. 1).

Significant marker:trait associations were discovered in eight of the 12 breeding pools tested, accounting for an average 38% difference in seed yield (Table 1). The diagnostic microsatellite, amplicon, and direction of allelic effect were not conserved across breeding pools. A pool-specific example of the marker:trait association effects are shown in Fig. 2, demonstrating a 33% change in seed yield potential on the basis of a single diagnostic PCR amplicon from a QTL-associated microsatellite marker in pool 2.

Table 1 Summary statistics and marker effects for seed yield and yield per inflorescence in 12 white clover breeding pools. Each pool was evaluated for a single year at Lincoln, New Zealand; under protocols utilised in our cultivar development programme. Plants were tested with up to three microsatellite markers. Marker effect values are changes in mean seed yield within pools among individuals with versus individuals without a yield-associated microsatellite amplicon. Different microsatellites and amplicons were associated with yield in different pools. Marker effects are for the polymorphism associated with the biggest change in mean values

Pool	n	Seed yield (g/plant)				Yield per inflorescence (g)				Marker Effect (%)
		Mean	SD	Min.	Max.	Mean	SD	Min.	Max.	
1	184	13.5	8.91	0.3	39.7	0.177	0.089	0.005	0.401	69
2	176	27.0	13.91	3.0	61.1	0.195	0.090	0.033	0.456	33
3	184	26.8	9.98	0.2	56.9	0.155	0.054	0.003	0.030	36
4	184	25.9	10.86	1.2	55.6	0.175	0.075	0.008	0.454	32
5	184	25.6	11.15	1.5	56.5	0.164	0.065	0.026	0.348	32
6	184	21.0	9.41	0.9	47.8	0.175	0.068	0.015	0.359	30
7	184	19.2	10.44	1.2	53.2	0.146	0.072	0.007	0.462	40
8	92	17.5	8.60	2.6	45.0	0.170	0.052	0.018	0.313	ns
9	175	18.9	10.15	0.5	55.6	0.162	0.065	0.007	0.510	ns
10	175	13.6	9.44	0.3	57.7	0.088	0.050	0.006	0.290	33
11	207	22.9	12.58	0.1	65.5	0.162	0.081	0.001	0.427	ns
12	184	10.6	8.51	0.2	38.5	0.114	0.069	0.009	0.294	ns

These data suggest that current marker technologies, QTL resolution, and breeding mechanisms are sufficient to begin realising the value of MAS in forages. While there is continued scope for development of new marker platforms and tighter marker:trait associations, investments in marker technology per se must be balanced against the value of alternate investments in defining the genetic architecture of additional traits, developing enhanced phenotyping capabilities, and the actual deployment of MAS to demonstrate and realise value from its potential to accelerate genetic gain in forages. Decisions about enhanced marker platforms, with an eye toward perfect (functionally associated) markers, must also be made recognising that a substantial portion of the variation for most economically significant quantitative traits is accounted for by epistasis (Carlborg and Haley 2004), and that we are breeding for dynamic environments, suggesting that some functionally associated markers may still not be perfectly transferable in an ever changing landscape of target outcomes, novel germplasm, and climatic conditions (Humphreys et al. 2006; Newton and Edwards 2006).

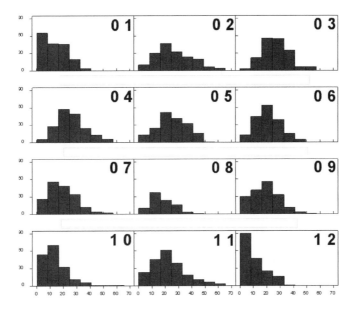

Fig. 1 Seed yield (g/plant) for 12 white clover breeding pools evaluated for a single season in either 2004, 2005, or 2006. Significant ($p < 0.01$) marker:trait associations for seed yield were observed in populations 1–7 and 10 when tested with microsatellite markers associated with QTL SY03-D2. Vertical axes are numbers of individuals, horizontal axes are seed yield (g/plant), numbers in chart are pool numbers

With suitable marker technology available, breeding systems optimised, and substantial opportunities to realise improved genetic gains, MAS may be poised to be broadly utilised in forage breeding. A realistic analysis of breeding targets, population composition, progression of the breeding cycle, trait genetic architecture, marker data collection and integration strategies, selection strategy, freedom to operate, and cost benefit ratios are necessary for success. The data we report suggest that value can be realised from the current investment in genomics for MAS in white clover, given QTL of moderate resolution, and widely used marker platforms such as microsatellites. Research investigating the change in rate of genetic gain associated with marker-informed crossing programmes, and investigation of QTL for high value traits with moderate or low heritability are necessary to more generally define the potential value of MAS in forages.

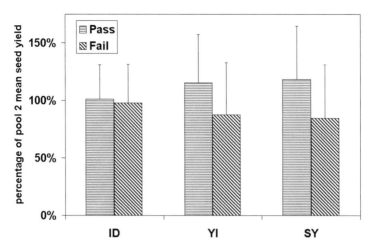

Fig. 2 The effect of the YI and SY associated microsatellite marker on the mean performance of those white clover individuals with (Pass) and without (Fail) the diagnostic amplicon in breeding pool 2, resulting in a 33% change in seed yield observed for those plants which passed versus those which failed a single PCR-based DNA marker test

References

Barrett B, Griffiths A, Schreiber M, Ellison N, Mercer C, Bouton JH, Ong B, Forster J, Sawbridge T, Spangenburg G, Bryan GJ, Woodfield DR (2004) A microsatellite map of white clover. Theor Appl Genet 109:596–608

Barrett B, Mercer C, Woodfield DR (2005a) Genetic mapping of a root-knot nematode resistance locus in *Trifolium*. Euphytica 143:85–92

Barrett BA, Baird IJ, Woodfield DR (2005b) A QTL analysis of white clover seed production. Crop Sci 45:1844–1850

Carlborg O, Haley CS (2004) Epistasis: too often neglected in complex trait studies? Nat Rev Genet 5:618–625

Chapman DF, Mackay AD, Devantier BP, Dymock N (1993) Impact of white clover cultivars on nitrogen fixation and livestock production in New Zealand hill pasture. Proc Int Grassl Cong 17:420–421

Cogan NOI, Abberton MT, Smith KF, Kearney G, Marshall AH, Williams A, Michaelson-Yeates TPT, Bowen C, Jones ES, Vecchies AC, Forster JW (2006a) Individual and multi-environment combined analyses identify QTLs for morphogenetic and reproductive development traits in white clover (*Trifolium repens* L.). Theor Appl Genet 112:1401–1415

Cogan NOI, Ponting RC, Vecchies AC, Drayton MC, George J, Dracatos PM, Dobrowolski MP, Sawbridge TI, Smith KF, Spangenberg GC, Forster JW

(2006b) Gene-associated single nucleotide polymorphism discovery in perennial ryegrass (*Lolium perenne* L.). Mol Genet Genomics 276:101–112

Cogan NOI, Drayton MC, Ponting RC, Vecchies AC, Bannan NR, Sawbridge TI, Smith KF, Spangenberg GC, Forster JW (2007) Validation of in silico-predicted genic SNPs in white clover (*Trifolium repens* L.), an outbreeding allopolyploid species. Mol Genet Genomics 277:413–425

Crush JR, Woodward SL, Eerens JPJ, MacDonald KA (2006) Growth and milksolids production in pastures of older and more recent ryegrass and white clover cultivars under dairy grazing. N Z J Agric Res 49:119–135

Easton HS, Baird DB, Cameron NE, Kerr GA, Norriss M, Stewart AV (2001) Perennial ryegrass cultivars: herbage yield in multi-site plot trials. Proc NZ Grassl Assoc 63:183–188

Faville M, Vecchies A, Schreiber M, Drayton M, Hughes L, Jones E, Guthridge K, Smith K, Sawbridge T, Spangenberg G, Bryan G, Forster J (2004) Functionally associated molecular genetic marker map construction in perennial ryegrass (*Lolium perenne* L.). Theor Appl Genet 110:12–32

Francia E, Tacconi G, Crosatti C, Barabaschi D, Bulgarelli D, Dall'Aglio E, Vale G (2005) Marker assisted selection in crop plants. Plant Cell Tissue Organ Cult 82:317–342

Hayes B, He J, Moen T, Bennewitz J (2006) Use of molecular markers to maximise diversity of founder populations for aquaculture breeding programs. Aquaculture 255:573–578

Helentjaris T, King G, Slocum M, Siedenstrang C, Wegman S (1985) Restriction fragment length polymorphisms as probes for plant diversity and their development as tools for applied plant breeding. Plant Mol Biol 5:109–118

Humphreys MW, Yadav RS, Cairns AJ, Turner LB, Humphreys J, Skøt L (2006) A changing climate for grassland research. New Phytol 169:9–26

Jones ES, Dupal MP, Dumsday JL, Hughes LJ, Forster J (2002) An SSR-based genetic linkage map for perennial ryegrass (*Lolium perenne* L.). Theor Appl Genet 105:577–584

Kölliker R, Boller B, Widmer F (2005) Marker assisted polycross breeding to increase diversity and yield in perennial ryegrass (*Lolium perenne* L.). Euphytica 146:55–65

Newton PCD, Edwards GR (2006) Plant breeding for a changing environment. In: Newton PCD, Carran RA, Edwards GR, Niklaus PA (eds) Agroecosystems in a changing climate. CRC Press, London, pp 309–322

Powell RL, Norman HD (2006) Major advances in genetic evaluation techniques. J Dairy Sci 89:1337–1348

Sawbridge T, Ong E, Binnion C, Emmerling M, Meath K, Nunan K, O'Neill M, O'Toole F, Simmounds JK, Winkworth A, Spangenburg G (2003) Generation and analysis of expressed sequence tags in white clover (*Trifolium repens* L.). Plant Sci 165:1077–1087

Sonesson AK (2007) Within-family marker-assisted selection for aquaculture species. Genet Select Evol 39:301–317

Stendal C, Casler MD, Jung G (2006) Marker-assisted selection for neutral detergent fiber in smooth bromegrass. Crop Sci 46:303–311

Tamaki H, Yoshizawa A, Fujii H, Sato K (2007) Modified synthetic varieties: a breeding method for forage crops to exploit specific combining ability. Plant Breed 126:95–100

Widdup K, Woodfield DR, Baird I, Clifford P (2004) Response to selection for seed yield in six white clover cultivars. Proc N Z Grassl Assoc 66:103–110

Woodfield DR (1999) Genetic improvements in New Zealand forage cultivars. Proc NZ Grassl Assoc 61:3–7

Woodfield DR, Easton HS (2004) Advances in pasture plant breeding for animal productivity and health. N Z Vet J 52:300–310

Woodfield DR, Baird I, Clifford P (2004) Genetic improvement of white clover seed production. Proc NZ Grassl Assoc 66:111–117

Xiong YW, Fei SZ, Brummer EC, Moore KJ, Barker RE, Jung GW, Curley J, Warnke SE (2006) QTL analyses of fiber components and crude protein in an annual × perennial ryegrass interspecific hybrid population. Mol Breed 18: 327–340

Yamada T, Jones ES, Cogan NOI, Vecchies AC, Nomura T, Hisano H, Shimamoto Y, Smith KF, Hayward MD, Forster JW (2004) QTL analysis of morphological, developmental, and winter hardiness-associated traits in perennial ryegrass. Crop Sci 44:925–935

Zhang Y, Sledge MK, Bouton JH (2007) Genome mapping of white clover (*Trifolium repens* L.) and comparative analysis within the Trifolieae using cross-species SSR markers. Theor Appl Genet 114:1367–1378

Molecular Mapping of QTLs Associated with Important Forage Traits in Tall Fescue

Malay Saha[1,2], Francis Kirigwi[1], Konstantin Chekhovskiy[1], Jennifer Black[1] and Andy Hopkins[1]

[1]Forage Improvement Division, The Samuel Roberts Noble Foundation, Ardmore, OK, USA
[2]Corresponding author, mcsaha@noble.org

Abstract. Tall fescue (*Festuca arundinacea* Schreb.) is a major perennial forage crop in the temperate regions of the world. Genetic linkage maps are an essential tool for genome research and have been used for tagging important traits. Tall fescue- and conserved grass EST-SSRs, genomic SSRs from tall fescue and Festuca × Lolium hybrids, and gene-specific STS markers from tall fescue orthologs were used to construct parental linkage maps followed by bi-parental consensus maps. A two-way pseudo testcross mapping strategy was followed to construct the linkage groups. The majority of markers segregated from either parent and showed a 1:1 Mendelian segregation ratio, thus indicating that the loci were in a heterozygous state in one parent and in a homozygous recessive state in the other parent. Markers present in both parents and showing a 3:1 segregation ratio were used for identifying homologous groups between maps. A distinctly reduced level of recombination was observed in the male parent compared to the female parent. Markers in general were evenly distributed throughout the genome. However, clustering of markers in some regions and few gaps of >20 cM in some linkage groups (LGs) were also evident. The mapping population was evaluated in field experiments at Ardmore, OK, for three consecutive years. Data on morphological, reproductive and quality traits were collected. Quantitative trait loci (QTL) and markers associated with these traits were identified. Marker-assisted breeding was initiated with markers associated with forage digestibility.

T. Yamada and G. Spangenberg (eds.), *Molecular Breeding of Forage and Turf*,
doi: 10.1007/978-0-387-79144-9_23, © Springer Science + Business Media, LLC 2009

Introduction

The relatively high abundance of microsatellites in most genomes, their high rate of mutations, simple Mendelian inheritance, co-dominant nature, locus specificity and interspecies transferability make them a marker class of choice (Jones et al. 2002; La Rota et al. 2005; Zhang et al. 2005). Genetic linkage maps are an essential tool in identifying the genomic regions that control traits of agronomic interest. The 'pseudo test cross' strategy has been widely applied in the construction of genetic linkage maps for a number of species such as tall fescue (Saha et al. 2005), ryegrass (Warnke et al. 2004) and alfalfa (Echt et al. 1994). Markers segregating in 1:1 ratio are ideal for framework map construction and those segregating in a 3:1 ratio can act as bridging loci to align homologous chromosomes of male and female linkage maps (Maliepaard et al. 1998).

Tall fescue (*Festuca arundinacea* Schreb.) is the most important cool-season perennial forage grass species in the United States. Improved cultivars with enhanced forage quality and yield are the major objectives for forage breeders. The advent of molecular markers provides the opportunity to determine the number, position, and individual effects of loci showing quantitative inheritance (Studer et al. 2006). In tall fescue, the biological functions of genes involved in forage digestibility and biomass production remain unknown. Thus, the objectives of this study were to develop a microsatellite map of tall fescue and use the map to identify quantitative trait loci (QTL) associated with forage digestibility and biomass yield. Molecular markers associated with the traits of interest will be of great value to initiate marker-assisted selection (MAS).

Materials and Methods

The mapping population used in this study was derived from a cross between HD28-56 (♀) and R43-64 (♂). The HD28-56 is the high digestibility parent and was kindly provided by Dr. David Sleper, University of Missouri, USA. R43-64 is a genotype with good agronomic attributes and high persistence selected from accession 97TF1 collected from Woodward Co., OK. A pseudo F_1 population of 124 genotypes was constructed and evaluated for 3 years from 2002 to 2005 in replicated field experiments following a randomized complete block design at Ardmore, OK, USA. Maturity and in vitro dry matter digestibility (IVDMD) data were collected. IVDMD was determined from the harvested dry forages following the protocol suggested by Ankom Technology (Ankom Technology Corp., Fairport, NY). Genomic DNA was extracted from tender leaves using a DNeasy® DNA Extraction Kit (QIAGEN Inc., Valencia, CA, USA).

Tall fescue expressed sequence tag (EST)- and genomic simple sequence repeat (SSR), tall fescue sequence tag site (STS) markers, conserved grass EST-SSRs and *Festuca × Lolium* hybrid genomic SSRs were used to construct the genetic linkage maps. Forward primers were modified by M13 tail at the 5′ end which permitted concurrent fluorescence labeling of PCR products. A detailed genotyping protocol can be obtained from Saha et al. (2005). JOINMAP ver. 4.0 (http://www.kyazma.nl) was used for defining the parental linkage groups (LGs) using a LOD threshold of 4.0 or above. LGs were constructed by treating the segregation data as a cross pollinator (CP). Kosambi mapping function was applied for calculating the map distances. Digestibility data were first analyzed using single-factor analysis of variance with SAS followed by interval- and multiple QTL mapping (MQM) using MapQTL version 4.0 (Van Ooijen 2004). Significance threshold for declaring a QTL were confirmed by permutation tests with a chromosome- and genome wide level significance of $\alpha = 0.05$, $n = 1,000$.

Results

Molecular Markers

A total of 384 tall fescue EST-SSRs, 511 tall fescue genomic SSRs, 96 tall fescue STS markers, 101 conserved grass EST-SSRs and 60 *Festuca × Lolium* hybrid genomic SSRs were first screened and then the selected polymorphic markers were used for genotyping the whole population. The EST primers generated 201 polymorphic markers, with an average of 1.7 markers per primer. The genomic SSR primers generated 376 polymorphic markers with an average of 1.9 markers per primer. From a total of 577 loci scored for mapping, 90% segregated from either parent (genetic constitution of Aa·aa or aa·Aa) and the remaining 10% segregated from both parents (Aa·Aa). A very high level of segregation distortion was observed among the scored markers.

Map Construction

The HD28-56 map was constructed with 276 markers distributed among 21 LGs. All 21 LGs were covered by both EST- and genomic SSR loci except LG 5 which only comprised of genomic SSRs. Markers in the 21 LGs covered a total length of 1,580.4 cM with an average of 13 markers per LG. The average marker density was 5.73 cM/marker. The R43-64 map was constructed with a total of 235 markers. The total length covered by 22 LGs was 1,363.8 cM with an average length of 61.9 cM/LG. The average marker density was 5.8 cM/marker. Many of the distorted markers were concentrated in LG 1 of the R43-64 map (Fig. 1).

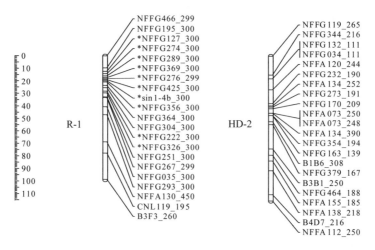

Fig. 1 Tall fescue parental [R43-64 (R) and HD28-56 (HD)] LGs. The *first letters* of the locus name represents the primer type (*NFFG* Noble genomic SSR; *NFFA* Tall fescue EST-SSR; *CNL* conserved grass EST-SSR; *B-* Festuca × Lolium genomic SSR). The *next three digits* represent primer number and the *last three digits* indicate the amplified fragment size in bp. Distorted markers (prefix *) are clustered on LG 1 of the male parental map (R)

QTL Analysis

A large variation was observed in the parents and progeny of the mapping population for the forage digestibility and morphological traits observed over 3 years. Here, we report the digestibility QTL identified in two parental maps. In general, higher digestibility was recorded from the fall than from the spring harvests. Both parents contributed significantly to forage digestibility. Thirteen linkage groups of each parent contributed significant QTLs associated with higher and lower digestibility. However,

QTL on HD28-56 parental LGs 1, 4, and 19 and R43-64 LGs 15 and 18, were consistently identified across several years. Figure 2 summarizes the QTL detected on HD28-56 parental LG 1. Step wise regression analysis indicated that only few markers mainly contributed to the overall expression of the trait (data not shown). We finally selected seven and six markers associated with high- and low digestibility, respectively, to initiate marker-assisted breeding (Table 1).

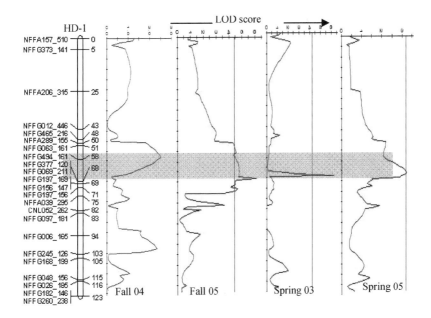

Fig. 2 Digestibility QTL identified on HD28-56 parental LG 1 (HD-1) identified across different environments (years, seasons). Significant QTL were detected in Fall 05, Spring 03 and Spring 05

Table 1 Marker loci associated with high- and low digestibility selected for marker-assisted selection

Locus	Effect[a]	LG/parent	Locus	Effect[a]	LG/parent
Positive effect			Negative effect		
NFFG494_161	4.4	HD-1	NFFG416_122	−4.3	R-15
NFFG388_140	3.8	HD-4	NFFA338_259	−2.3	R-18
NFFG147_128	4	HD-5	NFFG407_241	−3.3	R-18
NFFA016_222	2.5	HD-unmapped	fve2g_195	−9.5	R-unmapped
NFFG344_183	3.8	R-4	NFFG465_216	−2.6	HD-1
NFFG240_258	4.2	R-unmapped	NFFG192_277	−4.7	HD-19
NFFG299_112	10.1	R-unmapped			

[a]Additive regression coefficient (%A)

Discussion

Microsatellite markers can be developed from either genomic or EST sequences. EST-SSR markers are likely to be more conserved and thus more transferable across species than genomic-SSRs (Decroocq et al. 2003). Genomic SSRs tend to be widely distributed throughout the genome resulting in better map coverage (Warnke et al. 2004; La Rota et al. 2005). Using a combination of EST- and genomic-SSR markers is an important strategy for genetic linkage mapping. The double pseudo-testcross strategy was used to construct the parental maps (Alm et al. 2003; Echt et al 1994; Saha et al. 2005; Warnke et al. 2004). Tall fescue is an allohexaploid with disomic inheritance. Markers heterozygous in one parent (Aa) and recessive homozygous in the other parent (aa) segregate in a 1:1 ratio (similar to backcross) and markers heterozygous in both parents (Aa × Aa) segregate in a 3:1 ratio. Markers segregating in a 1:1 ratio were used to construct the framework map and, in a later attempt, 3:1 and distorted markers were added to the map. Marker density in both parental maps was similar but the female map spanned 217 cM more than the male map. Differences in map length can result from a variation in the number of recombination events in the two parents as well as variations in the numbers and locations of mapped loci.

Several QTL were found in this study with up to 4.4% effect on forage digestibility as estimated using IVDMD. Besides, several unmapped markers were also associated with the trait. It is estimated that a 1% increase in IVDMD can lead up to 3.2% increase in daily animal live-weight gains (Casler and Vogel 1999). It would be useful to select for alleles leading to increased digestibility while at the same time eliminating alleles contri-buting to decreased digestibility. The use of MAS with multiplexing for markers associated with high IVDMD QTL may lead to accelerated increases in forage digestibility. Three different populations are being developed i.e. (a) based on MAS, (b) based on phenotypic selection (IVDMD analysis), and (c) based on a combination of marker (MAS) and phenotypic selection, to assess the effectiveness of MAS to increase forage digestibility. These populations will be planted in the field from fall 2008.

Conclusion

We constructed a microsatellite map of tall fescue and used it to map QTL for IVDMD. Our analysis indicated that forage digestibility is controlled by several QTL. Alleles leading to increased or decreased IVDMD were identified. Markers identified in this study are being evaluated through several selection schemes to verify their effectiveness in increasing forage digestibility.

References

Alm V, Fang C, Busso CS, Devos KM, Vollan K, Grieg Z, Rognli OA (2003) A linkage map of meadow fescue (*Festuca pratensis* Huds.) and comparative mapping with other Poaceae species. Theor Appl Genet 108:25–40

Casler MD, Vogel KP (1999) Accomplishments and impact from breeding for increased forage nutritional value. Crop Sci 39:12–20

Decroocq V, Favé MG, Hagen L, Bordenave L, Decroocq S (2003) Development and transferability of apricot and grape EST microsatellite markers across taxa. Theor Appl Genet 106:912–922

Echt CS, Kidwell KK, Osborn TC, Knapp SJ, McCoy TJ (1994) Linkage mapping in diploid alfalfa (*Medicago sativa*). Genome 37:61–71

Jones ES, Dupal MP, Dumsday JL, Hughes LJ, Forster JW (2002) An SSR-based genetic linkage map for perennial ryegrass (*Lolium perenne* L.). Theor Appl Genet 105:577–584

La Rota M, Kantety RV, Yu JK, Sorrells ME (2005) Nonrandom distribution and frequencies of genomic and EST-derived microsatellite markers in rice wheat and barley. BMC Genomics 6:23–35

Maliepaard C, Alston FH, van Arkel G, Brown LM, Chevreau E, Dunemann F, Evans KM, Gardiner S, Guilford P, van Heusden AW, Janse J, Laurens F, et al. (1998) Aligning male and female linkage maps of apple (*Malus pumila* Mill) using multi-allelic markers. Theor Appl Genet 97:60–73

Saha MC, Mian MAR, Zwonitzer JC, Chekhovskiy K, Hopkins AA (2005) An SSR and AFLP based genetic linkage map of tall fescue. Theor Appl Genet 110:323–336

Studer B, Boller B, Herrmann D, Bauer E, Posselt UK, Widmer F, Kölliker R (2006) Genetic mapping reveals a single major QTL for bacterial wilt resistance in Italian ryegrass (*Lolium multiflorum*). Theor Appl Genet 113:661–671

Van Ooijen JW (2004) MapQTL®5, software for the mapping of quantitative trait loci in experimental populations. Kyazma B.V., Wageningen, the Netherlands

Warnke SE, Barker RE, Jung G, Sim SC, Mian MAR, Saha MC, Brilman LA, Dupal MP, Forster JW (2004) Genetic linkage mapping of an annual × perennial ryegrass population. Theor Appl Genet 109:294–304

Zhang LV, Bernard M, Leroy P, Feuillet C, Sourdille P (2005) High transferability of bread wheat EST-derived SSRs to other cereals. Theor Appl Genet 111:677–687

Utilizing Linkage Disequilibrium and Association Mapping to Implement Candidate Gene Based Markers in Perennial Ryegrass Breeding

Kevin F. Smith[1,3,4], Mark P. Dobrowolski[1,3], Noel O.I. Cogan[2,3], Germán C. Spangenberg[2,3] and John W. Forster[2,3]

[1]Primary Industries Research Victoria, Hamilton Centre, Hamilton, VIC 3300, Australia, kevin.f.smith@dpi.vic.gov.au
[2]Primary Industries Research Victoria, Victorian AgriBiosciences Centre, La Trobe Research and Development Park, Bundoora, VIC 3083, Australia
[3]Molecular Plant Breeding Cooperative Research Centre, Australia
[4]Corresponding author, kevin.f.smith@dpi.vic.gov.au

Abstract. Development of accurate high-throughput molecular marker systems such as SNPs permits evaluation and selection of favourable gene variants to accelerate elite varietal production. SNP discovery in perennial ryegrass has been based on PCR amplification and sequencing of multiple amplicons designed to scan all components of the transcriptional unit.

Full-length genes (with complete intron–exon structure and promoter information) corresponding to well-defined biochemical functions such as lignin biosynthesis and oligosaccharide metabolism are ideal for complete SNP haplotype determination. Multiple SNPs at regular intervals across the transcriptional unit were detected within and between the heterozygous parents and validated in the progeny of the $F_1(NA_6 \times AU_6)$ genetic mapping family. Haplotype structures in the parental genotypes were defined and haplotypic abundance, structure and variation were assessed in diverse germplasm sources. Decay of LD to r^2 values of c. 0.2 typically occurs over 500–3,000 bp, comparable with gene length and with little apparent variation between diverse, ecotypic and varietal population subgroups. Similar patterns were revealed as limited blocks of intragenic LD. The results are compatible with the reproductive biology of perennial ryegrass and the effects of large ancestral population size. This analysis provides crucial information to validate strategies for correlation of haplotypic diversity and phenotypic variation through association mapping.

T. Yamada and G. Spangenberg (eds.), *Molecular Breeding of Forage and Turf*, doi: 10.1007/978-0-387-79144-9_24, © Springer Science + Business Media, LLC 2009

Introduction

Functionally-associated genetic markers for use in molecular breeding of crop plants are based on nucleotide polymorphism located close to or within candidate genic sequences (Andersen and Lübberstedt 2003). Effective correlation of such sequence diversity with related phenotypic trait variation provides diagnostic markers for direct selection of superior allele content in germplasm improvement programs (Sorrells and Wilson 1997). Such systems have been proposed to be critically important for outbreeding forage species, which employ complex cultivar development strategies not ideally suited to linked marker allele-trait gene introgression (Forster et al. 2004; Spangenberg et al. 2005; Dobrowolski and Forster 2007). Single nucleotide polymorphisms (SNPs) are the most versatile class of functionally-associated genetic marker. Efficient methods for in vitro discovery of SNPs and characterisation of SNP haplotype structure have been described for perennial ryegrass (*Lolium perenne* L.) (Spangenberg et al. 2005; Cogan et al. 2006).

The ability to correlate genetic and phenotypic variation through association analysis is determined by the extent of linkage disequilibrium (LD), which is expected to be limited in allogamous pasture species, especially for long-established populations derived from a large number of parental genotypes (Mackay 2001; Forster et al. 2004; Dobrowolski and Forster 2007). In a previous study (Cogan et al. 2006), rapid decay of LD was observed within the perennial ryegrass abiotic stress-associated *Lp*ASRa2 gene, based on analysis of SNP variation in diverse germplasm. Proof-of-concept for association mapping in perennial ryegrass would be more readily obtained for genes related to well-characterised physiological processes with established agronomic outcomes. Full-length genes encoding enzymes involved in herbage nutritive quality traits fulfill both of these criteria.

Diagnostic genetic markers for herbage quality traits are of be of high value for implementation in pasture grass molecular breeding. The functional categories of lignin biosynthesis and oligosaccharide metabolism provide primary candidate genes (Forster et al. 2004; Cogan et al. 2006) for SNP discovery and association mapping of nutritive quality characters. *Lolium* cDNAs encoding enzymes associated with these pathways have been isolated and characterised (Chalmers et al. 2003, 2005; Gallagher and Pollock 1998; Gallagher et al. 2004; Heath et al. 1998, 2002; Lidgett et al. 2002; Lynch et al. 2002; Johnson et al. 2003; McInnes et al. 2002; Yamada et al. 2005). Full-length genomic clones have also been isolated for the lignin metabolism enzyme-encoding genes *Lp*CCR1 (McInnes et al. 2002), (cinnamoyl CoA-reductase) and *Lp*CAD2 (Lynch et al. 2002) (one of several cinnamyl alcohol dehydrogenases), as well as the oligosaccharide metabolism enzyme-encoding genes *Lp*FT1 (Lidgett et al. 2002) (putative sucrose:fructose 6-fructosyltransferase), and *Lp*1-SST (formerly *Lp*FT3: Chalmers et al. 2003) (sucrose:sucrose 1-fructosyltransferase). In addition to sequence annotation and functional genomics analysis, co-location of quantitative trait loci (QTLs) herbage quality components with the map locations of *Lp*CAD2 and *Lp*CCR1 (Cogan et al. 2005) provides support for the biological significance of these genes.

Full-length genomic sequences provide the opportunity for in vitro SNP discovery over substantial molecular distances across transcriptional units including promoter, untranslated region (UTR), exon and intron elements. Confirmation of SNP locus and haplotype structure in specific target genotypes, such as the parental genotypes of the $F_1(NA_6 \times AU_6)$ reference genetic mapping family (Faville et al. 2004; Cogan et al. 2006), may be followed by validation in larger customised germplasm samples to evaluate haplotype structure and LD (Cogan et al. 2006). This paper describes the application of the strategy to the full-length herbage quality genes and evaluation of haplotype structure, haplotype prevalence and gene-length LD. The results of this analysis are applicable to association mapping strategies for a broad range of economically important forage species.

Materials and Methods

Plant Materials

Genomic DNA was extracted from parents and progeny of the $F_1(NA_6 \times AU_6)$ genetic mapping population using methods described in Ponting et al. (2007). The association mapping panel (AMP) incorporated material from six distinct population sub-groups (Ponting et al. 2007), selected in order to permit evaluation of the effect of population complexity on haplotype structure and LD. The sub-group of diverse genotypes included individuals from a broad range of ecotypic and varietal sources including both pasture and turf types. The long-established Victorian ecotype is derived from introductions from Europe to Australia which have undergone local adaptation to lower rainfall regions of south-eastern Australia.

Ecotype Kangaroo Valley is adapted to a higher rainfall region of New South Wales, while Zürich Uplands is characterised by high levels of pseudostem water soluble carbohydrate (WSC) and was the germplasm source for development of variety Aurora. New Zealand-derived Cultivars A and B were both derived from the polycross of four distinct parental genotypes, and were included to determine the degree of 'founder effect' through imposition of population bottlenecks.

SNP Genotyping

SNP detection was performed using the single nucleotide primer extension (SNuPe) genotyping kit (GE Healthcare) and separation using the MegaBACE 1,000 and 4,000 platforms (GE Healthcare), as described by Cogan et al. (2006).

Linkage Disequilibrium Analysis

Haplotypes were reconstructed from unphased SNP data using the PHASE 2.1 Bayesian statistical analysis method (Stephens et al. 2001; Stephens and Donnelly 2003). From the most probable haplotypes derived from this analysis r^2 and $|D'|$ were calculated and visualised using Haploview (Barrett et al. 2005), excluding missing SNP data for which PHASE inserts random estimators. The decay of LD with distance in base pairs (bp) was evaluated using nonlinear regression using S-PLUS (version 6.1, Insightful Corporation, Seattle, USA) as described in Remington et al. (2001) using the expectation of r^2 described by Hill and Weir (1988) appropriate for low level of mutation and adjustment for sample size.

Results

In Vitro SNP Discovery and Validation

The *Lp*CAD1, *Lp*CCR1, *Lp*FT1 and *Lp*1-SST genes were predicted to contain 69, 198, 667 and 191 SNPs, respectively (Ponting et al. 2007)). Some amplicons generated multiple haplotypes indicative of gene dupli-cation effects, especially within the *Lp*CAD2 and *Lp*1-SST coding regions and the promoter-proximal regions of *Lp*CCR1 and (to a lesser extent) *Lp*FT1.

SNuPe-based SNP validation permitted characterisation of polymorphism and haplotype structure within and between parental genotypes. Visual assessment of structural affinities between haplotypes allowed tentative identification of haplogroups. Two of the three *Lp*CAD2 haplotypes and three of the four *Lp*FT1 haplotypes showed obvious structural similarities, but all *Lp*CCR1 and *Lp*1-SST haplotypes were distinct. The distances between the two most remote SNP loci for each gene were 3,309 bp (*Lp*CAD2), 4,210 bp (*Lp*CCR1), 7,140 bp (*Lp*FT1) and 4,269 bp (*Lp*1-SST), with a mean value of 4,732 bp (Ponting et al. 2007). These values define the maximum separation between genetic loci for estimation of LD decay, and are comparable with typical gene length in many higher plant species.

LD Analysis

SNP genotyping for each of the 71 validated SNP loci from the four full-length genes was performed across the AMP. Following most-probable haplotype reconstruction, the rate of decay of LD, as measured by decline to $r^2 = 0.2$, was evaluated as a function of distance for each gene across each genotypic sub-set. The general trend of LD decay is broadly similar for each population sub-group across each of the four genes. For *Lp*CAD2 (Fig. 1), the average distance for decline to $r^2 = 0.2$ varies from c. 500 bp to c. 2,000 bp.

This variation is not obviously correlated with immediate population complexity, as the most rapid rate of LD decline is observed for Cultivar B (derived from four parental genotypes) while the slowest rate of decline is observed for the ecotypes Kangaroo Valley and Victorian. Similar behaviour is observed for *Lp*CCR1, for which the average distance associated with LD decline to $r^2 = 0.2$ varies from c. 500 bp to c. 1,500 bp, with little obvious difference between sub-groups. For *Lp*FT1, more rapid LD decline is observed for the diverse genotypes ($r^2 = 0.2$ at c. 500 bp) as compared to the ecotypic sub-groups ($r^2 = 0.2$ at c. 1,000–3,000 bp), but differences are also observed between the cultivar samples, such that Cultivar A exhibits more rapid decline than Cultivar B. Minimal LD was detected between any sites within the *Lp*1-SST gene (Fig. 2), except for the Zürich Uplands ecotype.

Data was also analysed in the form of 'heat maps' using the $|D'|$ metric, to highlight blocks of elevated LD. (Ponting et al. 2007) For *Lp*CAD2, the most obvious regions of LD are within the 3'-region of the gene, from

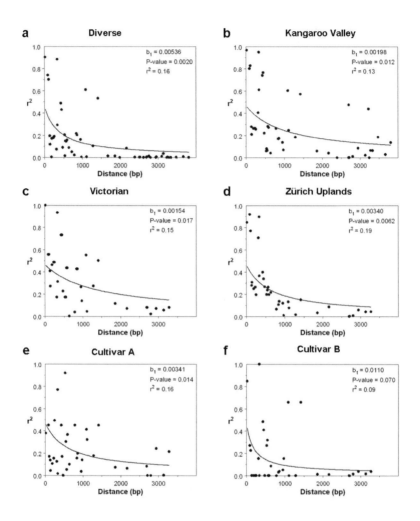

Fig. 1 Plots of LD, as measured by squared correlations of allele frequencies (r^2), against distance between SNP loci (base pairs) in the *Lp*CAD2 gene in diverse perennial ryegrass (**a**), the ecotypes Kangaroo Valley (**b**), Victorian (**c**), and Zürich Uplands (**d**), and the cultivars A (**e**) and B (**f**)

exon 2 onwards. This broad pattern is observed for each group, apart from Cultivar B, extending over 1–2 kb. For *Lp*CCR1 blocks of extensive LD are observed over c. 1 kb within intron 4 in the central portion of the gene, especially for the diverse genotypic sub-group, the Kangaroo Valley ecotype and Cultivar B. As observed for *Lp*CAD2, the position of blocks of elevated LD is biased towards the 3′-region of the gene. For *Lp*FT1

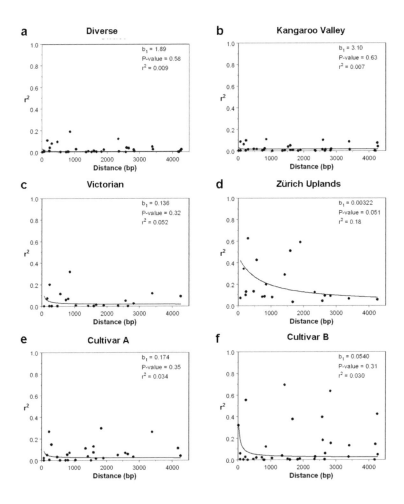

Fig. 2 Plots of LD, as measured by squared correlations of allele frequencies (r^2), against distance between SNP loci (base pairs) in the Lp1-SST gene. Details are as described in the legend for Fig. 1

elevated regions of LD are also observed predominantly in the 3′-region of the gene. The most obvious examples were for the Zürich Uplands ecotype (over c. 400 bp in the vicinity of exon 4) and Cultivar B (over c. 1,200 bp spanning intron 3 and exon 4). No obvious blocks of LD were observed for Lp1-SST (syn. LpFT3) consistent with the decay curves.

Haplotype Number and Diversity

The information for assessment of gene-length LD was based on recon-struction of most probable haplotypes. Numbers of reconstructed haplo-types with frequencies greater or equal to 1% in the relevant population sub-group were obtained, along with the proportions of total haplotype variation explained by the most common sets of all haplotypes. For the diverse and ecotypic population sub-groups, this was performed for the most common 5, 10 and 20 haplotypes. The majority of haplotype vari-ation was explained by up to ten most common haplotypes, except for the *Lp*FT1 gene. For the cultivars, the proportions explained by the eight most common haplotypes were determined, representing the theoretical maximum for these sub-groups. Whilst the number varied between 9 and 40 (Ponting et al. 2007) according to gene and cultivar, the general trend was for either eight or ten haplotypes to account for approximately 80–90% of the haplotypic variation.

Discussion

Efficiency of In Vitro SNP Discovery and Validation

The efficiency of the in vitro SNP discovery process was previously assessed for c. 100 genes from a broad range of functional categories, employing low, medium- and high-intensity activity streams (Cogan et al. 2006). The full-length genes in the present study were formerly included in the high-intensity activity stream, identifying 1,125 putative SNPs across the four genes. The incidence of predicted SNPs solely across the resequ-enced regions of the four genes was 1 per 28 bp, compared to 1 per 54 bp for the larger sample group. However, this value was biased by the high predicted SNP frequency in the *Lp*FT1 and *Lp*1-SST genes (1 per 13 bp and 1 per 21 bp, respectively), relative to *Lp*CAD2 and *Lp*CCR1 (1 per 104 bp and 1 per 58 bp, respectively). These larger predicted incidences may be related to high levels of intron variability, or, more likely, to the inclusion of paralogous amplification products in sequence contigs.

SNPs additional to those previously described (Cogan et al. 2006) were validated for the purpose of LD analysis. The validation efficiency was 45% across the four genes, with a range from 23% (*Lp*CCR1) to 84% (*Lp*CAD2). This proportion is lower than the value of 60% reported by Cogan et al. (2006), which excluded the effects of failed genotypic assays (presumably due to locus amplification primer site mutations and SNP clustering) and failure to detect polymorphism in F_1 progeny sets due to the detection of paralogous sequence variants (PSVs) arising between duplicated gene copies (Fredman et al. 2004). Both factors were evident for the target genes during amplicon sequence alignment.

Extent of Linkage Disequilibrium

The collective data for the four full-length genes demonstrates rapid decay of LD over the transcriptional unit, along with the frequent presence of limited block structure. As for LpASRa2, the data is compatible with a complex interplay of recombination, mutation and gene conversion events (Cogan et al. 2006). Although the numbers of genotypes analysed are larger than the 35 genotypes used to detect LD decay in LpASRa2, especially for the diverse sub-group, the sample size may still be limiting for accurate estimation. Underlying population structure arising from sub-division or admixture may also contribute to overestimation of LD (Mackay 2001; Flint-Garcia et al. 2003). However, considering the limited region over which LD is observed to extend in the full-length genes, this effect is not likely to be of major significance for perennial ryegrass.

Although three of the full-length genes (LpCAD2, LpCCR1 and LpFT1) show regular LD decay over distances from 500 to 3,000 bp, minimal LD is observed for the Lp1-SST gene. This may reflect a complex and extensive recombinational gene history. It is also possible that the validated SNPs have been obtained from closely linked duplicated gene copies. In this case, co-segregation in the $F_1(NA_6 \times AU_6)$ progeny set would be observed, but the molecular distances between paired loci may be larger than predicted and in excess of the average decay range. The resolution of genetic mapping studies is insufficient to address this possibility, which requires detailed physical characterisation of contiguous DNA sequences.

No consistent trend in terms of population sub-group behaviour, which could be related to population complexity or parental origin, was observed for either decay curve or heat map analysis. The reason for apparent bias in distribution of LD blocks towards 3'-genic regions is unclear. This may be an artefact of the set of genes sampled in this study, or possibly due to biased recombination frequency across the transcriptional unit reflecting the presence of 'hot-spots'. A number of the full-length genes possess introns of variable length in the 5'-region, which may contain non-coding recombination-promoting elements.

Previous studies of LD distribution in perennial ryegrass have been based on whole-genome scans with AFLP loci in the context of an association mapping study (Skøt et al. 2002, 2005b). Natural populations from Europe were selected on the basis of pre-existing genecological data (Sackville-Hamilton et al. 2002) for traits such as cold tolerance and flowering time variation. A small number of AFLP loci were identified as showing potential association both with one another and with genes controlling the flowering time trait, for which identified loci were inferred on the basis of co-migration to map in the vicinity of a major heading date variation QTL on LG7 (Armstead et al. 2004). However, significant LD was also observed between unlinked loci on different LGs, indicative of possible residual population structure effects. The observed associations between genetically-linked loci may hence also be spurious in nature (Dobrowolski and Forster 2007): alternatively, positive selection for flowering time variation locus on LG7 may have generated an extended LD block which is atypical in size. Subsequent studies based on SNP discovery across the 6.3 kb transcriptional unit of the Lp*AlkInv* alkaline invertase gene (Skøt et al. 2005a) revealed decay of LD to $r^2 = 0.1$ over 2–3 kb, comparable with the more extensive dataset reported here. Long-range LD in perennial ryegrass is likely to be limited on the basis of these studies, except in the instances of strong directional selection and narrow population bottlenecks.

The extent of LD observed across the diverse and ecotypic germplasm samples is compatible with that observed for other outbreeding species, such as land-race populations (Remington et al. 2001) and selected European elite samples (Zein et al. 2007) of maize, potato (Simko et al. 2006) and forestry species. Limited extent of LD is characteristic of species with outbreeding reproductive habits and large effective population sizes, either recently or in the ancestral past. Perennial ryegrass exhibits gametophytic self-incompatibility (Cornish et al. 1979) and high levels of intra- and inter-population genetic diversity (Guthridge et al. 2001), similar to the majority of other temperate forage species, and consistent with these characteristics. Although cytological studies of *Lolium* species have revealed limited and localised chiasma distributions, which also vary in frequency between populations on the basis of longevity and intensity of selection (Rees and Ahmad 1963; Jones and Rees 1966; Rees and Dale 1974), the cumulative effects of genetic recombination in large populations during species history would be expected to effectively dissipate long-range LD.

Haplotype Structure

Haplotype number data is biased by the relatively different numbers of individuals sampled for haplotype reconstruction (192 for the diverse set, 64 for each of the ecotypes, 48 for each of the cultivars). Haplotype number is consequently over-estimated for the cultivars, as very rare haplotypes may arise due to data errors or ambiguities in phase reconstruction and even a single occurrence will reach the >1% threshold. Nonetheless, the most noticeable observations for the diverse and ecotypic sub-groups are the relatively similar number of predicted haplotypes, and the large proportion of total haplotype variation for each sub-group explained by the ten most common haplotypes, with the partial exception of LpFT1. As many of these common haplotypes are apparently shared between the diverse and ecotypic sub-groups, the majority of genic sequence variation in perennial ryegrass may be moderately discrete in nature. The observed rapid decay of LD with distance is consistent with detection of multiple haplotypes.

For the cultivars, discrepancy between predicted and actual haplotype numbers is apparent. Both varieties are restricted-base synthetics (Guthridge et al. 2001) derived from four parental clones, with a maximum predicted haplotype number of eight in the absence of very rare intragenic recombination events. Most, but not all, of the total haplotype variation for each gene-cultivar combination was explained by the eight most common haplotypes. Although some of the excess haplotypes may indeed arise from analytical artefacts, the possibility remains that cultural practices during varietal development may have increased the haplotype pool of the cultivars. Gene flow by pollen dispersal and incorporation of adventitious maternal genotypes during seed multiplication are possible contributory mechanisms. Similar effects, especially those of gene flow, were inferred to have influenced estimates of allelic diversity at the S and Z self-incompatibility genes of perennial ryegrass (Devey et al. 1994).

Interpretation of the data on haplotype sharing is potentially complicated by some of the biasing factors described above, particularly the different numbers of individuals in each sub-group. The criterion of $p \geq 0.01$ for inclusion in the graphical data ensures that rare haplotypes within the diverse or ecotypic sub-groups which are common within the cultivar samples are not readily identified: such haplotypes are present and may be represented at proportions of up to 3 copies among the 384 possible haplotypes in the diverse set without inclusion in this analysis. Nonetheless, the degree of differentiation shown by the cultivar groups is striking and may have arisen from the imposition of a population bottleneck during varietal development. However, this interpretation is not so consistent with the excess haplotype numbers observed in the cultivars. A possible explanation for this apparent discrepancy is population admixture early in the history of the cultivated varieties, through the proposed mechanisms, but largely attributable to specific rather than generic germplasm sources. Such effects are compatible with the structure of many commercial breeding programs.

Implications for Association Genetics Studies

The rate of decay of LD in all of the population groups is comparable to gene length, matching both knowledge of species biology and the objective of diagnostic development for superior allele content (as assessed by haplotype structure) in target genes. The similar rates of LD decay observed for diverse genotypes, ecotypes and cultivars do not correlate with recent population histories, especially for restricted-base varieties. Estimated effective population sizes (N_e) for these distinct sub-groups during the recent past would be very different, leading to predictions of more extensive LD over longer physical and genetic distances. However, limited variation over shorter distance corresponding to gene-length LD is consistent with predicted large effective population sizes in the distant past (equal or greater than 5,000 generations ago, in this instance), assuming constant recombination rates, as demonstrated for human and dairy cattle populations (Hayes et al. 2003). Association mapping based on candidate gene variation is consequently likely to be useful for all of these population types, not solely for diverse and ecotypic sample sets. It is also possible that the reduced effective population sizes typical of ecotypic and varietal populations may contribute to superimposition of recently-generated long-range LD on an ancestral short-range pattern. In this case, whole-genome scans using SSR and SNP may provide identification of target genomic regions at low resolution, prior to candidate gene-based haplotype analysis.

The candidate gene-based approach may be potentially hindered by the more discrete haplotype distributions observed in cultivars, and to some extent in ecotypes, as haplotypes with significant phenotypic effects may have been excluded from these populations simply by random sampling. In addition, the power to detect significant associations with low frequency haplotypes can be limited for all population types, irrespective of functional effect. This problem may be addressed through hierarchical clustering of highly similar haplotypes, such as those differentiated by small number of synonymous or non-coding base variants, into affinity groups (haplogroups) (Olsen et al. 2004). Haplogroups may correspond to functionally-distinct units, and collectively provide the basis for association of higher frequency genetic variants with trait-specific variation. Comparative correlation analysis based on both predicted haplogroups and individual haplotypes provides an empirical test of these possibilities. Alternatively, individual SNP loci may show higher predictive capability and provide high resolution of detection (Neale and Savolainen 2003), although such SNPs would be required to be in strong LD with the relevant functional change or QTN (qualitative or quantitative trait nucleotide).

Conclusions

The results of this study demonstrate that extensive SNP locus data may be obtained for full-length gene sequences of perennial ryegrass, and used to determine SNP haplotype structure, prevalence and stability in various population types. Detailed studies of gene-length LD have been performed for the first time, revealing rapid decay and high nucleotide diversity, consistent with the biology of the species and appropriate to candidate gene-based approaches to association genetics. The results reported here are also relevant to other outbreeding pasture grasses such as Italian ryegrass (*Lolium multiflorum* Lam.), meadow fescue (*Festuca pratensis* Hud.), tall fescue (*Festuca arundinacea* Schreb.), cocksfoot (*Dactylis glomerata* L.) and timothy (*Phleum pratense* L.), as well as the forage legumes white clover (*Trifolium repens* L.), red clover (*Trifolium pratense* L.) and alfalfa (*Medicago sativa* L.).

Acknowledgements

This work was supported by funding from the Victorian Department of Primary Industries, Dairy Australia Ltd., the Geoffrey Gardiner Dairy Foundation, Meat and Livestock Australia Ltd. and the Molecular Plant Breeding Cooperative Research Centre. The authors thank Prof. Michael Hayward and Dr. Ben Hayes for careful critical assessment of the manuscript.

References

Andersen JR, Lübberstedt T (2003) Functional markers in plants. Trends Plant Sci 8: 554–560

Armstead IP, Turner LB, Farrell M, Skøt L, Gomez P, Montoya T, Donnison IS, King IP, Humphreys MO (2004) Synteny between a major heading-date QTL in perennial ryegrass (*Lolium perenne* L.) and the *Hd3* heading-date locus in rice. Theor Appl Genet 108: 822–828

Barrett JC, Fry B, Maller J, Daly MJ (2005) Haploview: analysis and visualisation of LD and haplotype maps. Bioinformatics 21: 263–265

Chalmers J, Johnson X, Lidgett A, Spangenberg GC (2003) Isolation and characterisation of a sucrose:sucrose 1-fructosyltransferase gene from perennial ryegrass (*Lolium perenne* L.). J Plant Physiol 160: 1385–1391

Chalmers J, Lidgett A, Johnson X, Jennings K, Cummings N, Forster J, Spangenberg G (2005) Molecular genetics of fructan metabolism in temperate grasses. Plant Biotechnol J 3: 459–474

Cogan NOI, Smith KF, Yamada T, Francki MG, Vecchies AC, Jones ES, Spangenberg GC, Forster JW (2005) QTL analysis and comparative genomics of herbage quality traits in perennial ryegrass (*Lolium perenne* L.). Theor Appl Genet 110: 364–380

Cogan NOI, Ponting RC, Vecchies AC, Drayton MC, George J, Dobrowolski MP, Sawbridge TI, Spangenberg GC, Smith KF, Forster JW (2006) Gene-associated single nucleotide polymorphism (SNP) discovery in perennial ryegrass (*Lolium perenne* L.). Mol Genet Genomics 276: 101–112

Cornish MA, Hayward MD, Lawrence MJ (1979) Self-incompatibility in ryegrass. I. Genetic control in diploid *Lolium perenne* L. Heredity 43: 95–106

Devey F, Fearon CH, Hayward MD, Lawrence MJ (1994) Self-incompatibility in ryegrass. 11. Number and frequency of alleles in a cultivar of *Lolium perenne* L. Heredity 73: 262–264

Dobrowolski MP, Forster JW (2007) Chapter 9: Linkage disequilibrium-based association mapping in forage species. In: Oraguzie NC, Rikkerink E, Gardiner SE, De Silva NH (eds.) Association mapping in plants. Springer, New York, pp. 197–209

Faville M, Vecchies AC, Schreiber M, Drayton MC, Hughes LJ, Jones ES, Guthridge KM, Smith KF, Sawbridge T, Spangenberg GC, Bryan GT, Forster JW (2004) Functionally-associated molecular genetic marker map construction in perennial ryegrass (*Lolium perenne* L.). Theor Appl Genet 110: 12–32

Flint-Garcia SA, Thornsberry JM, Buckler ESI (2003) Structure of linkage disequilibrium in plants. Annu Rev Plant Biol 54: 357–374

Forster JW, Jones ES, Batley J, Smith KF (2004) Molecular marker-based genetic analysis of pasture and turf grasses. In: Hopkins A, Wang Z-Y, Sledge M, Barker RE (eds.) Molecular breeding of forage and turf. Kluwer, Dordrecht, pp. 197–239

Fredman D, White SJ, Potter S, Eichler EE, Den Dunnen JT, Brookes AJ (2004) Complex SNP-related sequence variation in segmental genome duplications. Nat Genet 36: 861–866

Gallagher JA, Pollock CJ (1998) Isolation and characterisation of a cDNA clone from *Lolium temulentum* L. encoding for a sucrose hydrolytic enzyme which shows alkaline/neutral invertase activity. J Exp Bot 49: 789–795

Gallagher JA, Cairns AJ, Pollock CJ (2004) Cloning and characterisation of a putative fructosyltransferase and two putative invertase genes from the temperate grass *Lolium temulentum* L. J Exp Bot 55: 557–569

Guthridge KM, Dupal MD, Kölliker R, Jones ES, Smith KF, Forster JW (2001) AFLP analysis of genetic diversity within and between populations of perennial ryegrass (*Lolium perenne* L.). Euphytica 122: 191–201

Hayes BJ, Visscher PM, McPartlan HC, Goddard ME (2003) Novel multilocus measure of linkage disequilibrium to estimate past effective population size. Genome Res 13: 635–643

Heath R, Huxley H, Stone B, Spangenberg G (1998) cDNA cloning and differential expression of three caffeic acid *O*-methyltransferase homologues from perennial ryegrass (*Lolium perenne* L.). J Plant Physiol 153: 649–657

Heath R, McInnes R, Lidgett A, Huxley H, Lynch D, Jones ES, Mahoney NL, Spangenberg GC (2002) Isolation and characterisation of three 4-coumarate:CoA-ligase homologue cDNAs from perennial ryegrass (*Lolium perenne* L.). J Plant Physiol 159: 773–779

Hill WG, Weir BS (1988) Variances and covariances of squared linkage disequilibria in finite populations. Theor Popul Biol 33: 54–78

Johnson X, Lidgett A, Chalmers J, Guthridge K, Jones E, Spangenberg GC (2003) Isolation and characterisation of an invertase gene from perennial ryegrass (*Lolium perenne* L.). J Plant Physiol 160: 903–911

Jones RN, Rees H (1966) Chiasma frequencies and the potential genetic variability of *Lolium* populations. Nature 211: 432–433

Lidgett A, Jennings K, Johnson X, Guthridge K, Jones E, Spangenberg G (2002) Isolation and characterisation of fructosyltransferase gene from perennial ryegrass (*Lolium perenne*). J Plant Physiol 159: 415–422

Lynch D, Lidgett A, McInnes R, Huxley H, Jones E, Mahoney N, Spangenberg G (2002) Isolation and characterisation of three cinnamyl alcohol dehydrogenase homologue cDNAs from perennial ryegrass (*Lolium perenne* L.). J Plant Physiol 159: 653–660

Mackay TFC (2001) The genetic architecture of quantitative traits. Annu Rev Genet 35: 303–309

McInnes R, Lidgett A, Lynch D, Huxley H, Jones E, Mahoney N, Spangenberg G (2002) Isolation and characterisation of a cinnamoyl-CoA reductase gene from perennial ryegrass (*Lolium perenne*). J Plant Physiol 159: 415–422

Neale DB, Savolainen O (2004) Association of complex traits in conifers. Trends Plant Sci 9: 325–330

Olesen KM, Halldorsdottir SS, Stitchcombe JR, Weinig C, Schmitt J, Purugganan MD (2004) Linkage disequilibrium mapping of *Arabidopsis CRY2* flowering time alleles. Genetics 167: 1361–1367

Ponting RC, Drayton MD, Cogan NOI, Dobrowolski MP, Smith KF, Spangenberg GC, Forster JW (2007) SNP discovery, validation and haplotype structure in full-length herbage quality genes of perennial ryegrass (*Lolium perenne* L.). Mol Genet Genomics 278:585–597

Rees H, Ahmad K (1963) Chiasma frequencies in *Lolium* populations. Evolution 17: 575–579

Rees H, Dale PJ (1974) Chiasmata and variability in *Lolium* and *Festuca* populations. Chromosoma 47: 335–351

Remington DL, Thornsberry JM, Matsuoka Y, Wilson LM, Whitt SR, Doebley J, Kresovich S, Goodman MM, Buckler ES IV (2001) Structure of linkage disequilibrium and phenotypic associations in the maize genome. Proc Natl Acad Sci U S A 98: 11479–11484

Sackville-Hamilton NR, Skøt L, Chorlton KH, Thomas ID, Mizen S (2002) Molecular genecology of temperature response in *Lolium perenne*: 1. Preliminary analysis to reduce false positives. Mol Ecol 11: 1855–1863

Simko I, Haynes KG, Jones RW (2006) Assessment of linkage disequilibrium in potato genome with single nucleotide polymorphism markers. Genetics 173: 2237–2245

Skøt L, Sackville-Hamilton NR, Mizen S, Chorlton KH, Thomas ID (2002) Molecular genecology of temperature response in *Lolium perenne*: 2. Association of AFLP markers with ecogeography. Mol Ecol 11: 1865–1876

Skøt L, Humphreys J, Armstead IP, Humphreys MO, Gallagher JA, Thomas ID (2005a) Approaches for associating molecular polymorphisms with phenotypic traits based on linkage disequilibrium in natural populations of *Lolium perenne*. In: Humphreys MO (ed.) Molecular breeding of the genetic improvement of forage crops and turf. Wageningen Academic Publishers, the Netherlands, p 157

Skøt L, Humphreys MO, Armstead I, Heywood S, Skøt KP, Sanderson R, Thomas ID, Chorlton KH, Sackville-Hamilton NR (2005b) An association mapping approach to identify flowering time genes in natural populations of *Lolium perenne* (L.). Mol Breed 15: 233–245

Sorrells ME, Wilson WA (1997) Direct classification and selection of superior alleles for crop improvement. Crop Sci 37: 691–697

Spangenberg GC, Forster JW, Edwards D, John U, Mouradov A, Emmerling M, Batley J, Felitti S, Cogan NOI, Smith KF, Dobrowolski MP (2005) Future directions in the molecular breeding of forage and turf. In: Humphreys MO (ed.) Molecular breeding of the genetic improvement of forage crops and turf. Wageningen Academic Publishers, the Netherlands, pp. 83–97

Stephens M, Donnelly P (2003) A comparison of Bayesian methods for haplotype reconstruction. Am J Hum Genet 73: 1162–1169

Stephens M, Smith NJ, Donnelly P (2001) A new statistical method for haplotype reconstruction from population data. Am J Hum Genet 68: 978–989

Yamada T, Forster JW, Humphreys MW, Takamizo T (2005) Genetics and molecular breeding in the *Lolium/Festuca* pasture grass species complex. Grassland Sci 51: 89–106

Zein I, Wenzel G, Andersen JR, Lubberstedtm T (2007) Low level of linkage disequilibrium at the COMT (caffeic acid *O*-methyl transferase) locus in European maize (*Zea mays* L.). Genetic Res Crop Evol 54: 139–148

Genetic Diversity in Australasian Populations of the Crown Rust Pathogen of Ryegrasses (*Puccinia coronata* f.sp. *lolii*)

Peter Dracatos[1,3], Jeremy Dumsday[1,3], Alan Stewart[4],
Mark Dobrowolski[1,3], Noel Cogan[1,3], Kevin Smith[2,3] and John Forster[1,3,5]

[1]Department of Primary Industries, Biosciences Research Division, Victorian
AgriBiosciences Centre, 1 Park Drive, La Trobe University Research and
Development Park, Bundoora, VIC 3083, Australia
[2]Department of Primary Industries, Biosciences Research Division, Hamilton
Centre, Mount Napier Road, Hamilton, VIC 3330, Australia
[3]Molecular Plant Breeding Cooperative Research Centre, Australia
[4]PGG Wrightson Seeds Lincoln Christchurch 7640, New Zealand
[5]Corresponding author, john.forster@dpi.vic.gov.au

Abstract. Crown rust fungus, *Puccinia coronata* f.sp. *lolii* is an obligate
biotrophic pathogen of ryegrasses which causes significant reductions of herbage
yield, palatability and digestibility. Genetic diversity in virulence has been
reported in all the major temperate regions of the world and is therefore a major
problem for pasture and turf breeders developing varieties with durable resistance
for crown rust. Knowledge of the genetic variation present both within and
between Australasian crown rust populations is essential for the efficient
production of resistant varieties. A total of 11 efficient simple sequence repeat
(SSR) markers developed from a urediniospore-derived expressed sequence tag
(EST) resource have been used for Australasian intraspecific genetic diversity
analysis. Seventy-two single pustule samples comprising three main populations
from both North and South Islands of New Zealand and from south-eastern
Australia were genotyped. The analysis identified 59 distinct genotypes, high
levels of genetic diversity being detected both within and between populations. All
methods of analysis detected no significant difference between isolates from the
North and South Islands of New Zealand ($p > 0.05$), and high intrapopulation
diversity between Victorian isolates. However high population differentiation
($p < 0.001$) was detected between Victorian isolates and those from the South
Island (PhiPT [estimate of genetic variability = 0.101) and especially the North
Islands (PhiPT = 0.162) of New Zealand. Genetic dissection of crown rust

T. Yamada and G. Spangenberg (eds.), *Molecular Breeding of Forage and Turf*,
doi: 10.1007/978-0-387-79144-9_25, © Springer Science + Business Media, LLC 2009

population structure within Australasia will inform the magnitude of gene pyramiding required for the development of varieties with durable resistance.

Introduction

Puccinia coronata f.sp. *lolii* , the causal agent of crown rust, is the most serious foliar pathogen infecting perennial ryegrass as it causes significant damage for both forage and turf applications (Price 1987). Genetic resistance to the crown rust pathogen of ryegrasses is well established. More recently, trait-dissection studies of crown rust resistance in ryegrasses have been performed using host genome-specific molecular marker systems, and both major genes and quantitative trait loci (QTLs) for resistance have been defined (Dumsday et al. 2003; Studer et al. 2007; Muylle et al. 2005). However, the potential race-specificity of such resistance genes and QTLs determinants are currently unknown, and will restrict the suitability of such genes for varietal improvement. The variable nature of genetic control both within and between varieties may be at least partially attributable to pathotype diversity and adaptive potential. It is unclear whether the differences between quantitative and qualitative host responses are due to genetic variation within and between pathogen populations, and differences between host resistance status. Furthermore, it is not known how many races of the pathogen cause a single field infection and the magnitude of genetic diversity present both within and between crown rust populations derived from Australasian temperate regions.

The recent development of expressed sequence tag (EST)-derived simple sequence repeat (SSR) markers for *P. coronata* f.sp. *lolii* provides the means to assess intraspecific genetic diversity within and between geographic locations and to define aspects of population structure (Dracatos et al. 2006). SSR markers have previously been used to assess genetic diversity in the wheat stripe rust pathogen (*P. striiformis* f.sp. *tritici*) (Keiper et al. 2003) and the *Neotyphodium* endophyte (van Zijll de Jong et al. 2004) in addition to many other crop, pasture and animal species. Here we report the use of EST-SSR markers to test the hypothesis that crown rust genetic diversity is highly complex and that large proportions of intraspecific variation arise within and to some extent between populations.

Assessment of Crown Rust Genetic Diversity

Background

The crown rust pathogen has originated from a diverse sexual lineage (Browning and Frey 1969). Recent evidence suggests that populations of plant pathogenic fungi within a single location can be genetically diverse due to de novo mutations (Steele et al. 2001), spore migration (Brown and Hovmøller 2002) and heterozygosity due to either previous sexual recombination or mitotic recombination (Murphy 1935). Given that urediniospores of the crown rust pathogen are dikaryotic, heterokaryosis can only occur following recombination through meiosis or mitosis creating the possibility of heterozygosity at each virulence locus. The presence of sexual recombination on the secondary host (*Rhamnus* spp.) within the crown rust pathogen has been reported in many locations throughout the Northern hemisphere. It is thought that basidiospores produced during the sexual cycle are absent within Australasia due to the ability of asexual urediniospores to survive the relatively benign environmental conditions and to over-winter on the graminaceous host. It is consequently also unknown whether it is possible for the crown rust pathogen to undergo sexual recombination within Australasia and whether such processes have had an impact on race diversity (reviewed by Kimbeng 1999).

PCR Based-EST SSR Genotyping of Single Pustule Isolates of *P. coronata* f.sp *lolii*

A unique set of 55 EST-SSR molecular markers for the crown rust pathogen was developed from a urediniospore-derived cDNA collection. Each of the EST-SSR loci were labelled with the fluorochromes (HEX or FAM) for size discrimination on the MegaBACE 4000 capillary sequencer (GE Healthcare). Genomic DNA template derived from single pustule samples was extracted using Chelex (Biorad) resin which yielded picogram quantities of DNA that were unsuitable for use in PCR reactions. Whole genome amplification techniques such as multiple displacement amplification (MDA) commercialised as GenomiPhi (GE Healthcare) were used to produce microgram quantities of high molecular weight product suitable for efficient PCR amplification using DNA samples from a single pustule (Dean et al. 2002). Further characterisation of 55 SSR loci yielded a subset of 12 highly efficient and polymorphic markers suitable for the genetic analysis of single pustule isolates of *P. coronata* f.sp. *lolii*. Initial genetic diversity analysis using 16 single-pustule isolates from geographically diverse temperate regions (New Zealand, UK, Japan and Australia) identified 15 genetically distinct genotypes. A preliminary analysis using UPGMA generated a dendrogram showing a lack of correlation between clustering and geographic origin of isolates, which may indicate early evolution of population complexity and diversity (Dracatos et al. 2006).

Assessment of Australasian Genetic Diversity Within *P. coronata* f.sp *lolii* Populations

Genetic diversity analysis of 72 single pustule isolates from the North and South Islands of New Zealand and from temperate south-eastern Australia (listed in Table 1) was conducted using 11 selected *Puccinia coronata* EST-SSR locus (PCESTSSR) markers (Dracatos et al. 2006). All 11 SSR markers were polymorphic across the entire dataset detecting high genetic diversity among and between the populations of *P. coronata* f.sp *lolii* isolates used in this study. A UPGMA dendrogram was constructed which detected 59 distinct genotypes separated by genetic distance (Fig. 1). There was a significant correlation between geographic origin and clustering within the dendrogram (Fig. 1). Isolates from Victoria were predominantly found within different major clusters and were in most instances genetically dissimilar to isolates from both the North and South Islands of New Zealand. However, genetic diversity was observed both within and between geographic regions within Victoria, as two subclusters each containing isolates from all regions were found in separate main clusters. Isolates from the North and South Islands of New Zealand were located within the same main cluster and failed to differentiate apart from a small number of individuals from the south island which were genetically similar to Victorian isolates. Isolates from Taupo in the South were similar to that from Blenheim in the North Island as were isolates from Christchurch and Hamilton.

Analysis of molecular variance (AMOVA) was performed to determine any significant differences between and within crown rust populations from New Zealand and Australia (Table 2). A total of 8% of genetic variability was detected between populations ($p > 0.001$), which indicates that most of the variation occurs within populations. This may be expected as the number of clonal genotypes detected through genotypic analysis in this study was comparatively lower than that of the symbiotic *Neotyphodium* endophyte (van Zijll de Jong et al. 2004) and bacterial wilt pathogen (Kölliker et al. 2004) which are known to function as asexual lineages with.

Table 1 Summary of single pustule *P. coronata* f. sp. *lolii* samples used for genetic analysis

Sample number	Sample identification	Geographic origin	Source location
1	NZs5WAITCIT	New Zealand	North Island (Waitakere city)
2	NZ7HAM	New Zealand	North Island (Hamilton)
3	NZ8HAM	New Zealand	North Island (Hamilton)
4	NZ9HAM	New Zealand	North Island (Hamilton)
5	NZ10HAM	New Zealand	North Island (Hamilton)
6	NZ11HAM	New Zealand	North Island (Hamilton)
7	NZ12HAM	New Zealand	North Island (Hamilton)
8	NZ13HAM	New Zealand	North Island (Hamilton)
9	NZ14HAM	New Zealand	North Island (Hamilton)
10	NZ16HAM	New Zealand	North Island (Hamilton)
11	NZ18ROT	New Zealand	North Island (Rotorua)
12	NZ19ROT	New Zealand	North Island (Rotorua)
13	NZ20ROT	New Zealand	North Island (Rotorua)
14	NZ21ROT	New Zealand	North Island (Rotorua)
15	NZ22TAUP	New Zealand	North Island (Taupo)
16	NZ23TAUP	New Zealand	North Island (Taupo)
17	NZ24TAUP	New Zealand	North Island (Taupo)
18	NZ25TAUP	New Zealand	North Island (Taupo)
19	NZ26TAUP	New Zealand	North Island (Taupo)
20	NZ27HAST	New Zealand	North Island (Hastings)
21	NZ29HAST	New Zealand	North Island (Hastings)
22	NZ30PN	New Zealand	North Island (Palmerston)
23	NZ31PN	New Zealand	North Island (Palmerston)
24	NZ32PN	New Zealand	North Island (Palmerston)
25	NZ33WofB	New Zealand	North Island (West of Bulls)
26	NZ34WofB	New Zealand	North Island (West of Bulls)
27	NZ35WofB	New Zealand	North Island (West of Bulls)
28	NZ36WELL	New Zealand	North Island (Wellington)
29	NZ37BLEN	New Zealand	South Island (Blenheim)
30	NZ38BLEN	New Zealand	South Island (Blenheim)
31	NZ39BLEN	New Zealand	South Island (Blenheim)
32	NZBLEM	New Zealand	South Island (Blenheim)
33	NZ41BLEN	New Zealand	South Island (Blenheim)
34	NZ42BLEN	New Zealand	South Island (Blenheim)
35	NZ43BLEN	New Zealand	South Island (Blenheim)
36	NZ44BLEN	New Zealand	South Island (Blenheim)
37	NZ45BLEN	New Zealand	South Island (Blenheim)
38	NZ46BLEN	New Zealand	South Island (Blenheim)
39	NZ47REEF	New Zealand	South Island (Reefton)
40	NZ48REEF	New Zealand	South Island (Reefton)
41	NZ49REEF	New Zealand	South Island (Reefton)
42	NZ51ROSS	New Zealand	South Island (Ross)
43	NZ52ROSS	New Zealand	South Island (Ross)
44	NZ53FJ	New Zealand	South Island (Franz Josep)
45	NZ54FJ	New Zealand	South Island (Franz Josep)
46	NZ55FJ	New Zealand	South Island (Franz Josep)
47	NZ56CHRIST	New Zealand	South Island (Christchurch)
48	NZ57CHRIST	New Zealand	South Island (Christchurch)
49	NZ58CHRIST	New Zealand	South Island (Christchurch)
50	NZ59CHRIST	New Zealand	South Island (Christchurch)
51	NZ60CHRIST	New Zealand	South Island (Christchurch)
52	NZ61CHRIST	New Zealand	South Island (Christchurch)
53	NZ62CHRIST	New Zealand	South Island (Christchurch)
54	NZ66DUN	New Zealand	South Island (Christchurch)
55	NZ67TIM	New Zealand	South Island (Timaru)
56	AUS2	Australia	Hamilton, Victoria
57	AUS1	Australia	Hamilton, Victoria
58	VIC-Ham3	Australia	Hamilton, Victoria
59	VIC-Ham6	Australia	Hamilton, Victoria
60	VIC-How1	Australia	Howlong Victoria
61	VIC-How4	Australia	Howlong Victoria
62	VIC-How5	Australia	Howlong Victoria
63	VIC-Garv	Australia	Garvoc Vicroria
64	VIC-Mort1	Australia	Mortlake Victoria
65	VIC-Mort4	Australia	Mortlake Victoria
66	VIC-Mort6	Australia	Mortlake Victoria
67	VIC-Mort7	Australia	Mortlake Victoria
68	VIC-Mort8	Australia	Mortlake Victoria
69	VIC-Beeac3	Australia	Beeac Victoria
70	VIC-Beeac4	Australia	Beeac Victoria
71	VIC-Beeac6	Australia	Beeac Victoria
72	VIC-Beeac11	Australia	Beeac Victoria

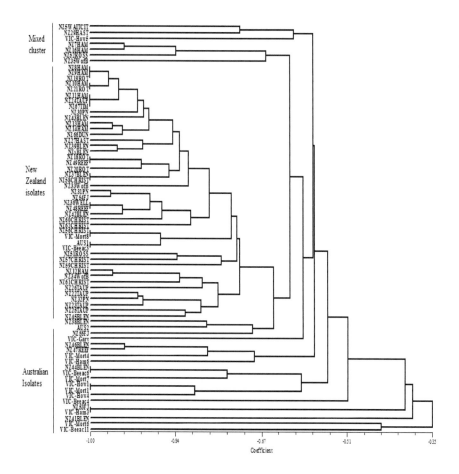

Fig. 1 UPGMA dendrogram cluster analysis for 72 *P. coronata* f.sp. *lolii* isolates based on analysis with 11 PCESTSSR markers. Isolates are labelled with a VIC (Victorian) or NZ (New Zealand) prefix followed by isolate number and location of sampling as described in Table 1

extensive linkage disequilibrium. The capacity to overwinter and undergo sexual recombination on its alternate host may account for the magnitude of genetic diversity observed in this study (as reviewed by Simons 1970).

It has however been reported that host specialisation within the *Rhamnus* genus is broader than that within the Poaceae, and as many different *Rhamnus* species are presented in both New Zealand and Australia, possible role of sexual recombination cannot be eliminated as a major source of genetic diversity.

Table 2 Differences between and within crown rust fungi populations from Australia and New Zealand. Probability values were deemed significant (*) at the 0.05 confidence interval and highly significant at (**) at $p < 0.01$ and (***) at $p < 0.001$

Source	df	SS	MS	Estimated variation	%	Statistic	Value	p-Value
Among pops	2	101.445	50.722	1.446	8			
Within pops	69	1,156.166	16.756	16.756	92	PhiPT	0.079	0.000***
Total	71	1,257.611	67.756	18.202				

To attribute genetic variation detected between individual populations a pair-wise population test was performed by calculating the PhiPT value for genetic variability in conjunction with principle coordinate analysis (PCoA) (Fig. 2). Both forms of analysis detected significant ($p < 0.001$) population differentiation (PhiPT values) between Australian and New Zealand isolates, but gene diversity was larger between isolates from Australia and the North Island of New Zealand (Table 3). No significant population differentiation ($p > 0.05$) was observed between isolates from the North and South Islands of New Zealand supporting the clustering patterns within the dendrogram and PCoA.

Clustering of isolates using PCoA detected three distinct groups of Victorian isolates, while isolates from the North and South Islands of New Zealand clustered together in two main groups (Fig. 2). The three groups

Table 3 Pairwise population PhiPT values and estimates of the number of migrants per generation between three populations of *Puccinia coronata* f.sp. *lolii* isolates from two islands of New Zealand (NZ, North Island and South Island) and south eastern Victoria in Australia (AUS Victoria). Probabilities for pairwise comparison were deemed significant (*) at the 0.05 confidence interval and highly significant at (**) at $p < 0.01$ and (***) at $p < 0.001$

Population 1	Population 2	PhiPT	Nm	#Pop 1	#Pop 2	p value
NZ North Island	NZ South Island	0.000	0.000	28	27	0.428 ns
NZ North Island	AUS Victoria	0.162	1.293	28	17	0.000***
NZ South Islands	AUS Victoria	0.101	2.226	27	17	0.000***

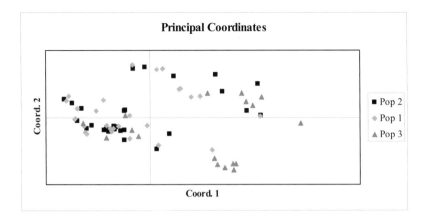

Fig. 2 Principle coordinate analysis clustering of single-pustule isolates of *Puccinia coronata* f.sp. *lolii* in Australasia, based on data from PCESTSSR analysis. Population 1: from North Island in New Zealand, Population 2: from South Island in New Zealand and Population 3: from Victoria in Australia

from Victoria were either genetically distinct from all other isolates, similar to isolates from the South Island or similar to a large cluster of isolates from both the North and South Islands. Population differentiation of this kind may be partially due to directed gene flow by means of wind-aided migration. Although there is no documented evidence for *P. coronata*, such effects have been observed for *P. striiformis* f.sp. *tritici* and *P. graminis* (Reviewed in Brown and Hovmøller 2002). Here we present preliminary evidence for spore migration of *P. coronata* between the South Island of New Zealand and Australia, which may partially account for the closer relationship between some Victorian isolates and those from the South Island.

Conclusions

Previous studies have identified diversity in virulence within Australia for both *P. coronata* f.sp. *lolii* and *avenae* through pathotype analysis (Aldaoud et al. 2004; Brake et al. 2001). All forms of genetic analysis in this study have detected population differentiation between isolates from Australia and New Zealand and genetic similarity of isolates between the North and South Islands of New Zealand. The genetic diversity detected within populations was far greater within Australia than New Zealand, which is likely to reflect host adaptation and the genetic diversity of perennial ryegrass varieties used in both countries. The observed separation of all isolates into three main groups may be due to wind-assisted spore migration between countries possibly resulting in the introduction of inoculum which has undergone previous sexual cycles. Further analysis to confirm the extent of linkage disequilibrium within each of three populations may confirm whether sexual recombination may explain the observed levels intra-population diversity observed in this study. The crown rust pathogen is highly genetically diverse, especially within Australia. This observation is likely to affect the durability of resistance genes deployed in elite *L. perenne* germplasms, emphasising the requirement for molecular marker development for both major and minor resistance genes for gene pyramiding.

References

Aldaoud R, Anderson MW, Reed KFM, Smith KF (2004) Evidence of pathotypes among Australian isolates of crown rust infecting perennial ryegrass. Plant Breed 123:395–397

Brake VM, Irwin JAG, Park RF (2001) Genetic variability in Australian isolates of *Puccinia coronata* f.sp. *avenae* assessed with molecular and pathogenicity markers. Australas Plant Pathol 30:259–266

Brown JKM, Hovmøller MS (2002) Aerial dispersal of pathogens on the global and continental scales and its impact on plant disease. Science 297:537–541

Browning JA, Frey KJ (1969) Multiline cultivars as a means for disease control. Annu Rev Phytopathol 7:355–382

Dean FB, Hosono S, Fang L, Wu X, Faruqi F, Bray-Ward P, Sun Z, Zong Q, Du Y, Du J, Driscoll M, Song W, Kingsmore SF, Egholm M, Lasken RS (2002) Comprehensive human genome amplification using multiple displacement amplification. Proc Natl Acad Sci USA 99:5261–5266

Dracatos PM, Dumsday JL, Olle RS, Cogan NOI, Dobrowolski MP, Fujimori M, Roderick H, Stewart AV, Smith KF, Forster JF (2006) Development and characterisation of EST-SSR markers for the crown rust pathogen of ryegrass (*Puccinia coronata* f.sp. *lolii*). Genome 49:572–583

Dumsday JL, Smith KF, Forster JW, Jones ES (2003) SSR-based genetic linkage analysis of resistance to crown rust (*Puccinia coronata* f.sp. *lolii*) in perennial ryegrass (*Lolium perenne*). Plant Pathol 52:628–637

Keiper FJ, Hayden MJ, Park RF, Wellings CR (2003) Molecular genetic variability of Australian isolates of five cereal rust pathogens. Mycol Res 107:545–556

Kimbeng CA (1999) Genetic basis of crown rust resistance in perennial ryegrass, breeding strategies, and genetic variation amongst pathogen populations: a review. Aus J Exp Agric 39:361–378

Kölliker R, Krahenbuhl R, Schubiger FX, Widmer F (2004) Genetic diversity and pathogenicity of the grass pathogen *Xanthomonas translucens* pv.*graminis*.

In: Hopkins A, Wang Z-Y, Sledge M, Barker RE (eds) Molecular breeding of forage and turf, Kluwer, Dordrecht, the Netherlands, pp. 53–59

Murphy HC (1935). Physiological specialisation in *Puccinia coronata* avenae. USDA Tech Bull 433:1–48

Muylle H, Baert J, Van Bockstaele E, Pertijs J, Roldan-Ruiz I (2005). Four QTLs determine crown rust (*Puccinia coronata* f. sp. *lolii*) resistance in a perennial ryegrass (*Lolium perenne*) population. Heredity 95:348–357

Price T (1987) Ryegrass rust in Victoria. Plant Prot Q 2:189

Simons MD (1970) Crown rust of oats and grasses. Monograph No. 5. The Heffernan Press, Worcester, MA

Steele KA, Humphreys E, Wellings CR, Dickinson MJ (2001). Support for a stepwise mutation model for pathogen evolution in Australasian *Puccinia striiformis* f.sp. *tritici* by use of molecular markers. Plant Pathol 50(2):174–180

Studer B, Boller B, Bauer E, Posselt U, Widmer F, Kölliker R (2007) Consistent detection of QTLs for crown rust resistance in Italian ryegrass (*Lolium multiflorum* Lam.) across environments and phenotyping methods. Theor Appl Genet 115:9–17

van Zijll de Jong E, Bannan NR, Batley J, Guthridge KM, Spangenberg GC, Smith KF, Forster JW (2004) Genetic diversity in the perennial ryegrass fungal endophyte *Neotyphodium lolii*. In: Hopkins A, Wang Z-Y, Sledge M, Barker RE (eds) Molecular breeding of forage and turf, Kluwer, Dordrecht, the Netherlands, pp. 155–164

Fungal Endophytes in *Lolium* and *Festuca* Species

Christopher L. Schardl

Department of Plant Pathology, University of Kentucky, Lexington,
KY 40546-0312, USA, schardl@uky.edu

Abstract. The epichloë endophytes (fungal genera *Epichloë* and *Neotyphodium*) are symbionts of many temperate pasture grasses, and are particularly important in *Festuca* and *Lolium* species. Many epichloae are known to protect host grasses against a wide range of insects, but some of the anti-insect alkaloids that they produce can negatively affect the health and reproduction of livestock. Here I review the evolutionary relationships and origins of *Epichloë* and *Neotyphodium* species, and what is known of the biosynthetic pathways and genetics of protective alkaloids that they produce.

Epichloë Relationships and Host Interactions

Clavicipitaceae, as currently recognized by most mycologists, is a fungal family within the order Hypocreales, and is dominated by parasites of invertebrate animals. Some Clavicipitaceae parasitize insects (such as scale insects) that feed on plants, then secondarily parasitize the plants (Sullivan et al. 2000). A group derived within the Clavicipitaceae is associated primarily with plants and have no requirement to first infect insect hosts. These include members of the type genus, *Claviceps* (ergot fungi), as well as the *Epichloë* species (anamorphs = *Neotyphodium* spp.) and several genera in the tribe Balansieae (Spatafora et al. 2007). Although many of the Balansieae were originally classified as *Epichloë* species, common usage now restricts genus *Epichloë* to those clavicipitaceous fungi which fruit by forming white stromata that surround host inflorescences, and which develop yellow, tan or orange perithecia following fertilization (Fig. 1). As

T. Yamada and G. Spangenberg (eds.), *Molecular Breeding of Forage and Turf*,
doi: 10.1007/978-0-387-79144-9_26, © Springer Science + Business Media, LLC 2009

such, the genus appears to be very coherent phylogenetically, and also appears to be restricted in host range to the plant subfamily Pooideae (Clay and Schardl 2002). The *Epichloë* species have given rise to numerous asexual, nonpathogenic plant symbionts that are vertically transmitted by seed infection, and conventionally classified as *Neotyphodium* species. Collectively, I will refer to the *Epichloë* species and their asexual derivatives as epichloae (singular, epichloë).

Fig. 1 *Epichloë* endophyte–grass host interactions. (a) An asymptomatic inflorescence (*left*) and a choked inflorescence (*right*) of *Agrostis perennans* symbiotic with *Epichloë amarillans*. Both inflorescences are systemically infected with the endophyte. The choked inflorescence has an ectophytic fungal stroma (*arrow*) required for its sexual cycle. (b) A fertilized stroma of *E. amarillans*. (c) stromata of *Epichloë typhina* on *Phleum pratense* growing on Hokkaido, Japan. (d) *Neotyphodium coenophialum* (stained with aniline blue) growing between host cells

Most of the *Epichloë* species are capable of transmitting in the seeds of their host plants, and do so at extremely high efficiency (Schardl et al.

2004). Microscopic studies of this process have only been conducted for asexual derivatives of *Epichloë* spp. Investigation of the process in ryegrasses indicates that *Neotyphodium occultans* (common in annual ryegrasses), and *Neotyphodium lolii* (in perennial ryegrass, *Lolium perenne*) colonize the embryo in the developing seed, and this appears to be the basis for transmission to the next generation (Freeman 1904; Philipson and Christey 1986). It seems likely that seed-transmissible *Epichloë* spp. utilize the same mechanism. Of the other Clavicipitaceae, only *Balansia hypoxylon* is known to be seed-transmitted, apparently by growth between leaf primordia in the embryos of cleistogamous seeds (Clay 1994). Highly efficient vertical transmission in open pollinated seeds seems to be peculiar to epichloae, and may have originated with the genus *Epichloë*.

Host Specificity of *Epichloë* Species

Most of the 11 described *Epichloë* species are associated with a narrow taxonomic range of hosts (Schardl and Leuchtmann 2005) (Fig. 2). In particular, *Epichloë amarillans* and *Epichloë baconii* are associated with tribe Agrostideae (=Aveneae), *Epichloë festucae* is almost exclusively associated with *Lolium* and *Festuca* spp., tribe Poeae (but has also been found in *Koeleria* sp., tribe Agrostideae), *Epichloë elymi* with the Hordeeae, *E. bromicola* with *Bromus* spp. (tribe Bromeae), *Epichloë sylvatica* with *Brachypodium sylvatica* (tribe Brachypodieae), *Epichloë glyceriae* with *Glyceria* spp. (tribe Meliceae), and *Epichloë brachyelytri* with the genus *Brachyelytrum* (tribe Brachyelytreae). *Epichloë yangzii* has been described from *Roegneria kamoji* (tribe Hordeeae), but has a very close phylogenetic relationship with *E. bromicola* (Li et al. 2006), and tests of possible mating compatibility with *E. bromicola* have not yet been reported. *Epichloë clarkii* has only been reported on *Holcus lanatus* (tribe Poeae) (Schardl and Leuchtmann 2005). The host range of *Epichloë typhina* is particularly broad, and includes genera *Dactylis*, *Poa*, *Puccinellia*, and *Lolium* from tribe Poeae, genera *Anthoxanthum*, *Arrhenatherum* and *Phleum* from tribe Agrostideae, and *Brachypodium pinnatum* from tribe Brachypodieae. Also, *E. typhina* is exceptionally diverse with respect to gene sequences. Interestingly, *E. typhina* genotypes found on *Lolium perenne*, *Poa trivialis*, and *Dactylis glomerata* are closely related and interfertile with *E. clarkii*, and *E. typhina* genotypes on *Bp. pinnatum* are closely related but apparently not interfertile with *E. sylvatica*.

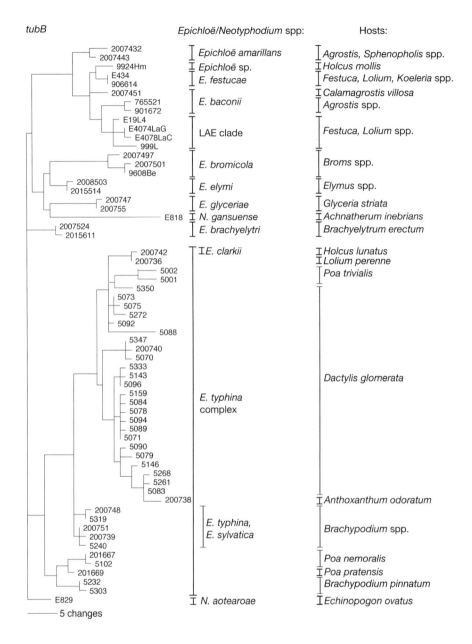

Fig. 2 Phylogenetic relationships of *Epichloë* and *Neotyphodium* species, with emphasis on the *Epichloë typhina* complex. The phylogram was inferred by maximum parsimony search on intron sequences from the β-tubulin genes (*tubB*)

The possibility that *E. typhina* represents a complex of host-specialized populations was tested by sampling populations from four hosts – *D. glomerata*, *Poa trivialis*, *Poa nemoralis*, and *Brachypodium pinnatum* – and sequencing segments of their genes for β-tubulin (*tubB*) (Fig. 2) and translation elongation factor 1-α (*tefA*) (Schardl CL, Leuchtmann A and McDonald BA, unpublished data). Although isolates from any one host often shared identical *tubB* or *tefA* sequences (haplotypes), there were no shared haplotypes between isolates from different hosts. This result indicated that there are barriers to productive mating between *E. typhina* on different host species. Whether such barriers are prezygotic or postzygotic is unknown. Symbiotic flies (genus *Botanophila*) mediate spermatial transfers, and are therefore necessary for *E. typhina* mating. It is possible that each fly may be selective for stromata on a single host, thereby restricting the range of fungal matings. Indeed, there is evidence for such preference, but it is not absolute (Bultman and Leuchtmann 2003). Another possibility is that any hybrids forming between *E. typhina* genotypes on different host species may have no compatible host or reduced compatibility (relative to their parents) on available hosts. A study of *E. typhina* hybrids between isolates from *D. glomerata* and isolates from *L. perenne* provides some support for this, although at least one such hybrid was capable of completing its life cycle on *D. glomerata* (Chung et al. 1997). Therefore, although both of these factors seem to contribute, neither appears to explain fully the observed specificity of interbreeding populations of *E. typhina*.

The *Neotyphodium* Species

In accordance with botanical nomenclature codes, asexual ascomycetes or basidiomycetes are classified in the Fungi Imperfecti, and given distinct form genus names. This system reflects the abundance of asexual fungus species, and the difficulty in establishing on morphological grounds the link between asexual and sexual fungi. With molecular phylogenetic methods, these links are often established unambiguously, but the naming convention remains. A clear link has been established between the described *Neotyphodium* species and *Epichloë* spp. (Moon et al. 2004). This was expected given the similarity in asexual spore states and, especially, similarity in host range (grass subfamily Pooideae) and endophytic lifestyle. The *Neotyphodium* species differ from the *Epichloë* species in being incapable of producing a stroma (or, on rare occasions, producing a stroma that fails to develop and produce the meiotically derived ascospores). Also, some *Neotyphodium* species have spores that are substantially larger than those of *Epichloë* spp., apparently due to their polyploid or heteroploid nature (more on this momentarily).

Phylogenetic analysis has identified very close links between certain *Neotyphodium* species and their apparently ancestral *Epichloë* species (Moon et al. 2004) (Fig. 3). For example, *N. lolii* (from *L. perenne*) as well as an endophyte of *Festuca obtusa* appear to be asexual derivatives of *E. festucae*. Similarly, *Neotyphodium huerfanum* (from *Festuca arizonica*) and *Neotyphodium typhinum* var. *canariense* (from *Lolium edwardii*) appear to be derived from *E. typhina*, and are particularly closely related to those *E. typhina* genotypes associated with *P. nemoralis*.

In addition to the described *Epichloë* species, *Neotyphodium* species have phylogenetic relationships suggesting their derivation from *Epichloë* species that have not been identified or are extinct (Moon et al. 2004; Figs. 2 and 3). For example, *Neotyphodium aotearoae* appears to constitute the sister clade to the *E. typhina*/*E. sylvatica*/*E. clarkii* clade. Also, *Neotyphodium gansuense* is not closely related to any described *Epichloë* species (Moon et al. 2007).

Many *Neotyphodium* species show clear genetic indicators of hybrid origin, combining some or most of the genomes of two or more ancestral *Epichloë* species (Moon et al. 2004). Of particular interest are the endophytes in *Lolium* and *Festuca* spp. (Fig. 3). *Neotyphodium coenophialum*, the common endophyte of tall fescue (*Festuca arundinacea* = *Lolium arundinaceum*) in Northern Europe (and transplanted to North America, New Zealand and Australia) has three ancestors: *E. typhina* (specifically, a genotype related to those on *P. nemoralis*), *E. festucae*, and an unknown or extinct *Epichloë* species with gene sequences in the "*Lolium*-associated endophyte" (LAE) clade. (This clade is so named because each endophyte occurs in a species of genus *Lolium* as defined by Darbyshire (1993) on phylogenetic grounds, and it is the preference of this author to use Darbyshire's treatment.) *N. occultans*, the common symbiont of annual ryegrasses, possesses an LAE clade genome along with a genome from *E. bromicola*. Certain Mediterranean tall fescue relatives also have endophytes with LAE genomes. One of these is a hybrid with *E. festucae* and LAE genomes, the other is a different hybrid possessing LAE and *E. typhina* genomes. Their hybrid origins explain the large genome sizes, which in turn explain the larger conidium sizes relative to sexual *Epichloë* species.

Alkaloids

A well-documented characteristic of epichloae in *Lolium* and *Festuca* species is defense of the plants against herbivory (Schardl et al. 2004).

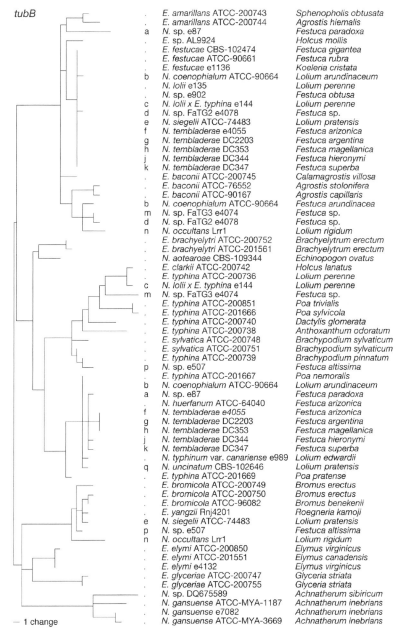

Fig. 3 Phylogenetic relationships of *Epichloë* and *Neotyphodium* species, with emphasis on the endophytes of *Lolium* and *Festuca* species. The phylogram was inferred by maximum parsimony search on intron sequences from the β-tubulin genes (*tubB*). *Letters* indicate interspecific hybrids, most of which have multiple *tubB* haplotypes. Endophyte species and strain designations are followed by host species

Several of the well-known forage grasses in these genera bear epichloë strains that produce any of four classes of alkaloids: ergot alkaloids, indole-diterpenes, peramine and lolines (Fig. 4). All four classes include anti-insect compounds, which either deter or kill herbivorous insects (Clay 1990; Riedell et al. 1991; Rowan 1993; Tanaka et al. 2005). In fact, the presence of a peramine producing endophyte, such as *N. lolii*, is essential for survival of *L. perenne* in much of New Zealand, where the grass would otherwise be eliminated by the Argentine stem weevil. Such protection is not as critical in the natural range of *L. perenne* in Europe, where many populations have a majority of endophyte-free plants (Lewis et al. 1997).

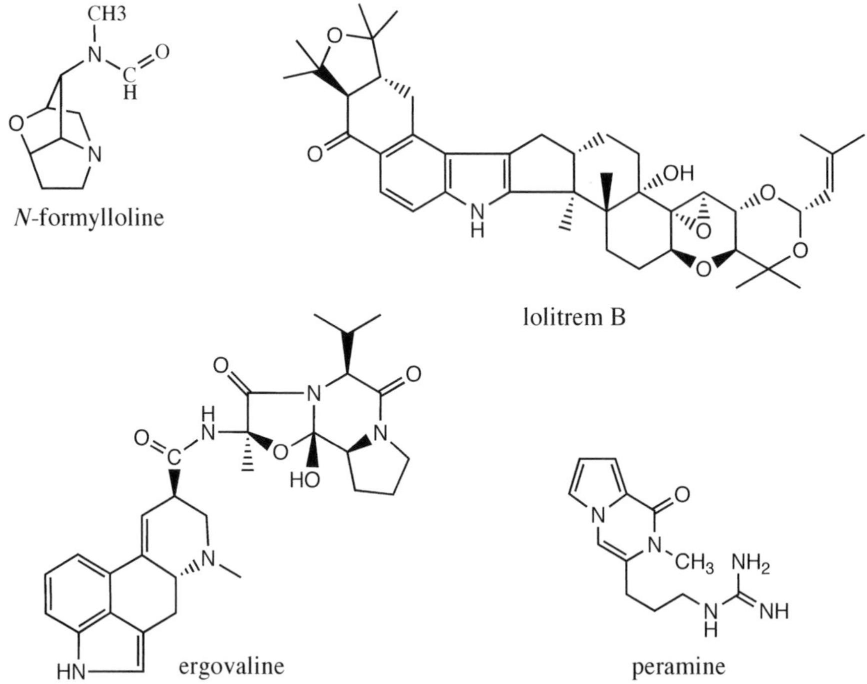

Fig. 4 Major alkaloids produced by endophytes of *Lolium* and *Festuca* species. Most *N. coenophialum* strains in tall fescue produce the loline alkaloid *N*-formylloline, the ergot alkaloid ergovaline, and peramine. Most *N. lolii* strains in perennial ryegrass produce the indole-diterpene alkaloid, lolitrem B, as well as ergovaline and peramine

Peramine is an insect feeding deterrent, and is currently the only known representative of the alkaloid class, pyrrolopyrazine. Synthesis of peramine requires a nonribosomal peptide synthetase, encoded by the *perA* gene identified in *Epichloë festucae* strain Fl1 (Tanaka et al. 2005). The seven domains of *perA* – designated in order as A, T, C, A, M, T, and R – have signature sequences indicative of key enzyme activities; namely, activation (A) and thiolation (T) of each precursor amino acid, and their condensation (C), as well as *N*-methyltransferase (M), and reductase (R). The two amino acid precursors have been suggested to be Δ1-pyrroline-5-carboxylate and L-arginine.

Another group of anti-insect alkaloids, the lolines (Schardl et al. 2007), was first characterized from the annual ryegrass species, *Lolium temulentum*. Structurally, lolines are *exo*-1-aminopyrrolizidines with an oxygen bridge between carbons 2 and 7, and often with methyl, formyl or acyl groups on the 1-amine. Lolines have also been found in other annual ryegrasses in association with *N. occultans*, tall fescue with *N. coenophialum*, and meadow fescue (*Festuca pratensis* = *Lolium pratense*) with *Neotyphodium uncinatum* or *Neotyphodium siegelii*. Production of lolines by cultured *N. uncinatum* supports the hypothesis that, in all of these symbioses it is the endophyte that produces these alkaloids. The lolines have broad insecticidal activities, are inducible by damage to the plant, and can reach levels near 2% of the plant dry mass.

L-Proline (Pro) and L-homoserine (Hse) are loline alkaloid precursors (Blankenship et al. 2005). Although the fungus is capable of producing these precursors from simple nitrogen and organic carbon sources, it is unknown to what degree the plant provides the precursors and what influence precursor availability may have on loline levels in planta.

Much of the loline alkaloid biosynthesis pathway has been elucidated by applying isotopically labeled putative precursors and intermediates to *N. uncinatum* cultures and analyzing the labeling patterns in *N*-formylloline (Faulkner et al. 2006). The first step seems likely to be an unusual γ-substitution reaction condensing Pro and Hse. Less understood, but equally interesting, is the addition of an unusual ether linkage between bridgehead carbons 2 and 7 of the pyrrolizidine ring system.

The ergot alkaloids (Schardl et al. 2006) are toxins of major concern to livestock producers. These alkaloids can be particularly high in the resting structures (ergots) of *Claviceps* species, and the ergots can contaminate grain fed to livestock. Mechanical means of screening out ergots has reduced this problem. However, some of the epichloae produce similar ergot alkaloids and, if these alkaloids are sufficiently abundant in the forage, and especially if the animals are under heat stress, symptoms of ergot alkaloid poisoning become apparent. Problems can range from poor weight gain to low conception rates, agalactia, stillbirths, and dry gangrene.

The enzyme, dimethylallyltryptophan synthase, which catalyzes the initial ergot alkaloid biosynthesis step, was purified from *Claviceps fusiformis* SD58. The gene *dmaW* was then identified from *C. fusiformis* and subsequently from *C. purpurea, Balansia obtecta,* and *Neotyphodium* species (Schardl et al. 2006; Wang et al. 2004). Sequence analysis demonstrated that the enzyme represents a novel class, unrelated to other known prenyltransferases. Homologs of *dmaW* are present in many sequenced genomes of Pezizomycotina (filamentous, ascomycetous fungi). Elimination of *dmaW* in a ryegrass endophyte eliminated all clavine and ergot alkaloids, demonstrating that dimethylallyltryptophan synthase is indeed required for ergot alkaloid biosynthesis.

Some *Claviceps* species or strains produce simple ergolene compounds, such as agroclavine and elymoclavine (typical of *C. fusiformis*), whereas others produce lysergic acid amides, often including the complex ergopeptines (Fig. 4) (Gröger and Floss 1998; Schardl et al. 2006). The initial pathway steps are prenylation of tryptophan at the aromatic position 4, followed by methylation of the alpha amine, and oxidative cyclization coupled with decarboxylation. This gives chanoclavine I, which is further oxidized to the aldehyde, then acted on by probably two distinct enzymes of the "chanoclavine cyclase" step to give agroclavine. Oxidation of the methyl group of agroclavine gives elymoclavine. Further oxidation gives paspalic acid, which may spontaneously isomerize to lysergic acid.

The ergopeptine alkaloids are built from lysergic acid and three amino acids, the nature of which can vary, but which generally have hydrophobic side chains (Haarmann et al. 2005; Panaccione et al. 2001; Schardl et al. 2006). The ergopeptine commonly associated with grass endophytes is ergovaline, which is made by a series of peptide bonds linking lysergic acid to L-alanine (Ala), L-valine (Val), and Pro. The enzyme responsible, lysergyl peptide synthetase, consists of two subunits. One of these (LPS2) activates lysergic acid and generates a thiol ester with 4'-phosphopantetheine. The other subunit (LPS1) has three modules containing domains for activation, thiolation and condensation of Ala, Val and Pro, and a final domain for cyclization of the Val-Pro peptide to give the ergopeptide lactam. A separate enzyme is thought to hydroxylate the alpha carbon of the Ala moiety, allowing formation of the cyclol ring that characterizes the ergopeptines.

Genes for ergot alkaloid biosynthesis are clustered in *C. purpurea* (Haarmann et al. 2005; Schardl et al. 2006), *C. fusiformis* (Lorenz et al.

2007), *Aspergillus fumigatus* (Coyle and Panaccione 2005) and *N. lolii* (Fleetwood et al. 2007).

The fourth known class of epichloë alkaloids are the indole-diterpenes, exemplified by lolitrem B, produced by strains of *N. lolii, E. festucae*, and *Neotyphodium melicicola* (among others). These are tremorgenic alkaloids, and livestock that ingest *L. perenne* with lolitrem-producing strains of *N. lolii* can show symptoms of ryegrass staggers. The indole-diterpenes are produced from indole glycerol and geranylgeranyl diphosphate. Clusters of genes for indole-diterpene alkaloids have been identified first in *Penicillium paxilli*, then *N. lolii* and *E. festucae* (McMillan et al. 2003; Young et al. 2006). The common *N. lolii* genotypes in *L. perenne* in New Zealand pastures produce these alkaloids, but can now be replaced with *L. perenne–N. lolii* cultivars lacking these alkaloids (Bluett et al. 2004).

Practical Uses of Endophytes

The *N. coenophialum* strain most commonly present in North American tall fescue pastures sometimes produces ergot alkaloids at levels toxic to grazing livestock. Likewise, perennial ryegrass in Australia and New Zealand, as well as Oregon, can have toxic levels of lolitrems produced by common strains of *N. lolii*. Such strains also produce ergot alkaloids, which conceivably might synergize with lolitrems (this, to my knowledge, remains untested). Simple removal of the endophytes is possible, but not desirable because they can also provide protection from nematodes (Panaccione et al. 2006), increased drought tolerance, and other benefits (Malinowski and Belesky 2000). Alternatives are to use forage grasses that have no endophyte, or that have endophytes that produce no ergot alkaloids or indole-diterpenes. The latter include meadow fescue with *N. uncinatum*, and annual ryegrasses with *N. occultans*, both of which produce lolines but not the anti-livestock alkaloids (Schardl and Leuchtmann 2005).

Novel strains of endophytes have been identified and put into tall fescue and perennial ryegrass, generating cultivars with little or none of ergot alkaloids or lolitrems, and these appear promising for pasture and forage (Bluett et al. 2005; Bouton et al. 2002). An alternative approach, to knock out genes for the determinant steps in the respective pathways, is also being explored (Wang et al. 2004). As the problems with endophytes are addressed, the research focus is shifting to address the extent and mechanisms of the benefits these endophytes provide to their hosts. The

Epichloë festucae genome and transcriptome sequencing efforts nearing completion will greatly enhance these research efforts.

Acknowledgements

Resent research by the author's group has been supported by grants EF-0523661 and MCB-0213217 from the U.S. National Science Foundation, and grants 200403171013, 200506271031, 2005-35319-16141, and 2005-35318-16184 from the United States Department of Agriculture.

References

Blankenship JD, Houseknecht JB, Pal S, Bush LP, Grossman RB, Schardl CL (2005) Biosynthetic precursors of fungal pyrrolizidines, the loline alkaloids. Chembiochem 6:1016–1022

Bluett SJ, Thom ER, Dow BW, Burggraaf VT, Hume DE, Davies E, Tapper BA (2004) Effects of natural reseeding and establishment method on contamination of a novel endophyte-infected perennial ryegrass dairy pasture with other ryegrass/endophyte associations. N Z J Agric Res 47:333–344

Bluett SJ, Thom ER, Clark DA, Macdonald KA, Minnee EMK (2005) Effects of perennial ryegrass infected with either AR1 or wild endophyte on dairy production in the Waikato. N Z J Agric Res 48:197–212

Bouton JH, Latch GCM, Hill NS, Hoveland CS, McCann MA, Watson RH, Parish JA, Hawkins LL, Thompson FN (2002) Reinfection of tall fescue cultivars with non-ergot alkaloid-producing endophytes. Agron J 94:567–574

Bultman TL, Leuchtmann A (2003) A test of host specialization by insect vectors as a mechanism for reproductive isolation among entomophilous fungal species. Oikos 103:681–687

Chung KR, Hollin W, Siegel MR, Schardl CL (1997) Genetics of host specificity in *Epichloë typhina*. Phytopathology 87:599–605

Clay K (1990) Fungal endophytes of grasses. Annu Rev Ecol Syst 21:275–295

Clay K (1994) Hereditary symbiosis in the grass genus *Danthonia*. New Phytol 126:223–231

Clay K, Schardl C (2002) Evolutionary origins and ecological consequences of endophyte symbiosis with grasses. Am Nat 160 Suppl.:S99–S127

Coyle CM, Panaccione DG (2005) An ergot alkaloid biosynthesis gene and clustered hypothetical genes from *Aspergillus fumigatus*. Appl Environ Microbiol 71:3112–3118

Darbyshire SJ (1993) Realignment of *Festuca* subgenus *Schedonorus* with the genus *Lolium* (Poaceae). Novon 3:239–243

Faulkner JR, Hussaini SR, Blankenship JD, Pal S, Branan BM, Grossman RB, Schardl CL (2006) On the sequence of bond formation in loline alkaloid biosynthesis. Chembiochem 7:1078–1088

Fleetwood DJ, Scott B, Lane GA, Tanaka A, Johnson RD (2007) A complex ergovaline gene cluster in Epichloë endophytes of grasses. Appl Environ Microbiol 73:2571–2579

Freeman EM (1904) The seed fungus of Lolium temulentum L., the darnel. Philos Trans R Soc Lond B 196:1–27

Gröger D, Floss HG (1998) Biochemistry of ergot alkaloids – achievements and challenges. In Alkaloids. Vol. 50. Cordell GA (ed). New York: Academic Press, pp. 171–218

Haarmann T, Machado C, Lübbe Y, Correia T, Schardl CL, Panaccione DG, Tudzynski P (2005) The ergot alkaloid gene cluster in Claviceps purpurea: extension of the cluster sequence and intra species evolution. Phytochemistry 66:1312–1320

Lewis GC, Ravel C, Naffaa W, Astier C, Charmet G (1997) Occurrence of Acremonium endophytes in wild populations of Lolium spp. in European countries and a relationship between level of infection and climate in France. Annu Appl Biol 130:227–238

Li W, Ji Y-l, Yu H-S, Wang Z-W (2006) A new species of Epichloë symbiotic with Chinese grasses. Mycologia 98:560–570

Lorenz N, Wilson EV, Machado C, Schardl C, Tudzynski P (2007) Comparison of ergot alkaloid biosynthesis gene clusters in Claviceps species indicate loss of late pathway steps in evolution of C. fusiformis. Appl Environ Microbiol 73:7185–7191

Malinowski DP, Belesky DP (2000) Adaptations of endophyte-infected cool-season grasses to environmental stresses: mechanisms of drought and mineral stress tolerance. Crop Sci 40:923–940

McMillan LK, Carr RL, Young CA, Astin JW, Lowe RGT, Parker EJ, Jameson GB, Finch SC, Miles CO, McManus OB, Schmalhofer WA, Garcia ML, Kaczorowski GJ, Goetz M, Tkacz JS, Scott B (2003) Molecular analysis of two cytochrome P450 monooxygenase genes required for paxilline biosynthesis in Penicillium paxilli, and effects of paxilline intermediates on mammalian maxi-K ion channels. Mol Genet Genomics 270:9–23

Moon CD, Craven KD, Leuchtmann A, Clement SL, Schardl CL (2004) Prevalence of interspecific hybrids amongst asexual fungal endophytes of grasses. Mol Ecol 13:1455–1467

Moon CD, Guillaumin J-J, Ravel C, Li C, Craven KD, Schardl CL (2007) New Neotyphodium endophyte species from the grass tribes Stipeae and Meliceae. Mycologia 99:895–905

Panaccione DG, Johnson RD, Wang JH, Young CA, Damrongkool P, Scott B, Schardl CL (2001) Elimination of ergovaline from a grass-Neotyphodium endophyte symbiosis by genetic modification of the endophyte. Proc Natl Acad Sci USA 98:12820–12825

Panaccione DG, Kotcon JB, Schardl CL, Johnson RD, Morton JB (2006) Ergot alkaloids are not essential for endophytic fungus-associated population suppression of the lesion nematode, *Pratylenchus scribneri*, on perennial ryegrass. Nematology 8:583–590

Philipson MN, Christey MC (1986) The relationship of host and endophyte during flowering, seed formation, and germination of *Lolium perenne*. N Z J Bot 24:125–134

Riedell WE, Kieckhefer RE, Petroski RJ, Powell RG (1991) Naturally occurring and synthetic loline alkaloid derivatives: insect feeding behavior modification and toxicity. J Entomol Sci 26:122–129

Rowan DD (1993) Lolitrems, paxilline and peramine: mycotoxins of the ryegrass/endophyte interaction. Agric Ecosyst Environ 44:103–122

Schardl CL, Leuchtmann A (2005) The epichloë endophytes of grasses and the symbiotic continuum. In: The Fungal Community: its organization and role in the ecosystem. Vol. 23. Dighton J, White JF, Oudemans P (eds). Boca Raton, FL: CRC Press, pp. 475–503

Schardl CL, Leuchtmann A, Spiering MJ (2004) Symbioses of grasses with seedborne fungal endophytes. Annu Rev Plant Biol 55:315–340

Schardl CL, Panaccione DG, Tudzynski P (2006) Ergot alkaloids-biology and molecular biology. Alkaloids Chem Biol 63:45–86

Schardl CL, Grossman RB, Nagabhyru P, Faulkner JR, Mallik UP (2007) Loline alkaloids: currencies of mutualism. Phytochemistry 68:980–996

Spatafora JW, Sung GH, Sung JM, Hywel-Jones NL, White JF (2007) Phylogenetic evidence for an animal pathogen origin of ergot and the grass endophytes. Mol Ecol 16:1701–1711

Sullivan RF, Bills GF, Hywel-Jones NL, White JF Jr (2000) *Hyperdermium*: a new clavicipitalean genus for some tropical epibionts of dicotyledonous plants. Mycologia 92:908–918

Tanaka A, Tapper BA, Popay A, Parker EJ, Scott B (2005) A symbiosis expressed non-ribosomal peptide synthetase from a mutualistic fungal endophyte of perennial ryegrass confers protection to the symbiotum from insect herbivory. Mol Microbiol 57:1036–1050

Wang J, Machado C, Panaccione DG, Tsai H-F, Schardl CL (2004) The determinant step in ergot alkaloid biosynthesis by an endophyte of perennial ryegrass. Fungal Genet Biol 41:189–198

Young CA, Felitti S, Shields K, Spangenberg G, Johnson RD, Bryan GT, Saikia S, Scott B (2006) A complex gene cluster for indole-diterpene biosynthesis in the grass endophyte *Neotyphodium lolii*. Fungal Genet Biol 43:679–693

Seed Transmission of Endophytic Fungus, *Neotyphodium occultans*, in Cross Breeding of Italian Ryegrass (*Lolium multiflorum*) Using Detached Panicle Culture, and Comparison with Situations in Interspecific/Intergeneric Crossings Including *Festuca* Species

Koya Sugawara[1,3], Akira Arakawa[1], Takuya Shiba[1], Hiroto Ohkubo[2] and Takao Tsukiboshi[1]

[1]National Institute of Livestock and Grassland Science, Nasushiobara, Tochigi 329-2793, Japan
[2]Japan International Research Center for Agricultural Sciences, 1-1, Ohwashi, Tsukuba, Ibaraki 305-8686, Japan
[3]Corresponding author, skoya@affrc.go.jp

Abstract. *Neotyphodium* species are seed-transmitted endophytic fungi that form mutualistic (symbiotic) associations with grasses of the subfamily Pooideae, and their presence can increase stress tolerance of host grasses. Italian (annual) ryegrass (*Lolium multiflorum*) clones infected with *N. occultans* were pollinated by pollen from annual ryegrass and perennial ryegrass (*L. perenne*) using detached panicles, and from tall fescue (*Festuca arundinacea*) by conventional bagging, as part of studies on the use of this endophytic fungus to enhance productivity of this important forage grass and also of interspecific/intergeneric hybrids. In mating involving *L. multiflorum*, 64–100% seed transmission ratios of the fungus were observed among 13 cross-combinations, along with one exceptional case of no infection. In mating with *L. perenne* 53–100% transmission was observed, whereas less than 30% transmission was observed in crosses involving *F. arundinacea*. The results indicated that the symbiont can be seed transmitted through mating using detached panicles, and the possibility of poor compatibility between *N. occultans* and some *L. multiflorum* genotypes as well as with *L. perenne* and *F. arundinacea*.

Introduction

Neotyphodium species (formerly known as *Acremonium* species, sect. *Albolanosa*) are seed transmitted endophytic fungi belonging to the family Clavicipitaceae which form mutualistic (symbiotic) associations with grasses of the subfamily Pooideae. They are ecologically and agriculturally important, producing chemicals that can adversely affect grazing animals and insect pests (reviewed in Clay and Schardl 2002). Recent surveys revealed that naturalized Italian (annual) ryegrasses (*Lolium multiflorum*) infected with the endophytic fungus, *N. occultans* (Moon et al. 2000) is prevalent in the western part of Japan and its possible use for providing insect resistance in *L. multiflorum* is of interest (Sugawara et al. 2005, 2006). The fungus is assumed to deter some insect pests from feeding on young host seedlings (Stewart 1987) and influencing not only insect population dynamics of an ecosystem (Omacini et al. 2001), but also the condition of soil (Omacini et al. 2004).

This paper reports on studies with *L. multiflorum* infected with *N. occultans*, to improve the ease of carrying out mating by utilizing detached panicles, and to determine if the fungus can form associations with hybrids formed with perennial ryegrass (*L. perenne*) and tall fescue (*Fesutuca arundinacea*), two grass species that are not natural hosts of this endophytic fungus.

Materials and Methods

Plants

Seed parents (maternal plants) were chosen from a sister group derived from an open pollination of a naturalized *L. multiflorum* clone infected with *N. occultans*, collected in Shizuoka prefecture, Japan (54-10-4-4, Sugawara et al. 2006). Commercially available cultivars were used as pollen parents; *L. multiflorum* cv. Hataaoba, cv. Nioudachi, cv. Sachiaoba; *L. perenne* cv. Kiyosato; *F. arundinacea* cv. Nanryo. All plants were grown in glasshouses of the National Institute of Livestock and Grassland Science, Nasushiobara, Tochigi, Japan, using a potting mixture (Kureha Engei-baido, Kureha Chemical Industry Co., Ltd, Tokyo, Japan).

Mating of Plants

Mating was done following the method described by Shimizu and Ohsugi (1977). In brief, panicles in which the first flowers were about to open were excised from the both parental plants and inserted into bottles with water solution containing 3% sucrose and 0.015% sulfurous acid (H_2SO_3). The desired combinations of panicles were put together in a paper bag (approx. 12 cm wide × 40 cm high), 3–10 tillers from each parent per bag, and were kept in an air conditioned room (22°C) under weak fluorescent light (16L:8D). The panicles were shaken occasionally during the first 17 days to facilitate pollination. Seeds were harvested after 37 days. For crossing between *F. arundinacea* and *N. occultans*-infected *L. multiflorum*, conventional mating with the same paper bags using intact panicles (not detached) was used to overcome a likely lower fertility of the cross combination.

Detection of Endophytic Fungi in Plants

Endophyte infection of offspring plants was confirmed by microscopic observation of plant tissue from young seedlings after they generated four or more tillers, using the method described in Sugawara et al. (2007). In short, plant tissues around the apical meristem were dissected, taking two or more tillers form a seedling, and observed under a differential interference contrast microscope after mounting on glass slides with a solution made by mixing lactic acid, glycerol, and distilled water in the ratio of 1:2:1 (v/v/v) and stored until tissues became clear enough for mycelia to be seen (Fig. 1, left). For seeds from crossing with *L. multiflorum* cultivars, content of *N*-formylloline, an insect repelling chemical secreted by the fungus, was also checked prior to the microscopical observations, as an indicator of the existence of the fungus and its activity for insect pests, using gas chromatography by taking 30 mg of subsamples from each seed sample (Shiba et al. 2007).

Results

Sufficient amounts of seeds to check both *N*-formylloline content and endophyte infection were generated from 35 mating combinations with *L. multiflorum* cultivars, out of 40 mating trials. The content of *N*-formylloline varied from less than 50 ppm (below the threshold of detection) to 1,879 ppm (w/w) (Shiba et al. 2007). Due to the smaller number of seeds available from the pollination with *L. perenne* and *F. arundinacea*,

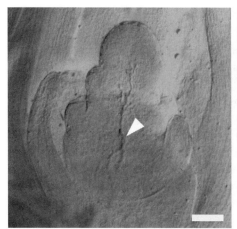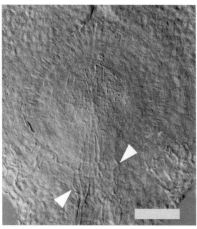

Fig.1 Endophytic fungus, *Neotyphodium occultans* in host plant, Italian ryegrass (*Lolium multiflorum*). Fungal mycelia are indicated by *arrowheads*. *Left*: an apical meristem differentiating into a panicle, *right*: an immature ovule of the plant (differential interference contrast microscopy of chemically cleared tissue, bars = 50 μm)

we could not check the amount of the chemical. Among the seed samples from mating within the species, seeds from 13 combinations were sown and presence of the endophytic fungus in the growing seedlings was checked microscopically. In a sample from which *N*-formylloline was not detected, no infections by *N. occultans* were observed. In the other 12 samples from which the chemical was detected, infection ratios varying from 64 to 100% were observed (Table 1). From mating with perennial ryegrass (by detached panicles) and tall fescue (in conventional bagging), seeds were generated from five out of five, and four out of five mating combinations respectively, and 53–100% infection was observed from the former combination, whereas less than 30% infection was observed with the latter combination (Tables 2 and 3). Although *L. multiflorum* is considered to be an outbreeder (Jauhar 1993), DNA content of the nuclei was checked by flow cytometry and only offspring plants considered to have chromosome composition of 2n = 28 (tetraploid) were considered hybrids in mating with *F. arundinacea*.

Discussion

The feasibility of plant pollination using detached panicle culture was recognized some 30 years ago (Shimizu and Ohsugi 1977; Wafford et al. 1986), but the suitability of the method for grasses infected with *Neotyphodium* endophytes had not been tested. The fact that the fungi are already present in apical meristems in very early stages of differentiation into panicles and, subsequently, the immature ovules of the plants (Fig. 1; Sugawara et al. 2004, 2006), implies the feasibility of the application, and we confirmed that with this current study.

Evolved from plant pathogenic *Epichloë* species, *Neotyphodium* endophytes are known to have natural host ranges (Clay and Schardl 2002), although it is difficult to evaluate if the host range can be extended through controlled mating. In this study it was shown that the infection rates differed between maternal genotypes with reduced or even no infection occurring even with mating involving a natural host species of *N. occultans*. There seemed to be a tendency that the further the genetic distance of parents, the lower the infection ratio of the symbiont to the hybrids (*L. multiflorum* > *L. perenne* > *F. arundinacea*). Further investigation is needed to see whether infected plants observed among population from these interspecific and intergeneric crosses can maintain stable associations.

Being seed transmitted, *Neotyphodium* endophytes can act just like maternally inherited genetic traits of the host plants, and add traits plants can not generate, such as insect repellence by chemicals secreted by the fungi. However, limited ways to introduce them into new plant–fungus associations (Latch and Christensen 1985) and problems with host compatibility (Koga et al. 1993; Stewart 1997, also indicated in this study), provide obstacles for their wide practical applications for broader ranges of grasses. To elucidate the mechanisms of this host compatibility/specificity, mating using detached panicles allowing the testing of many cross combinations within small amount of space with a low workload, may provide a useful experimental tool. The method is especially useful for *N. occultans* which we can not culture on media (Moon et al. 2000) and are unable to inoculate into other grasses for assessment of its host specificity and compatibility. In the current study, maintenance of *N. occultans* seemed to be negatively influenced by the hybridization of the host genome with another genome(s) from *Festuca*. This might be an obstacle for the practical use of the fungus in hybrid plants (*Festulolium* species), but also may be a key to study mechanisms for host specificity/compatibility of the symbiosis.

Table 1 Content of *N*-formylloline (NFL) in offspring seeds from crosses of Italian ryegrass (*Lolium multiflorum*) made by detached panicle culture, and infection status of *Neotyphodium occultans* among plants grown from them

Seed parent[b]	Pollen parent[c]	NFL content (ppm)[d]	Infection status of *Neotyphodium occultans*[a]			
			Offspring plants grown	Number of checked plants	Number of infected plants	(%)
D1-C3	Sachiaoba 06-11	1,879	12	12	11	91.7
Ful-D1	Sachiaoba 06-12	1,560	27	27	22	81.5
NC2-E4	Hataaoba 06-18	1,549	33	29	26	89.7
NC2-A2	Hataaoba 06-39	1,546	32	29	26	89.7
NC2-E5	Nioudachi 06-29	1,466	21	20	19	95.0
Fu2-B1	Hataaoba 06-41	1,422	33	33	30	90.9
DC2-D4	Hataaoba 06-5	1,384	31	31	27	87.1
NC2-B5	Hataaoba 06-19	1,299	34	34	34	100.0
Fu4-A7	Hataaoba 06-26	1,208	35	35	33	94.3
NC2-B1	Hataaoba 06-16	1,202	34	34	29	85.3
Fu3-C3	Nioudachi 06-24	1,120	31	30	20	66.7
Fu3-A1	Hataaoba 06-1	1,118	31	31	20	64.5
Fu3-B3	Nioudachi 06-21	ND	29	26	0	0.0

[a] Checked by microscopic observations of basal part of tillers
[b] Sister clones from open pollination of a *L. multiflorum* clone infected with *N. occultans* (54-10-4, Sugawara et al. 2006)
[c] Chosen from established cultivars available in Japan
[d] Checked by a gas chromatography using ca.30 mg seeds. ND: Lower than detectable threshold (50 ppm)

Table 2 Infection status of *Neotyphodium occultans* among plants from cross pollination of infected annual ryegrass (*Lolium multiflorum*) with perennial ryegrasses (*L. perenne*) made using detached panicle culture

		Infection status of *Neotyphodium occultans*[a]			
Seed parent[b]	Pollen parent[c]	Offspring plants grown	Number of checked plants	Number of infected plants	(%)
Fu1-B5	ki04-20	2	2	2	100.0
Fu3-A7	ki04-11	7	7	7	100.0
Fu1-A5	ki04-6	20	19	17	89.5
Fu4-A6	ki04-14	4	4	3	75.0
NC-A1	ki04-19	15	15	8	53.3

[a] Checked by microscopical observations of basal part of tillers
[b] Sister clones from open pollination of a *L. multiflorum* clone infected with *N. occultans* (54-10-4-4, Sugawara et al. 2006)
[c] Chosen from established *L. perenne* cultivar available in Japan, cv. Kiyosato

Table 3 Infection status of *Neotyphodium occultans* among plants from cross pollination of infected annual ryegrass (*Lolium multiflorum*, 2n = 14, diploid) with tall fescue (*Festuca arundinacea*, 2n = 42, hexaploid)

		Infection status of *Neotyphodium occultans*[a]			
Seed parent[b]	Pollen parent[c]	Offspring plants grown[d]	Number of checked plants	Number of infected plants	(%)
Nc2-B1	Na2	23	23	6	26.1
Fu3-E3	Na4	35	35	8	22.9
D1-A2	Na5	13	13	2	15.4
Fu1-C5	Na3	7	7	0	0.0

[a] Checked by microscopical observations of basal part of tillers
[b] Sister clones from open pollination of a *L. multiflorum* clone infected with *N. occultans* (54-10-4-4, Sugawara et al. 2006)
[c] Chosen from established *F. arundinacea* cultivar available in Japan, cv. Nanryo
[d] Plants considered to have chromosome composition of 2n = 28 (tetraploid) were chosen by measuring DNA content of the nuclear by flow cytometry and used

Acknowledgements

The authors are grateful for helpful advice given by Alan V. Stewart, PGG Wrightson Seeds, Christchurch, New Zealand and Michael. J. Christensen, AgResearch Grassland Research Center, Palmerston North, New Zealand, for our research. This work was supported in part by a research project of the Ministry of Agriculture, Forestry and Fisheries of Japan (Integrated research for developing Japanese-style forage feeding system to increase forage self-support ratio, No. 14004) and Grants-in-Aid for Scientific Research from the Japan Society for the Promotion of Science (JSPS) (No. 18380156).

References

Clay K, Schardl CL (2002) Evolutionary origins and ecological consequences of endophyte symbiosis with grasses. Am Nat 160 (Suppl): S99–S127

Jauhar PP (1993) Cytogenetics of the *Festuca–Loium* complex: relevance to breeding. Springer, Berlin Heidelberg, Germany

Koga H, Christensen MJ, Bennett RJ (1993) Incompatibility of some grass-*Acremonium* endophyte associations. Mycol Res 97: 1237–1244

Latch GCM, Christensen MJ (1985) Artificial infection of grass with endophytes. Ann Appl Biol 107: 17–24

Moon CD, Schardl CL, Christensen MJ (2000) The evolutionary origins of *Epichloë* endophytes from annual ryegrass. Mycologia 92: 1103–1118

Omacini M, Chaneton EJ, Ghersa CM, Müller CB (2001) Symbiotic fungal endophytes control insect host–parasite interaction webs. Nature 406: 78–81

Omacini M, Chaneton EJ, Ghersa CM, Otero P (2004) Do foliar endophytes affect grass litter decomposition? A microcosm approach using *Lolium multiflorum*. Oikos 104: 581–590

Shiba T, Sasaki T, Sugawara K, Arakawa A, Kanda K (2007) N-formylloline, a chemical produced within endophyte infected pasture grasses, can disrupt the growth of rice leaf bug, *Trigonotylus caelestialum*. Grassl Sci 53 (Suppl.): 358–359 (In Japanese. Translated title by the present authors)

Shimizu N, Ohsugi R (1977) Development and germination behaviour of seed produced by culture of cut ear in pasture plants. II. The optimum culture condition on Italian and perennial ryegrasses. Bull Natl Grassl Res Inst 11: 34–46

Sugawara K, Ohkubo H, Yamashita M, Mikoshiba Y (2004) Flowers for *Neotyphodium* endophytes detection: a new method for detection from grasses. Mycoscience 45: 222–226

Sugawara K, Shiba T, Yamashita M (2005) *Neotyphodium* research and application in Japan. In: Roberts CA, West CP, Spiers DE (eds), *Neotyphodium* in Cool-Season Grasses: Current Research and Applications. Blackwell, Ames, IA, pp. 55–63

Sugawara K, Inoue T, Yamashita M, Ohkubo H (2006) Distribution of the endophytic fungus, *Neotyphodium occultans* in naturalized Italian ryegrass in western Japan and its production of bioactive alkaloids known to repel insect pests. Grassl Sci 52: 147–154

Sugawara K, Ohkubo H, Tsukiboshi T (2007) Distribution and morphology of *Neotyphodium* endophytes in apical meristems of host plants as observed using differential interference contrast microscopy. Proceedings of the 6th International Symposium on Fungal Endophytes of Grasses, pp. 139–141

Stewart AV (1987) Plant breeding aspects of ryegrasses (*Lolium* sp.) infected with endophytic fungi. PhD thesis, Lincoln College

Stewart AV (1997) Observations on maintaining endophyte during back-crossing of endophyte from perennial ryegrass to annual ryegrasses. In: Bacon CW, Hill NS (eds), *Neotyphodium*/Grass Interactions (Proceedings of the 3rd International Symposium on *Acremonium*/grass Interactions). Plenum Press, New York, NY, p. 279

Wafford DS, Frakes RV, Chilcote DO (1986) A detached culm technique for seed production of tall fescue in isolation from foreign pollen sources. Crop Sci 26: 193–195

Colonial Bentgrass Genetic Linkage Mapping

David Rotter[1,2], Keenan Amundsen[3], Stacy A. Bonos[1], William A. Meyer[1], Scott E. Warnke[3] and Faith C. Belanger[1,2,4]

[1]Department of Plant Biology and Pathology and [2]The Biotechnology Center for Agriculture & the Environment, Rutgers, The State University of New Jersey, New Brunswick, NJ, USA
[3]United States Department of Agriculture, Agricultural Research Service, Beltsville, MD, USA
[4]Corresponding author, belanger@aesop.rutgers.edu

Abstract. Colonial bentgrass (*Agrostis capillaris*) is an important turfgrass species used on golf courses in temperate regions, although the related species, creeping bentgrass (*A. stolonifera*), is often preferred. One of the major management problems for creeping bentgrass is the fungal disease called dollar spot. Colonial bentgrass as a species has good resistance to dollar spot and may be a source of novel genes or alleles that could be used in the improvement of creeping bentgrass. As one approach to ultimately identifying the genes in colonial bentgrass that confer dollar spot resistance, we are developing a genetic linkage map of colonial bentgrass. To provide tools for mapping genes, we have generated expressed sequenced tag (EST) resources for both colonial and creeping bentgrass, and have developed a new approach to gene-based marker development.

Introduction

Turfgrasses are important for providing surfaces for recreation in public parks, sports fields, golf courses, and home lawns. With urbanization and the popularity of field sports, lawn and recreational grasses have become important commodities in the United States, as well as other parts of the world. Numerous academic and private breeding programs are engaged in the improvement of several grass species used for turf (Meyer and Funk 1989; Bonos et al. 2006). Creeping bentgrass (*Agrostis stolonifera* L.) and

T. Yamada and G. Spangenberg (eds.), *Molecular Breeding of Forage and Turf*,
doi: 10.1007/978-0-387-79144-9_28, © Springer Science + Business Media, LLC 2009

the related species colonial bentgrass (*A. capillaris* L.) are important turfgrass species that are used extensively on golf courses in temperate regions (Warnke 2003; Ruemmele 2003). Generally creeping bentgrass is preferred and is planted on approximately 140,757 acres of golf course fairways and greens (Golf Course Superintendents Association of America, 2007). Colonial bentgrass does not have the desirable aggressive stoloniferous growth habit of creeping bentgrass, which aids in repair of the turf from the damage incurred during play.

Although creeping bentgrass is generally preferred for use on golf courses, it can be very susceptible to the fungal pathogen that causes the disease dollar spot (Walsh et al. 1999). In the literature the pathogen is referred to as *Sclerotinia homoeocarpa* F.T. Bennett, although it is recognized that it should be reclassified to one of the genera *Rutstroemia, Lanzia, Moellerodiscus,* or *Poculum* (Carbone and Kohn 1993; Holst-Jensen et al. 1997). The current method of disease control relies heavily on the use of fungicides (Watson et al. 1992).

Because dollar spot is such a major management problem for creeping bentgrass, turfgrass breeders are working to develop improved cultivars (Meyer and Belanger 1997; Bonos et al. 2006). No creeping bentgrass cultivars are completely resistant to dollar spot but some creeping bentgrass germplasm selections with better dollar spot resistance have been identified and are being used for cultivar development (Bonos et al. 2004). Genetic linkage maps of creeping bentgrass have been developed and quantitative trait loci (QTL) for dollar spot resistance have been identified (Chakraborty et al. 2005, 2006; Bonos et al. 2005). In the future, marker-assisted selection may be possible in creeping bentgrass breeding.

Another approach to the development of improved cultivars of creeping bentgrass may be interspecific hybridization with colonial bentgrass, which as a species has good resistance to dollar spot (Plumley et al. 2000). Interspecific hybridization has been used by breeders of many crops to introduce beneficial traits from related species into crop species. It has been successfully utilized in many important agronomic crops such as coffee (Lashermes et al. 2000), grapes (Lodhi et al. 1995), garlic (Yanagino et al. 2003), tomato (Bernatzky and Tanksley 1986), and wheat (Rabinovich 1998). However, interspecific hybridization has not yet been used in bentgrass breeding (Brilman 2001). The ultimate goal of such an approach would be to develop bentgrass cultivars with the stoloniferous growth habit of creeping bentgrass combined with the dollar spot resistance of colonial bentgrass. We have been working on this approach and our results to date are the subject of this chapter. We believe that combining the approaches of selecting for superior creeping bentgrass germplasm with introgression of dollar spot resistance genes from colonial bentgrass may provide the highest level of dollar spot resistance to new cultivars.

Genome Organization of Creeping and Colonial Bentgrasses

For interspecific hybridization to be a useful breeding strategy there must be compatibility between the genomes of the two species involved. An extensive cytological investigation into some *Agrostis* spp. and their interspecific hybrids was reported by Jones (1956a–c). Hybrids between creeping and colonial bentgrass were recovered and were examined for chromosome pairing at metaphase 1 of meiosis (Jones 1956b). He concluded that creeping and colonial bentgrass were both allotetraploids having one ancestral genome in common and a 2N chromosome number of 28. The genome designations for colonial bentgrass were $A_1A_1A_2A_2$ and for creeping bentgrass were $A_2A_2A_3A_3$. The chromosome pairing in the hybrids indicated complete pairing between the common A_2 genomes and partial pairing between the A_1 and A_3 genomes. Recently Zhao et al. (2007) reported a similar study on the interspecific hybrids and confirmed the observations of Jones (1956b).

Interspecific Hybridization of Creeping and Colonial Bentgrasses

In 1998, colonial bentgrass x creeping bentgrass hybrids were generated in projects investigating the potential outcrossing of transgenic creeping bentgrass (Belanger et al. 2003a,b). From controlled crosses it was determined that the frequency of interspecific hybridization was low, lower than the frequency of selfing which was estimated to be about 0.5% (Belanger et al. 2003b). The hybrids were found to be fertile, both through the pollen and the egg.

Knowing that colonial bentgrass as a species has good dollar spot resistance led to the evaluation of these hybrids as a potential source of genetic resistance to dollar spot for future crossing with creeping bentgrass. In 2001 and 2002, the colonial x creeping bentgrass interspecfic hybrids were field tested for dollar spot resistance and some of the hybrids exhibited excellent resistance (Belanger et al. 2004).

In the summer of 2002 an attractive hybrid having excellent dollar spot resistance was backcrossed to a creeping bentgrass plant. The resulting progeny from the cross were evaluated in field tests in 2003–2006. The population exhibited large variation with respect to dollar spot resistance, growth habit, and leaf morphology. Since some of the backcross progeny did have excellent dollar spot resistance (Belanger et al. 2005), this population will be ideal for determining the genetic basis of the colonial bentgrass contribution to the observed resistance. Overall, about 11% of the backcross progeny exhibited good dollar spot resistance, suggesting three colonial bentgrass genes may be involved in conferring resistance. We have found that the primary hybrids and their backcross progeny are generally fertile, which is promising regarding the potential for introgression of the colonial bentgrass dollar spot genes into creeping bentgrass.

Colonial Bentgrass Genetic Linkage Mapping

Our approach to identification of the colonial bentgrass genes responsible for dollar spot resistance is to develop a genetic linkage map of colonial bentgrass (Rotter et al. 2006). There is currently no linkage map available for colonial bentgrass. Our approach is to map the colonial bentgrass portion of the hybrid genome by using the backcross population described above as the mapping population. This is an unconventional mapping population structure but is being used to ultimately identify the colonial bentgrass genomic regions conferring dollar spot resistance in the backcross population. Colonial bentgrass as a species is very resistant to dollar spot, so populations segregating for resistance cannot be generated from intraspecific crosses. We are mapping colonial bentgrass by identifying polymorphic markers present in the interspecific hybrid but absent in the creeping bentgrass grandparent and parent of the mapping population. These polymorphic markers originated from the colonial bentgrass grandparent of the population and can therefore be used to map the colonial bentgrass genome. For any particular colonial bentgrass gene, half of the mapping population will have inherited that gene.

We are combining amplified fragment length polymorphism (AFLP) markers with gene-based markers to develop the linkage map. The AFLP markers are extremely useful for quickly establishing linkage groups providing a framework for placement of the gene-based markers. We have used several approaches to develop gene-based markers, which are described below. To provide the sequences from which to develop gene-based markers, we generated expressed sequence tag (EST) sequences from both colonial and creeping bentgrass (Rotter et al. 2007a). For colonial bentgrass 7,528 sequences were obtained with an average length of 745 bases (accession numbers DV852741-DV860268). For creeping bentgrass 8,470 sequences were obtained with an average length of 567 bases (accession numbers DV860269-DV868738). The EST sequences are from the 3′ ends of the cDNAs and so contain the 3′ untranslated regions of the genes. It is well known that the 3′ untranslated regions of genes are more variable than the coding sequences, and therefore more useful for developing gene-based molecular markers (Bhattramakki et al. 2002; Brady et al. 1997).

PCR Fragment Presence or Length Polymorphisms

We are using the sequence variations in the 3′ untranslated regions of the ESTs to develop colonial specific markers for linkage mapping in our population. We are using three different methods for polymorphism detection. One method is to identify +/− polymorphisms or length polymorphisms in PCR fragments generated from the 3′ ends of the genes. Through sequence alignments of colonial and creeping bentgrass ESTs, we can often design oligonucleotide primers that will selectively amplify only the colonial bentgrass gene, or that will amplify fragments of different sizes from the similar colonial and creeping bentgrass genes. The PCR fragments are visualized after separation on high-resolution agarose gels. Bands unique to the hybrid originated from the colonial bentgrass portion of the hybrid genome and can thus be used as markers for mapping colonial bentgrass.

Minisequencing

We have also used the colonial and creeping bentgrass EST sequence alignments to develop single nucleotide polymorphism (SNP) markers that are detected by minisequencing. In cases where it is not possible to develop colonial specific oligonucleotides, sometimes a SNP difference is present. SNPs are the most common type of polymorphism between individuals within a species and are being widely used as markers in many species (Rafalski 2002). However, the development of SNP markers generally requires considerable sequence information from multiple genotypes within the species. Because of the structure of our mapping population, instead of looking for SNP differences between individuals within a species, we are looking for SNP differences between two species, colonial and creeping bentgrass.

The minisequencing technique used for marker development was modified from that described by Carvalho and Pena (2005). Based on sequence alignments, SNPs between colonial and creeping bentgrass are identified. PCR primers are then designed to amplify the region surrounding the SNP. Unincorporated nucleotides and primers are then inactivated and a chain elongation reaction is performed using a labeling primer designed to end one base pair upstream of the identified SNP. Instead of traditional nucleotides, a single fluorescently labeled dideoxy nucleotide of the specific SNP is used in the reaction. If the individual contains the SNP in question then the result is a single base extension of the labeling primer with the fluorescent dideoxy nucleotide. If the individual does not contain the SNP then there is a mismatch at the SNP site and no label is incorporated. The products are then separated on a polyacrylamide sequencing gel and visualized with a phosphorimager.

Dideoxy Polymorphism Scanning

We have also developed a new method for marker development called dideoxy polymorphism scanning (ddPS), which can reveal SNPs and indels without prior knowledge of a sequence polymorphism (Rotter et al. 2007b). Minisequencing is limited to one specific SNP site that is targeted based on a known sequence variation revealed through sequence alignments. The new ddPS method is proving very useful for us since our EST database is relatively small and we do not always have enough sequence data to accurately identify useful SNPs.

The approach for ddPS marker development is diagramed in Fig. 1. PCR primers are designed to amplify an approximately 200 bp fragment from the 3′ end of a selected EST sequence. The forward PCR primer has an M13 sequence extension added to the 5′ end. For PCR primer design, we first align all the similar sequences for a particular gene available in our EST database. When multiple sequences are available it is often possible to identify sequence "types" based on variations in the 3′ untranslated regions. As discussed above, colonial bentgrass is an allotetraploid and so homoeologous genes are expected. Also numerous plant genes are known to be members of gene families. Whenever possible, PCR primers are

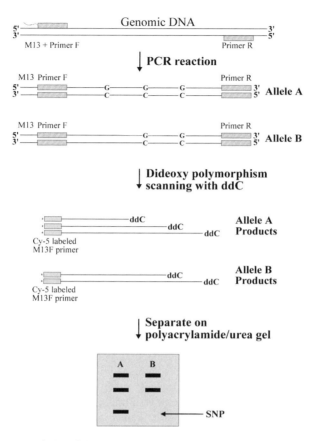

Fig. 1 Diagram of the dideoxy polymorphism scanning (ddPS) method (from Rotter et al. 2007b)

designed to amplify only one sequence type. This simplifies the ddPS pattern making polymorphism detection easier. The PCR products are treated with Exonuclease I to degrade the unused primers and shrimp alkaline phosphatase to dephosphorylate the unused deoxynucleotides. The PCR products are then subjected to a standard sequencing reaction using a *single* dideoxy nucleotide and the M13 sequencing primer, which has been labeled with the fluorescent dye Cy5 at the 5′ end. The dideoxy terminated reaction products are separated on a sequencing polyacrylamide gel and the fragments detected with a phosphorimager. Polymorphisms are revealed as different sized dideoxy terminated fragments, which result from SNPs and indels between the samples being compared. An example of ddPS detection of polymorphisms is shown in Fig. 2.

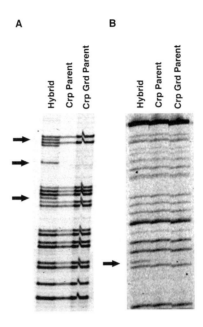

Fig. 2 Two examples of ddPS screens of PCR products from the hybrid, and the creeping parent and grandparent of the mapping population. (**a**) An example of a case where the PCR product from the interspecific hybrid was a mixture of a colonial bentgrass allele and a creeping bentgrass allele, as determined by cloning the PCR product and sequencing several resulting plasmids (Rotter et al. 2007b). With the ddPS method, hybrid specific bands originating from the colonial bentgrass allele were detected and found to be segregating in the mapping population. (**b**) An example where a single sequence variation between the hybrid and the creeping bentgrass parental plants was found and could be used to map the colonial bentgrass gene

It is important to emphasize that direct sequencing of the PCR products would not be a more informative substitute for the ddPS method of marker development. Mapping of heterozygous outcrossing species presents challenges not encountered with homozygous diploid species such as maize and *Arabidopsis* where mapping can be done with recombinant inbred lines. With obligate outcrossing species such as the bentgrasses, even single band PCR products are generally heterogeneous due to the presence of heterozygous alleles. In our case, the PCR products from the interspecific hybrid plant may contain both colonial and creeping bentgrass alleles as seen in Fig. 2a. Sequencing of such products is often impossible since a single base indel between the alleles will generate mixed bases at all positions of the sequence. With ddPS, where only a single dideoxy terminator is used, terminated fragments from a mixed PCR product are visualized as discreet bands. The allelic variation used for marker development is revealed as the presence or absence of one or more terminated fragments from the reaction. The ddPS method can be implemented by using either sequencing polyacrylamide gels or high throughput capillary sequencing machines (Rotter et al. 2007b).

Similarity of Colonial Bentgrass Linkage Groups to Wheat and Rice

Genome comparisons among many of the important cereal grasses have revealed considerable synteny in gene order (Devos 2005). Detailed comparisons of the rice and wheat chromosomes have been reported (Sorrells et al. 2003; LaRota and Sorrells 2004). Although there are numerous exceptions, the wheat chromosomes are largely derived from specific rice chromosomal regions (LaRota and Sorrells 2004). In general, similarities in chromosomal organization are found among members of the Pooideae and we expect the same for colonial bentgrass. A linkage map of creeping bentgrass using AFLP, random amplified polymorphic DNA (RAPD), and restriction fragment length polymorphism (RFLP) markers has been published (Chakraborty et al. 2005). Similarity of the creeping bentgrass linkage groups to the linkage groups of wheat was observed based on mapping some conserved sequences (Chakraborty et al. 2005). Not surprisingly, we can also see similarities of the colonial bentgrass linkage groups to those of wheat, an allohexaploid with a base chromosome number of 7 like the bentgrasses.

A comparison of colonial bentgrass linkage group 4A1 with rice chromosomes 3 and 11 is shown in Fig. 3. We are making colonial bentgrass linkage group designations based on the designations used for wheat. The wheat linkage groups 4A, 4B, and 4D are largely comprised of genes found on rice chromosomes 3 and 11 (LaRota and Sorrells 2004). The colonial bentgrass linkage group shown in Fig. 3 is designated as linkage group 4 based on having genes found on rice chromosomes 3 and 11. It is designated as from the A_1 genome of colonial bentgrass based on sequence comparisons. From sequence comparisons of the bentgrass ESTs, we can often make predictions of the subgenome since the colonial and creeping bentgrass A_2 subgenomes are quite similar to each other (Rotter et al. 2007a). We are using the rice-wheat chromosomal relationships and sequence comparisons among the colonial and creeping bentgrass ESTs to make assignments of the 14 colonial bentgrass linkage groups.

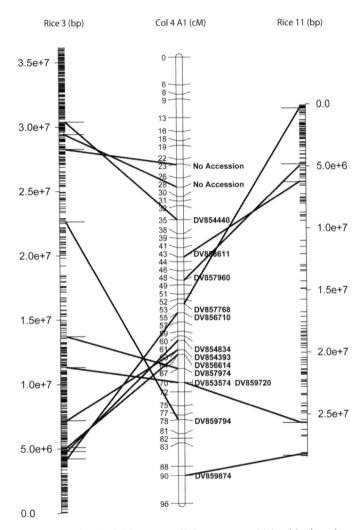

Fig. 3 Comparison of colonial bentgrass linkage group 4A1 with rice chromosomes 3 and 11. The *dashes* along the rice chromosome diagrams represent positions of matches of colonial bentgrass ESTs at e^{-20}. The *unlabeled dashes* along the colonial bentgrass linkage group represent AFLP markers

Conclusion

For the past several years we have been investigating the possibility of using interspecific hybridization with colonial bentgrass as an approach to improving the dollar spot resistance of creeping bentgrass. We have found that the interspecific hybrids can have good dollar spot resistance and that the resistance can be transmitted to progeny from a backcross to creeping bentgrass. These results are encouraging regarding the possibility of transferring the genes for dollar spot resistance from colonial bentgrass into creeping bentgrass. As an approach to developing markers for marker-assisted selection and to ultimately identify the genes underlying colonial bentgrass dollar spot resistance, we are currently working on a genetic linkage map of colonial bentgrass. To facilitate mapping, we have developed EST sequence resources and developed a new approach to marker development, dideoxy polymorphism scanning. We anticipate that this method will be useful for mapping in other species, especially for heterozygous species where direct sequencing of PCR products is not an option.

Acknowledgements

This research was supported with funds provided by the United States Golf Association, the Rutgers Center for Turfgrass Science, and the United States Department of Agriculture.

References

Belanger FC, Meagher TR, Day PR, Plumley K, Meyer WA (2003a) Interspecific hybridization between *Agrostis stolonifera* and related *Agrostis* species under field conditions. Crop Sci 43:240–246

Belanger FC, Plumley K, Day PR, Meyer WA (2003b) Interspecific hybridization as a potential method for improvement of *Agrostis* species. Crop Sci 43:2172–2176

Belanger FC, Bonos S, Meyer WA (2004) Dollar spot resistant hybrids between creeping bentgrass and colonial bentgrass. Crop Sci 44:581–586

Belanger FC, Bonos SA, Meyer WA (2005) Interspecific hybridization as a new approach to improving dollar spot resistance in creeping bentgrass. USGA Turfgrass Environ Res Online 4:1–5

Bernatzky R, Tanksley S (1986) Toward a saturated linkage map in tomato based on isozymes and random cDNA sequences. Genetics 112:887–898

Bhattramakki D, Dolan M, Hanafey M, Wineland R, Vaske D, Register JC III, Tingey SV, Rafalski A (2002) Insertion-deletion polymorphisms in 3′ regions of maize genes occur frequently and can be used as highly informative genetic markers. Plant Mol Biol 48:539–547

Bonos SA, Casler MD, Meyer WA (2004) Plant responses and characteristics associated with dollar spot resistance in creeping bentgrass. Crop Sci 44:1763–1769

Bonos SA, Honig J, Kubik C (2005) The identification of quantitative trait loci for dollar spot resistance in creeping bentgrass using simple sequence repeats. The ASA-CSSA-SSSA International Annual Meeting, Salt Lake City

Bonos SA, Clarke BB, Meyer WA (2006) Breeding for disease resistance in the major cool-season turfgrasses. Annu Rev Phytopathol 44:213–234

Brady KP, Rowe LB, Her H, Stevens TJ, Eppig J, Sussman DJ, Sikela J, Beier DR (1997) Genetic mapping of 262 loci derived from expressed sequences in a murine interspecific cross using single-strand conformational polymorphism analysis. Genome Res 7:1085–1093

Brilman LA (2001) Utilization of interspecific crosses for turfgrass improvement. Int Turfgrass Soc Res J 9:157–161

Carbone I, Kohn LM (1993) Ribosomal DNA sequence divergence within internal transcribed spacer 1 of the Sclerotiniaceae. Mycologia 85:415–427

Carvalho CMB, Pena SDJ (2005) Optimization of a multiplex minisequencing protocol for population studies and medical genetics. Genet Mol Res 4:115–125

Chakraborty N, Bae J, Warnke S, Chang T, Jung G (2005) Linkage map construction in allotetraploid creeping bentgrass (*Agrostis stolonifera* L.). Theor Appl Genet 111:795–803

Chakraborty N, Curley J, Warnke S, Casler MD, Jung G (2006) Mapping QTL for dollar spot resistance in creeping bentgrass (*Agrostis stolonifera* L.). Theor Appl Genet 113:1421–1435

Devos KM (2005) Updating the "crop circle". Curr Opin Plant Biol 8:155–162

Golf Course Superintendents Association of America (2007) Golf course environmental profile. www.eifg.org

Holst-Jensen A, Kohn LM, Schumacher T (1997) Nuclear rDNA phylogeny of the Sclerotiniaceae. Mycologia 89:885–899

Jones K (1956a) Species determination in *Agrostis*. Part I. Cytological relationships in *Agrostis canina* L. J Genet 54:370–376

Jones K (1956b) Species differentiation in *Agrostis*. Part II. The significance of chromosome pairing in the tetraploid hybrids of *Agrostis canina* subsp. *montana* Hartmn., *A. tenuis* Sibth. and *A. stolonifera* L. J Genet 54:377–393

Jones K (1956c) Species determination in *Agrostis*. Part III. *Agrostis gigantea* Roth. and its hybrids with *A. tenuis* Sibth. and *A. stolonifera* L. J Genet 54:394–399

LaRota M, Sorrells ME (2004) Comparative DNA sequence analysis of mapped wheat ESTs reveals the complexity of genome relationships between rice and wheat. Funct Integr Genomics 4:34–46

Lashermes P, Andrzejewski S, Bertrand B, Combes MC, Dussert S, Graziosi G, Trouslot P, Anthony F (2000) Molecular analysis of introgressive breeding in coffee (*Coffea arabica* L.). Theor Appl Genet 100:139–146

Lodhi MA, Daly MJ, Ye GN, Weeden NF, Reisch BI (1995) A molecular marker based linkage map of *Vitis*. Genome 38:786–794

Meyer WA, Belanger FC (1997) The role of conventional breeding and biotechnical approaches to improve disease resistance in cool-season turfgrasses. Int Turfgrass Soc Res J 8:777–790

Meyer WA, Funk CR (1989) Progress and benefits to humanity from breeding cool-season grasses for turf. In: Sleeper DA, Asay KH, Pederson JF (eds) Contributions from breeding forage and turfgrasses. Crop Science Society of America Special Publication Number 15. Madison, WI, pp 31–48

Plumley KA, Meyer WA, Murphy JA, Clarke BB, Bonos SA, Dickson WK, Clark JB, Smith DA (2000) Performance of bentgrass cultivars and selections in New Jersey turf trials. Rutgers Turfgrass Proc 32:1–21

Rabinovich SV (1998) Importance of wheat-rye translocations for breeding modern cultivars of *Triticum aestivum* L. Euphytica 100:323–340

Rafalski JA (2002) Novel genetic mapping tools in plants: SNPs and LD-based approaches. Plant Sci 162:329–333

Rotter D, Bonos SA, Meyer WA, Warnke S, Belanger FC (2006) Colonial bentgrass genetic linkage mapping. The ASA-CSSA-SSSA International Annual Meeting, Indianapolis

Rotter D, Bharti AK, Li HM, Luo C, Bonos SA, Bughrara S, Jung G, Messing J, Meyer WA, Rudd S, Warnke SE, Belanger FC (2007a) Analysis of EST sequences suggests recent origin of allotetraploid colonial and creeping bentgrasses. Mol Genet Genomics 278:197–209

Rotter D, Warnke SE, Belanger FC (2007b) Dideoxy polymorphism scanning, a gene-based method for marker development for genetic linkage mapping. Mol Breed 19:267–274

Ruemmele BA (2003) *Agrostis capillaris* (*Agrostis tenuis* Sibth.) colonial bentgrass. In: Casler MD, Duncan RR (eds) Turfgrass biology, genetics, and breeding. Wiley, Hoboken, NJ, pp 187–200

Sorrells ME et al. (2003) Comparative DNA sequence analysis of wheat and rice genomes. Genome Res 13:1818–1827

Walsh B, Ikeda SS, Boland GJ (1999) Biology and management of dollar spot (*Sclerotinia homoeocarpa*); an important disease of turfgrass. HortScience 34:13–21

Warnke SE (2003) Creeping bentgrass (*Agrostis stolonifera* L.). In: Casler MD, Duncan RR (eds) Turfgrass biology, genetics, and breeding. Wiley, Hoboken, NJ, pp 175–185

Watson JR, Kaeerwer HE, Martin DP (1992) The turfgrass industry. In: Waddington DV, Carrow RN, Shearman RC (eds) Turfgrass. ASA, Madison, WI, pp 29–88

Yanagino T, Sugawara E, Watanabe M, Takahata Y (2003) Production and characterization of an interspecific hybrid between leek and garlic. Theor Appl Genet 107:1–5

Zhao H, Bughrara S, Wang Y (2007) Cytology and pollen grain fertility in creeping bentgrass interspecific and intergeneric hybrids. Euphytica 156:227–235

Advances and Prospects in the Molecular Breeding of *Zoysia*

Makoto Yaneshita

Japan Turfgrass II Incorp., 344-1, Nase-cho, Totsuka-ku, Yokohama-City
245-0051, Japan, makoto.yaneshita@sakura.taisei.co.jp

Abstract. Zoysiagrasses are warm-season grasses distributed throughout tropical and temperate zones of eastern Asia and Oceania, mainly in Japan, Korea and China. In Japan, there are wide variations in morphological and physiological traits among natural populations of zoysiagrass, for example, in color-retention in winter, salinity tolerance, and growth rate of stolons. In order to facilitate the utilization of diversity in zoysiagrasses, we have applied DNA marker techniques into the breeding program of zoysiagrasses. In this paper, we review the genetic analyses of *Zoysia* using DNA markers, including (1) the investigation of genetic variations among *Z. japonica* accessions collected from several populations in Japan and (2) efforts to identify quantitative trait loci related to winter leaf color. Moreover, we have incorporated transformation approaches into the breeding program. In this paper, we review transformation experiments of zoysiagrass with functional genes such as herbicide resistance genes and a new trial to improve the forage quality of zoysiagrass by a new gene silencing technique.

Introduction-Distribution of *Zoysia* Species in Japan

Turfgrasses perform important roles in our lives such as producing ornamental areas in public gardens and protecting soil erosion from slope areas. Recently, demands for turfgrasses on recreational areas including golf courses and sport-fields have been increased. In Japan, *Zoysia* species, which are distributed in tropical and temperate zones of eastern Asia and Oceania, have been extensively used as turfs for a long time. Historical records indicate that zoysiagrasses were planted in temple gardens in the Heian era (Kondoh 1991). Since the 1700s, the utilization of zoysiagrasses for turfs has been popular (Nakamura 1980). Also, *Zoysia* species have

T. Yamada and G. Spangenberg (eds.), *Molecular Breeding of Forage and Turf,*
doi: 10.1007/978-0-387-79144-9_29, © Springer Science + Business Media, LLC 2009

been important forage grasses, because they can survive on infertile soils with heavy traffic stress. Hirayoshi and Matumura (1958) indicated that semi-natural grasslands of zoysiagrasses were the most productive grassland for the grazing of cows.

In Japan, five species of *Zoysia*, i.e. *Z. japonica*, *Z. matrella*, *Z. tenuifolia*, *Z. sinica*, and *Z. macrostachya*, out of 16 *Zoysia* species in the world, are observed (Shoji 1983). *Z. japonica* is distributed from the southern part of Hokkaido (43°N) to south-western Kyushu (30°N). Of the 16 *Zoysia* species, *Z. japonica* is adapted for growth at the highest latitude. Natural populations of *Z. matrella* and *Z. tenuifolia* are mainly observed along seashores in the subtropical zone from the west parts of Kyushu to the southwest islands of Japan. *Z. macrostachya* and *Z. sinica* had grown in sands of seashores from the temperate zone to the subtropical zone. But, habitats of those two species have been restricted because of the loss of natural seashores. Our sampling locations of five *Zoysia* species are shown in Fig. 1.

There are several reports for wide variations of morphological and physiological traits among zoysiagrasses indicating that we have a large genetic resource of *Zoysia* with a high breeding potential. These variations could be interesting for breeding, in photosynthetic abilities at low temperature (Okawara and Kaneko 1997), color-retention in winter, and salinity tolerance (Ikeda and Hoshino 1978). However, there have been few attempts to produce a new variety of zoysiagrass using variations in natural populations. Rather, ecotypes selected from natural populations have been used as domestic varieties.

Reasons for the limitation of systematic breeding in zoysiagrasses are due to (1) propagation systems of *Zoysia* species and (2) an insufficiency of genetic information and markers to evaluate variations among wild populations. Essentially, *Zoysia* species propagate vegetatively, but they also produce seeds by an out-crossing because of the protogynous nature of flowering. However, germination rates in natural conditions are very low because of hard seed coats and a requirement of high temperature. Therefore, it takes a long time to produce a genetic population of zoysiagrasses through cross-hybridizations.

Zoysia species have been classified based on morphological characteristics of leaf-blades and spikelets. However, Kitamura (1970) suggested from the investigation of morphological characteristics in natural populations that the classification criteria of *Zoysia* should be reconsidered, because they varied continuously among species. Chromosome numbers of five *Zoysia* species are 2n = 40 with no difference in karyotypes of chromosome (Kitamura 1970). Furthermore, Forbes (1952) indicated that the chromosome pairing at metaphase I in interspecific hybrids among three species of *Zoysia* produced by a hand-pollination showed normal configuration of twenty bivalents. Flavonoid analysis with paper-chromatography (Nakamura and Nakamae 1973) and isozyme analysis of peroxidase (Yamada and Fukuoka 1984) were applied to evaluate genetic variations among natural populations of *Zoysia*. However, information obtained from these investigations is insufficient for an overall understanding of classification and genetic variations in *Zoysia*.

In this paper, several genetic experiments conducted with the aim to facilitating the breeding program of *Zoysia* are reviewed. Firstly, genetic variations among *Z. japonica* accessions collected throughout Japan were

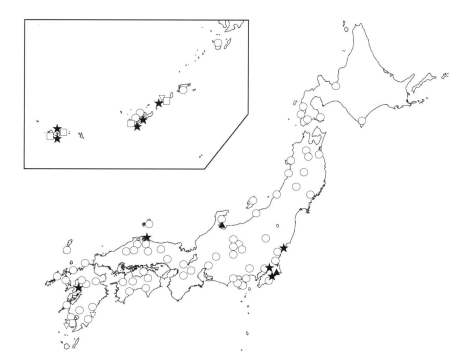

Fig. 1 Sampling locations of *Zoysia* accessions

○ : *Z. japonica*, ★ : *Z. matrella*, □ : *Z.tenuifolia*, ▲ : *Z. macrostacha*, ▽ : *Z. sinica*

evaluated by restriction fragment length polymorphism (RFLP) analysis. Secondly, a genetic linkage map was constructed and quantitative trait loci for leaf color in winter were identified. Finally, genetic transformation techniques were applied for the improvement of turf and forage quality.

Genetic Variation Among *Z. japonica* Accessions Revealed by the RFLP Analysis

Morphological Characteristics of *Z. japonica* Accessions

Z. japonica is the most popular species of zoysiagrass in Japan. In order to evaluate the breeding potential of wild populations of *Z. japonica* distributed

Table 1 Morphological data of *Z. japonica* accessions

Accession no.	Sampling location		Spikelet width (mm)	Spikelet length	Spikelet number per spike	Spike length (cm)	Leaf width (mm)	Leaf length (mm)	Color score in winter*
1	(seeded variety)		1.27	3.15	46.1	4.10	4.94	11.8	1.4
2	Erimo	Hokkaido	1.12	3.06	32.7	3.15	5.3	10.5	1.0
3	Zenibako-1	Hokkaido	1.25	2.89	57.7	3.62	5.14	11.0	1.0
4	Zenibako-2	Hokkaido	1.17	2.92	49.5	3.09	4.76	10.1	2.0
5	Shiraoi	Hokkaido	1.11	2.81	35.0	3.18	4.2	10.4	1.2
6	Kamiiso	Hokkaido	1.27	2.99	41.9	3.39	4.6	9.1	2.0
7	Niyama	Hokkaido	1.22	2.97	35.4	3.21	4.38	10.0	1.3
8	Esashi	Hokkaido	1.28	2.67	42.9	2.95	4.27	7.4	1.8
9	Hakkouda	Aomori	1.10	2.52	31.5	2.92	3.98	10.7	1.8
10	Asamushi	Aomori	1.27	3.38	43.9	3.87	4.51	12.5	1.5
11	Tanesashi	Iwate	1.30	2.91	41.1	3.79	4.97	12.8	1.5
12	Tazawa	Akita	1.28	3.46	56.3	4.28	4.71	9.6	0.7
13	Chokai	Akita	1.23	3.04	53.2	3.88	4.97	9.5	1.5
14	Sado	Niigata	1.15	2.66	39.1	2.96	4.63	8.2	1.8
15	Iitsuna	Nagano	1.26	3.23	44.7	3.48	4.56	9.5	1.2
16	Himi	Toyama	1.44	5.84	32.6	3.81	5.18	8.1	1.8
17	NIRGUS	Tochigi	1.31	3.46	46.2	4.12	5.36	11.0	1.3
18	Tanzawa	Kanagawa	1.27	3.03	33.3	3.44	5.54	11.0	2.0
19	Taito	Chiba	1.24	4.23	30.7	4.10	3.38	10.6	2.3
20	Choshi	Chiba	1.17	2.97	31.7	2.77	4.45	10.4	2.0
21	Manazuru	Kanagawa	1.15	3.06	39.5	3.09	4.64	11.7	2.0
22	Okuirou	Shizuoka	1.36	4.14	27.6	3.62	4.27	10.1	2.5
23	Irouzaki	Shizuoka	1.18	2.89	39.5	3.24	4.54	9.9	2.3
24	Nagakute	Aichi	1.06	2.80	38.0	3.18	5.17	11.2	2.5
25	Sencho	Gihu	1.16	3.19	49.9	3.70	4.67	9.2	2.5
26	Nomugi	Gihu	1.15	3.10	38.3	3.53	4.8	9.5	2.3
27	Inohana	Gihu	1.17	3.15	43.1	3.25	4.98	10.0	2.0
28	Wakakusa-yama	Nara	1.17	3.18	47.8	3.50	4.7	9.4	2.0
29	Iki	Tottori	1.41	4.93	31.8	4.74	4.45	11.0	1.5
30	Kagamiganari	Tottori	1.12	2.87	45.0	3.46	4.82	9.5	2.2
31	Dogo	Hiroshima	1.04	3.34	58.5	3.85	4.65	8.8	1.7
32	Shikoku-Noushi	Kagawa	1.16	2.77	35.7	3.99	4.76	11.1	2.5
33	Nomura	Ehime	1.08	2.89	38.8	3.17	3.92	9.1	2.5
33	Tosa-shouwa	Kochi	1.03	2.46	44.2	2.53	4.56	8.0	2.8
35	Goto	Hukuoka	1.13	2.76	34.2	2.89	5.19	9.5	2.5
36	Kashii	Hukuoka	1.02	2.69	29.7	2.92	4.07	11.2	2.5
37	Magarizaki	Kumamoto	1.13	4.18	30.9	3.71	3.5	7.9	2.5
38	Aso	Kumamoto	1.07	3.04	37.4	3.08	4.46	10.6	2.3
39	Miyazaki	Miyazaki	1.28	3.45	32.3	2.95	4.18	8.8	1.2
40	Hitotsuba	Miyazaki	1.17	2.73	32.4	2.86	4	9.9	2.3
41	Toi-cape	Miyazaki	1.00	3.26	27.6	2.83	3.94	8.1	2.3
42	Yaku island	Kagoshima	1.11	3.32	20.1	2.23	3.6	5.8	2.5
43	Amami island	Kagoshima	1.09	2.83	40.2	3.02	5.14	11.8	2.7
44	Naha-1	Okinawa	0.97	3.16	30.6	3.36	4.2	12.2	3.0
45	Naha-2	Okinawa	0.73	3.18	22.4	2.50	2.94	5.7	3.0
46	Shuri	Okinawa	1.00	3.33	39.3	4.22	4.08	10.9	3.8
47	(interspecific hybrid)		1.01	3.37	30.0	3.75	4.5	11.9	2.5
48	(*Z. matrella*)		0.95	3.69	17.5	2.49	1.94	6.7	3.9
	Average		1.16	3.28	38.4	3.37	4.5	9.7	2.1
	LSD(5%)		0.14	0.37	8.6	0.5	0.5	1.61	0.4

*0:dead – 5:dark green

in seashores and mountainous areas from the southern part of Hokkaido to the southwest islands, morphological characteristics and field performances of accessions were examined.

Forty-five accessions of *Z. japonica* were used in the analysis (Table 1). Also, a seeded variety of *Z. japonica*, a putative interspecific hybrid of zoysiagrass (Yaneshita et al. 1997) and one accession of *Z. matrella* were used as controls. Morphological characteristics in the classification key for *Zoysia* were measured (Table 1). Moreover, these accessions were planted in a test field with replications, and leaf color was scored visually from 0 (dead) to 5 (dark green) monthly.

Morphological data of the accessions are shown in Table 1. Differences in all of the characteristics observed among the accessions were significant at the 5% probability level. In addition, morphological characteristics varied continuously, not only among *Z. japonica* accessions, but also between species. For example, spikelet shapes of No. 19, 37, 45, 46 of *Z. japonica* accessions were similar to that of *Z. matrella*. In general, spikelets and leaf-blades of accessions collected from northern areas were wider than those of accessions from southern areas. In addition, a geographical differentiation was observed in the number of spikelets per spike. Accessions from northern areas produced more spikelets in a spike than southern accessions. Moreover, a significant difference of leaf color score measured in November was recognized among accessions. *Z. matrella* retained a green color of leaves longer than *Z. japonica*. In *Z. japonica*, accessions collected from Okinawa exhibited the same leaf color scores as *Z. matrella*, and southern accessions might be able to retain leaf color longer than northern accessions.

Genetic Variation Among *Z. japonica* Accessions

Since geographical differences in morphological characteristics and winter leaf color were observed in *Z. japonica* accessions, the genetic variation in *Z. japonica* accessions was examined by RFLP analysis in order to investigate the relationship between morphological and genetic variations.

Total DNA extracted from each accession was digested by *Hin*dIII or *Xba*I. Digested DNAs were hybridized with 37 genomic and cDNA markers. Labeling of probe DNAs and detection of hybridized fragments were carried out with the ECL system from Amersham Ltd. The forty-eight accessions could be distinguished from each other by the RFLP analysis, because the RFLP patterns obtained with 37 DNA markers varied among the accessions.

In order to investigate genetic relationships among *Z. japonica* accessions, principal component analysis (PCA) based on RFLP patterns was conducted. In PCA analysis, an averaged frequency of commonly shared RFLP fragments between a given accession and other accessions was calculated at each RFLP locus, and then applied as components. In Fig. 2, first principal factors and second principal factors of accessions were plotted on plane 1 and plane 2, respectively. Accessions were clustered into three groups according to sampling locations, indicated by three kinds of symbols in the figure. Also, plots of control accessions and unique accessions were indicated. Whereas accessions collected from Hokkaido and Tohoku were clustered with each other in the fourth quadrant, accessions from southern

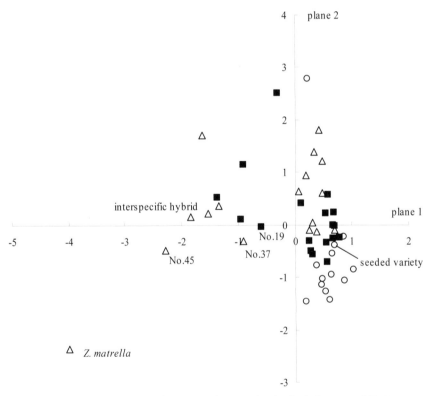

Fig. 2 Plots of first principal factors and second principal factors of *Z. japonica* accessions calculated from RFLP patterns on plane 1 and plane 2, respectively. ○ : accessions from Hokkaido and Tohoku areas, ■ : accessions from Kanto, Chubu and Kinki areas, △ : accessions from Shikoku, Kyushu and Okinawa areas

areas were widely distributed in the figure. *Z. japonica* accessions 19, 37 and 45, exhibiting the similar spikelet shapes to *Z. matrella*, were plotted in the third quadrant, as was *Z. matrella*. The putative interspecific hybrid was located adjacent to *Z. matrella* and unique accessions of *Z. japonica*. Since a coincidence between variations of morphological characteristics and clusters of accessions constructed with the RFLP analysis was recognized, geographical differentiations of morphological characteristics among *Z. japonica* accessions seemed to be genetically controlled. Furthermore, the putative interspecific hybrid and unique accessions in the third quadrant of figure retained leaf color in winter longer than other *Z. japonica* accessions.

Genetic Analysis of Winter Leaf Color of Zoysiagrass Facilitated by Molecular Markers

Variation of Winter Leaf Color in Self-pollinated Progenies from the Interspecific Hybrid

Based on investigations of morphological and genetic variations in *Z. japonica* accessions, it could be considered that the putative interspecific hybrid and unique accessions would harbor genes related to the color retention of leaf in winter from *Z. matrella* though an interspecific hybridization. In order to identify genetic regions related to the color retention of leaf in winter, quantitative trait loci (QTL) analysis of winter leaf color in self-pollinated progenies derived from the interspecific hybrid was conducted.

One hundred and five progenies were planted in 1 m^2 field test plots in May 1992. Rhizomes excised from each progeny were transplanted in pots filled with sand and grown in the green house for the DNA isolation. Leaf colors of the progenies were scored visually once a month with 0 (dead) ~5 (dark green) from April 1993 to December 1995. For the genetic analysis of winter leaf-color, color scores measured in November 1993, 1994, and 1995 were analyzed.

The mean color scores of 105 self-pollinated progenies in November 1993, 1994, and 1995 were 1.49 ± 0.73, 1.73 ± 0.76, and 1.51 ± 0.54, respectively. The color scores of progenies measured at three scoring periods were significantly correlated with each other (correlation coefficiency = 0.56 averaged in three combination, $P < 0.01$). For the statistic analysis, color scores at each scoring period were treated as replications for progenies. Averaged scores of progenies in three scoring periods were continuously distributed from 0.17 to 3.0, and were significantly different at the 5% probability level. Therefore, leaf color of *Zoysia* in winter might be a quantitative trait.

Genetic Analysis of Winter Leaf Color

We have constructed a RFLP linkage map of *Zoysia* with the self-pollinated progenies from the interspecific hybrid used in the field trial (Yaneshita et al. 1999). The linkage map consisted of 115 RFLP loci in 22 linkage groups ranging in size from 12.5 to 141.3 cM with a total map distance of 1,506 cM. In addition, the genome constitution of *Zoysia* was estimated to be allotetraploid based on a sharing of alignment of duplicated RFLP loci between pairs of linkage groups.

For the genetic analysis of winter leaf color, cDNA clones of wheat and tall fescue induced by environmental stresses and landmark clones of the rice genome were newly mapped on the linkage map of *Zoysia*, indicated by bold characters in Fig. 3. Out of ten cDNA clones of wheat induced by the salinity stress, which were given from Kihara Inst. Biol. Res., four clones; ESI3, WESR3, WESR4 and WESR5 detected allelic RFLP patterns in the self-pollinated progenies. TC11-2 clone out of four cDNA clones isolated from tall fescue after a freezing treatment was mapped on the linkage group of *Zoysia*. As for landmark clones of rice distributed from Society for Techno-innovation of Agriculture, Forestry and Fisheries (STAFF), cDNA clones mapped on the chromosome 8, 9 and 10 of rice were applied, because several QTL related to environmental stress tolerance have been identified on these linkage groups. In Fig. 3, the number in parenthesis of rice clones indicates the chromosome number of rice. Some genome synteny between rice and *Zoysia* could be recognized from the map positions of rice clones on the *Zoysia* map. For example, two rice clones on chromosome 9 were mapped on *Zoysia* linkage group IX.

For the genetic analysis of winter leaf color, the means of color scores in genotypes identified as *Z. japonica* type, hybrid type and *Z. matrella* type from the RFLP patterns were calculated at each RFLP locus. The one-way ANOVA was applied to test for differences among genotypic means. Genotypic means of color score at ZG3, ZG58, ZG133, ZCL62 and WESR5 were significantly different at the 1% probability level, and differences of genotypic means at ZG30, ZG60, ZG160, ZCL25, LHC, TC11-2 and

RC1454a were significant at the 5% level (Table 2). Based on these results, four genetic regions related to the winter leaf color of *Zoysia* were identified on linkage groups VI, VII, XII and XXIII. Mean scores of the *Z. matrella* type identified at loci on linkage group VII were higher than those of the *Z. japonica* type. For example, mean of *Z. japonica* type at ZG58 was 1.30 ± 0.28 (29 plants) and that of *Z. matrella* type was 1.89 ± 0.71 (18 plants). On the other hand, progenies of the *Z. japonica* type on group XII exhibited higher mean scores than the *Z. matrella* type.

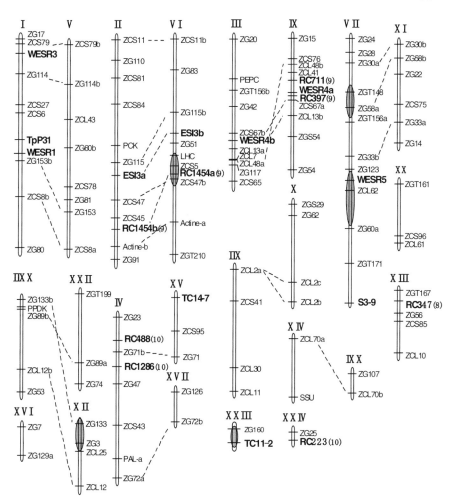

Fig. 3 RFLP linkage map of *Zoysia* indicating locations of cDNA clones from other species and genetic regions related to winter leaf color. *Bold* indicates cDNA clones from other species, *shaded areas* indicate the regions related to winter leaf color

It is interesting that some cDNA clones related to environmental stress tolerances of other plant species were mapped on genetic regions related to winter leaf color of *Zosyia*.

Possibility to Identify Functional Genes

Since a significant variation of winter leaf color was observed in the self-pollinated progenies from the interspecific hybrid, it could be considered that the winter leaf color of *Zoysia* is genetically controlled. Furthermore, *Z. matrella* might possess gene(s) related to the trait based on the field performance of *Zoysia* accessions.

According to the QTL analysis for winter leaf color of self-pollinated progenies, four genetic regions related to winter leaf color were identified. Whereas gene(s) related to the winter leaf color seemed to be introduced from *Z. matrella* into the interspecific hybrid, the population means of winter leaf color in the *Z. japonica* genotype at the genetic regions in linkage groups VI and XII were higher than those of the *Z. matrella* genotype. These data indicated that winter leaf color of *Zoysia* could be expressed

Table 2 Statistic data for the scores of winter leaf color in the self-pollinated progenies

Locus	Color score averaged in genotype[a]			F-ratio
	J	H	M	
Group VI				
LHC	1.71 (16)[b]	1.67 (59)	1.32 (24)	3.85*
RC1454a	1.54 (17)	1.64 (34)	1.01 (12)	5.80**
Group VII				
ZG30	1.40 (28)	1.58 (57)	1.85 (18)	3.54*
ZG58	1.30 (29)	1.62 (58)	1.89 (18)	6.70**
ZG60	1.44 (39)	1.55 (42)	1.90 (19)	4.64*
WESR5	1.31 (34)		1.71 (60)	11.85**
ZCL62	1.35 (38)	1.63 (58)	1.86 (18)	6.29**
Group XII				
ZG3	1.85 (14)	1.64 (58)	1.33 (29)	5.50**
ZG133	1.89 (15)	1.62 (64)	1.22 (21)	7.34**
ZCL25	1.85 (17)	1.62 (56)	1.37 (31)	4.31*
Group XXIII				
ZG160	1.31 (19)	1.58 (54)	1.73 (32)	3.36*
TC11-2	1.31 (18)		1.65 (67)	4.51*

*,**Indicate differences at 5% probability level and 1% level, respectively
[a]J: *Z. japonica* genotype, H: hybrid genotype, M: *Z. matrella* genotype
[b]Number of plants in each genotype indicated in *parentheses*

with several physiological responses to low temperature including photosynthetic activity and cold acclimation. Therefore, a fine genetic map of *Zoysia* should be constructed for the identification of functional genes. Cai et al. (2005) constructed a simple sequence repeat-amplified fragment length polymorphism (SSR-AFLP) based linkage map of *Zoysia*, which consisted of 540 markers and covered a total map length of 1,187 cM. In the future, the two *Zoysia* linkage maps should be combined in order to facilitate the isolation of genes associated with important phenotypic traits.

Genetic Transformation of *Zoysia*

Genetic transformation by exogenous genes has been recognized as a useful tool for the improvement of agronomic traits. We have established a transformation system of *Zoysia* by a polyethylene glycol (PEG)-mediated gene introduction into protoplasts (Inokuma et al. 1997). We have succeeded in producing transgenic plants containing an herbicide resistant gene with the protoplast system. In addition, the *Agrobacterium*-mediated transformation system is applicable to our reproducible cell lines derived from cv 'Miyako' of zoysiagrass (unpublished).

Recently, we have started a new transformation project of *Zoysia* in cooperation with the National Institute of Livestock and Grassland Science (NILGS), Advanced Industrial Science and Technology (AIST) and GreenSogna Inc., which is supported by a grant from Bio-oriented Technology Research Advancement Institution (BRAIN). The main objective of this project is to improve forage qualities of tall fescue and zoysiagrass through a reduction of lignin content by a new gene silencing system. The gene silencing system developed by the gene function team of AIST is to convert a transcription factor to a dominant gene repressor by the attachment of a repression domain consisting of a specific hexapeptide of DLELRL to the C terminus of the transcription factor. Mitsuda et al. (2006) have succeeded in producing male and female sterile plants of *Arabidopsis* and rice by expression of chimeric repressors constructed from transcription factors related to flower differentiation and the repression domain. In addition, they have isolated a transcription factor regulating secondary wall thickening in *Arabidopsis* (Mitsuda et al. 2005). We are attempting to reduce the lignin content in zoysiagrass by introducing the chimeric repressor constructed from this transcription factor.

In this paper, research activities applying molecular marker techniques to the breeding program of *Zoysia* are presented, focused on a series of investigations of winter leaf color of *Zoysia*. Although specific functional genes have not yet been identified, the genetic information obtained in these investigations will be applied further breeding programs and in the genetic transformation of *Zoysia*.

References

Cai HW, Inoue M, Yuyama N, Takahashi W, Hirata M, Sasaki T (2005) Isolation, characterization and mapping of simple sequence repeat markers in zoysiagrass (*Zoyisa* spp.). Theor Appl Genet 112:158–166

Forbes I Jr (1952) Chromosome numbers and hybrids in *Zoysia*. Agron J 44:147–151

Hirayoshi I, Matumura M (1958) Studies on *Zoysia* range. Grassl Sci 3:6–22 (in Japanese)

Ikeda H, Hoshino M (1978) Studies on ecotypes of *Zoysia japonica* collected from all over Japan. J Jpn Turfgrass Sci 7:27–34 (in Japanese)

Inokuma C, Sugiura K, Imaizumi N, Cho C, Kaneko S (1997) Transgenic Zoysia (*Zoysia japonica* Steud.) plants regenerated from protoplasts. Int Turfgrass Soc Res J 8:297–303

Kitamura F (1970) Studies on the horticultural classification and development of Japanese lawn grasses. Bull Kemigawa Arboretum Fac Agric Univ Tokyo 3:1–60 (in Japanese)

Kondoh K (1991) On historical and cultural backgrounds of lawns in Japanese ancient days. J Jpn Soc Turfgrass Sci 20:51–55 (in Japanese)

Mitsuda N, Seki M, Shinozaki K, Ohme-Takagi M (2005) The NAC transcription factors NST1 and NST2 of Arabidopsis regulate secondary wall thickenings and are required for anther dehiscence. Plant Cell 17:2993–3006

Mitsuda N, Hiratsu K, Todaka D, Nakashima K, Yamaguchi-Shinozaki K, Ohme-Takagi M (2006) Efficient production of male and female sterile plants by expression of a chimeric repressor in Arabidopsis and rice. Plant Biotechnol J 3:325–332

Nakamura N (1980) The improvement of *Zoysia* and other warm season grasses. J Jpn Soc Turfgrass Sci 9:113–117 (in Japanese)

Nakamura N, Nakamae H (1973) The studies on the taxonomy of grasses by the paper-chromatography. III. *Zoysia japonica*. Turfgrass Res Bull K G U Green Section Research Center 24:29–37 (in Japanese)

Okawara R, Kaneko S (1997) Effect of low growth temperature on photosynthetic O_2 evolution and chlorophyll florescence in zoysiagrasses. Grassl Sci 42:294–298

Shoji S (1983) Species ecology of *Zoysia* grass. J Jpn Soc Turfgrass Sci 2:105–110 (in Japanese)

Yamada T, Fukuoka H (1984) Variations in peroxidase of Japanese lawn grass (*Zoysia japonica* Steud.) populations in Japan. Jpn J Breed 34:431–438

Yaneshita M, Nagasawa R, Engelke MC, Sasakuma T (1997) Genetic variation and interspecific hybridization among natural populations of zoysiagrasses detected by RFLP analyses of chloroplast and nuclear DNA. Genes Genet Syst 72:173–179

Yaneshita M, Kanako S, Sasakuma T (1999) Allotetraploidy of Zoysia species with 2n = 40 based on a RFLP genetic map. Theor Appl Genet 98:751–756

Transgenesis in Forage Crops

Zeng-Yu Wang[1,2], Jeremey Bell[1], Xiaofei Cheng[1], Yaxin Ge[1],
Xuefeng Ma[1], Elane Wright[1], Yajun Xi[1], Xirong Xiao[1], Jiyi Zhang[1]
and Joseph Bouton[1]

[1]Forage Improvement Division, The Samuel Roberts Noble Foundation, 2510
Sam Noble Parkway, Ardmore, OK 73401, USA
[2]Corresponding author, zywang@noble.org

Abstract. We have established genetic transformation systems for a number of
important forage species including tall fescue, switchgrass, bermudagrass,
zoysiagrass, alfalfa, white clover and *Medicago truncatula*. The target agronomic
traits are forage quality, drought tolerance and phosphate uptake. This chapter
summarizes our efforts in improving major forage grasses and legumes by
transgenic approaches.

Introduction

Forages are consumed in many forms by domestic livestock and a wide
range of other animals. Most cultivated forages fit into two botanic
families, Poaceae (Gramineae), the grasses, and Fabaceae (Leguminosae),
the legumes. The commonly used forage grasses include tall fescue
(*Festuca arundinacea* Schreb.), a predominant cool-season species in the
USA, and bermudagrass (*Cynodon dactylon* (L.) Pers.), a productive
warm-season species. Switchgrass (*Panicum virgatum* L.) is a model
bioenergy crop that can also be used as forage. The most widely used
legume forages are alfalfa (*Medicago sativa* L.) and white clover
(*Trifolium repens* L.). These forage species exhibit gametophytic self-
incompatibility and hence require cross-pollinated breeding systems.
Darnel ryegrass (*Lolium temulentum* L.) is a self-fertile, diploid species

T. Yamada and G. Spangenberg (eds.), *Molecular Breeding of Forage and Turf,*
doi: 10.1007/978-0-387-79144-9_30, © Springer Science + Business Media, LLC 2009

with a short life cycle, it thus can be used as a suitable model for transgenic studies in grasses (Wang et al. 2005). *Medicago truncatula* is an annual forage species that has been developed into a model for legumes (Young et al. 2005).

Transgenesis refers to the introduction of heterologous or homologous DNA into a plant genome resulting in its stable integration and expression. As one of the key experimental methods in functional genomics, transgenesis allows not only the understanding of basic biological questions, but also the exploitation of genomic information for crop improvement (Dixon et al. 2007). Genetic improvement is one of the most effective ways to increase productivity of forages. Considering the genetic complexity and the associated difficulties encountered by conventional breeding methods, transgenic approaches offer many alternative and effective strategies to improve forages (Wang and Ge 2006).

Transgenesis in Grasses

We have established efficient plant regeneration and genetic transformation systems for a number of grass species, including tall fescue, switchgrass, Russian wildrye [*Psathyrostachys juncea* (Fisch.) Nevski], bermudagrass, zoysiagrass (*Zoysia japonica* Steud.), creeping bentgrass (*Agrostis stolonifera* L.) and *Lolium temulentum*. Transgenic grass plants were first obtained by biolistic transformation of embryogenic cell cultures (Chen et al. 2003, 2004; Wang et al. 2003a, 2004a). Later, we developed new protocols based on *Agrobacterium*-mediated transformation (Ge et al. 2006, 2007; Wang and Ge 2005a,b). Because biolistic transformation is a physical process involves only one biological system, it is a fairly repro-ducible procedure that can be easily adapted from one laboratory to another laboratory. The advantage of *Agrobacterium*-mediated transformation is that the method generally results in a lower copy number, fewer rearrangements and an improved stability of expression than the free DNA delivery methods.

A chimeric hygromycin phosphotransferase (*hph*) gene was used as a selectable marker for stable transformation of grasses (Ge et al. 2006, 2007; Wang et al. 2003a; Wang and Ge 2005b). For *Agrobacterium*-mediated transformation of tall fescue, switchgrass and darnel ryegrass, embryogenic calluses were co-cultivated with *Agrobacteria*, and hygromycin resistant calluses were obtained after hygromycin selection.

Transgenic plants were subsequently regenerated from the hygromycin resistant calluses (Ge et al. 2007; Wang and Ge 2005a). For bermudagrass, zoysiagrass and creeping bentgrass transformation, stolon nodes were co-cultivated with *Agrobacteria* and transgenic plants were obtained directly from the infected stolons (Ge et al. 2006; Wang and Ge 2005b). The transgenic nature of the regenerated plants was demonstrated by Southern and northern hybridization analyses, GUS staining and detection of GFP signals.

Progeny analyses revealed stable meiotic transmission of transgenes following Mendelian rules in transgenic tall fescue (Wang et al. 2003a; Wang and Ge 2005a). Replicated field trials showed similar agronomic performance between the progenies of transgenics (T1) and seed-derived plants (Wang et al. 2003b). In a separate study on pollen viability of transgenic and control plants, transgenic progenies (T1) showed similar pollen viability when compared with that of seed-derived plants (Wang et al. 2004b).

Forage digestibility is a limiting factor for animal productivity. Lignification of plant cell walls is largely responsible for lowering digestibility of forage tissues. We analyzed lignin deposition at different development stages of tall fescue and cloned cDNAs of major enzymes involved in lignin biosynthesis (Chen et al. 2002). Transgenic tall fescue plants were produced with gene constructs of cinnamyl alcohol dehydrogenase (CAD) and caffeic acid O-methyltransferase (COMT). Severely reduced mRNA levels and significantly decreased enzymatic activities were found in some transgenic lines. Chemical analyses showed that these transgenic tall fescue plants had reduced lignin content, altered lignin composition and increased in vitro dry matter digestibility (Chen et al. 2003, 2004).

The composition and structure of lignified cell walls has a dramatic impact on the technological value of biomass. Lignin also negatively affects the utilization of plant structural polysaccharides for ethanol production. The major obstacle for ethanol production from lignocellulose feedstocks is the relatively high cost of obtaining sugars (acid treatment, cellulase treatment) for fermentation. In addition, lignin degradation products are known inhibitors of ethanol fermentation and reduced lignin may correlate with significantly improved bioconversion of sugars to ethanol. We therefore believe that reducing lignin content will also reduce recalcitrance to saccharification in cellulosic bioenergy crops such as switchgrass. We have cloned major lignin biosynthetic genes from switchgrass and produced transgenic plants with RNAi gene constructs.

Transgenesis in Forage Legumes

Compared with grasses, *Agrobacterium*-mediated transformation is relatively easier in legume species. We have been able to routinely produce large numbers of transgenic alfalfa, white clover and *Medicago truncatula* plants (Crane et al. 2006; Wright et al. 2006; Xie et al. 2006; Zhang et al. 2005). Leaves and cotyledons have been used as explants for alfalfa and white clover transformation, respectively. Cotyledons, roots and leaves have been successfully used for *M. truncatula* transformation (Crane et al. 2006; Wright et al. 2006; Xie et al. 2006).

Drought stress is regular feature for perennial crops. We characterized two putative ERF transcription factor genes *WXP1* and its paralog *WXP2* from *Medicago truncatula*. Overexpression of *WXP1* in alfalfa led to a significant increase in cuticular wax loading on leaves of transgenic plants and resulted in improved drought tolerance in alfalfa (Zhang et al. 2005). Sequence comparison with *WXP1* revealed its homolog in *M. truncatula*, designated *WXP2*. Transgenic expression of both *WXP1* and *WXP2* in Arabidopsis resulted in improved drought tolerance, however, the transgenic plants were opposite in their freezing tolerance, with *WXP1* plants more tolerant and *WXP2* plants more sensitive to freezing stress (Zhang et al. 2006).

Phosphate is one of the least available macronutrients restricting crop production in many ecosystems. Improving phosphate uptake in plants is an economic way to increase forage production. We cloned and characterized a constitutive promoter, two root-specific promoters, a novel phytase gene and a purple acid phosphatase gene from *M. truncatula* (Xiao et al. 2005a,b; 2006a,b). Transgenic expression of the phytase gene or the purple acid phosphatase gene in Arabidopsis led to significant improvement in organic phosphorus uptake and plant growth (Xiao et al. 2005a, 2006a). Transgenic expression of the genes in white clover plants increased their abilities of utilizing organic phosphorus in response to P deficiency. The results showed that the genes have potentials for improving plant P nutrition and phytoremediation in commercially important crops.

References

Chen L, Auh C, Chen F, Cheng XF, Aljoe H, Dixon RA, Wang Z-Y (2002) Lignin deposition and associated changes in anatomy, enzyme activity, gene expression and ruminal degradability in stems of tall fescue at different developmental stages. J Agric Food Chem 50: 5558–5565

Chen L, Auh C, Dowling P, Bell J, Chen F, Hopkins A, Dixon RA, Wang Z-Y (2003) Improved forage digestibility of tall fescue (*Festuca arundinacea*) by transgenic down-regulation of cinnamyl alcohol dehydrogenase. Plant Biotechnol J 1: 437–449

Chen L, Auh C, Dowling P, Bell J, Lehmann D, Wang Z-Y (2004) Transgenic down-regulation of caffeic acid O-methyltransferase (COMT) led to improved digestibility in tall fescue (*Festuca arundinacea*). Funct Plant Biol 31: 235–245

Crane C, Wright E, Dixon RA, Wang Z-Y (2006) Transgenic *Medicago truncatula* plants obtained from *Agrobacterium tumefaciens*-transformed roots and *Agrobacterium rhizogenes*-transformed hairy roots. Planta 223: 1344–1354

Dixon RA, Bouton JH, Narasimhamoorthy B, Saha M, Wang Z-Y, May GD (2007) Beyond structural genomics for plant science. Adv Agron 95: 77–161

Ge Y, Norton T, Wang Z-Y (2006) Transgenic zoysiagrass (*Zoysia japonica*) plants obtained by *Agrobacterium*-mediated transformation. Plant Cell Rep 25: 792–798

Ge Y, Cheng X-F, Hopkins A, Wang Z-Y (2007) Generation of transgenic *Lolium temulentum* plants by *Agrobacterium tumefaciens*-mediated transformation. Plant Cell Rep 26: 783–789

Wang Z-Y, Ge Y (2005a) *Agrobacterium*-mediated high efficiency transformation of tall fescue (*Festuca arundinacea* Schreb.). J Plant Physiol 162: 103–113

Wang Z-Y, Ge Y (2005b) Rapid and efficient production of transgenic bermudagrass and creeping bentgrass bypassing the callus formation phase. Funct Plant Biol 32: 769–776

Wang Z-Y, Ge Y (2006) Recent advances in genetic transformation of forage and turf grasses. In Vitro Cell Dev Biol Plant 42: 1–18

Wang Z-Y, Bell J, Ge YX, Lehmann D (2003a) Inheritance of transgenes in transgenic tall fescue (*Festuca arundinacea* Schreb.). In Vitro Cell Dev Biol Plant 39: 277–282

Wang Z-Y, Scott M, Bell J, Hopkins A, Lehmann D (2003b) Field performance of transgenic tall fescue (*Festuca arundinacea* Schreb.) plants and their progenies. Theor Appl Genet 107: 406–412

Wang Z-Y, Bell J, Lehmann D (2004a) Transgenic Russian wildrye (*Psathyrostachys juncea*) plants obtained by biolistic transformation of embryogenic suspension cells. Plant Cell Rep 22: 903–909

Wang Z-Y, Ge YX, Scott M, Spangenberg G (2004b) Viability and longevity of pollen from transgenic and non-transgenic tall fescue (*Festuca arundinacea*) (Poaceae) plants. Am J Bot 91: 523–530

Wang Z-Y, Ge Y, Mian R, Baker J (2005) Development of highly tissue culture responsive lines of *Lolium temulentum* by anther culture. Plant Sci 168: 203–211

Wright E, Dixon RA, Wang Z-Y (2006) *Medicago truncatula* transformation using cotyledon explants. In: Wang K (ed) *Agrobacterium* protocols (2nd edition), Humana Press: Totowa, NJ, pp 129–135

Xiao K, Harrison M, Wang Z-Y (2005a) Transgenic expression of a novel *M. truncatula* phytase gene results in improved acquisition of organic phosphorus by *Arabidopsis*. Planta 222: 27–36

Xiao K, Zhang C, Harrison M, Wang Z-Y (2005b) Isolation and characterization of a novel plant promoter that directs strong constitutive expression of transgenes in plants. Mol Breed 15: 221–231

Xiao K, Katagi H, Harrison M, Wang Z-Y (2006a) Improved phosphorus acquisition and biomass production in *Arabidopsis* by transgenic expression of a purple acid phosphatase gene from *M. truncatula*. Plant Sci 170: 191–202

Xiao K, Liu J, Dewbre G, Harrison M, Wang Z-Y (2006b) Isolation and characterization of root-specific phosphate transporter promoters from *Medicago truncatula*. Plant Biol 8: 439–449

Xie D-Y, Sharma SB, Wright E, Wang Z-Y, Dixon RA (2006) Metabolic engineering of proanthocyanidins through co-expression of anthocyanidin reductase and the PAP1 MYB transcription factor. Plant J 45: 895–907

Young ND, Cannon SB, Sato S, Kim D, Cook DR, Town CD, Roe BA, Tabata S (2005) Sequencing the genespaces of *Medicago truncatula* and *Lotus japonicus*. Plant Physiol 137: 1174–1181

Zhang J-Y, Broeckling CD, Blancaflor EB, Sledge M, Sumner LW, Wang Z-Y (2005) Overexpression of *WXP1*, a putative *Medicago truncatula* AP2 domain-containing transcription factor gene, increases cuticular wax accumulation and enhances drought tolerance in transgenic alfalfa (*Medicago sativa*) Plant J, 42: 689–707

Zhang J-Y, Broeckling C, Sumner LW, Wang Z-Y (2006) Heterologous expression of two putative *Medicago truncatula* ERF transcription factor genes, *WXP1* and *WXP2*, in Arabidopsis led to increased leaf wax accumulation and improved drought tolerance, but differential response in freezing tolerance. Plant Mol Biol 64: 265–278

Evaluating the Role of Habitat Quality on Establishment of GM *Agrostis stolonifera* Plants in Non-agronomic Settings

Lidia S. Watrud[1,3], Mike Bollman[1], Marjorie Storm[2], George King[2], Jay R. Reichman[1], Connie Burdick[1] and E. Henry Lee[1]

[1]US Environmental Protection Agency, Office of Research and Development, National Health and Environmental Effects Research Laboratory, Western Ecology Division, 200 SW 35th Street, Corvallis, OR 97333, USA
[2]Dynamac Corporation, 200 SW 35th Street, Corvallis, OR 97333, USA
[3]Corresponding author, watrud.lidia@epa.gov

Abstract. We compared soil chemistry and plant community data at non-agronomic mesic locations that either did or did not contain genetically modified (GM) *Agrostis stolonifera*. The best two-variable logistic regression model included soil Mn content and *A. stolonifera* cover and explained 90% of the variance in the probability of a site having GM *A. stolonifera*. Inclusion of NH_4 as a third predictor variable increased the variance explained by the logistic model to 100%. Soils at GM locations were characterized by significantly lower ($P < 0.05$) Mn, *A. stolonifera* cover, and NH_4. Pairwise comparisons indicated that sites in which the GM plants became established had a significantly higher % of bare ground and significantly lower *A. stolonifera* cover, Mn, organic matter, and carbon ($P < 0.05$). The pH of soil at GM plant locations varied from 5.9 to 9.5. Our results suggest potential roles of soil disturbance and nutrient status in the establishment of *Agrostis* in mesic habitats. Additional research is needed to evaluate the ecological consequences of gene flow of GM *Agrostis* to non-agronomic plant communities.

Introduction

Agrostis stolonifera (creeping bentgrass) is a highly outcrossing, wind-pollinated, perennial cool season grass. It also reproduces asexually via

T. Yamada and G. Spangenberg (eds.), *Molecular Breeding of Forage and Turf,*
doi: 10.1007/978-0-387-79144-9_31, © Springer Science + Business Media, LLC 2009

stolons. Its small seeds can be disseminated by wind, water, wildlife, machinery and humans. *Agrostis* is a cosmopolitan genus that occupies habitats ranging from seaside dunes, agronomic areas, roadsides, riparian and residential areas to rangelands and mountain meadows. The habitats preferred by *A. stolonifera* overlap with those of other *Agrostis* species and there is potential for transgene flow from genetically modified (GM) *Agrostis stolonifera* into wild populations of sexually compatible species. The presentation below describes our efforts to begin to address the question "Do soil or plant community factors affect establishment of GM plants?"

Over 70% of the world's grass seed supply is grown in the Willamette Valley of western Oregon, USA. To address the concerns of grass seed and other growers about possible GM contamination of their crops in western OR, and following discussions between Scotts Company, Monsanto, the US Department of Agriculture Animal and Plant Health Inspection Service (USDA APHIS), the Oregon Department of Agriculture (ODA), permission was granted for the Scotts Company to field test GM creeping bentgrass in central OR. Under terms of the USDA APHIS permit, creeping bentgrass engineered with a gene (*CP4 EPSPS*) that made the GM plants resistant to Roundup ® (glyphosate) herbicide, could be grown in a control district defined by the ODA (Oregon Administrative Rules 2002). The 4,453 ha control district, in which GM creeping bentgrass but not conventional bentgrass could be grown, was located just north of Madras, OR to the east of the Cascade Mountains. Although Madras and surrounding areas receive little rainfall, irrigation provides water to grow diverse types of agronomic crops. The Deschutes and Crooked Rivers, several creeks and numerous irrigation canals, ponds, roadside ditches and natural drainages in the area each provide mesic habitats for resident populations of compatible grasses including *A. stolonifera* and *A. gigantea*. In late 2002, nine fields, totaling 162 ha, were planted in the control district and the GM fields flowered for the first time in 2003. Based on two simple assumptions (1) prevailing winds at the time of anthesis would be approximately 10 km/h from the northwest and (2) pollen would remain alive for 2–3 h, we designed a sampling plan that included the use of "sentinel plants" and compatible wild resident plants over a landscape area that extended up to 21 km beyond the perimeter of the control district. The sentinel plants were pots of *A. stolonifera* that were placed in all directions, but mostly to the south and southeast of the control district, for a period of 6 weeks. The six week period coincided with the expected time of flowering of the GM fields; it also allowed for maturation of seeds that might form on the plants. After 6 weeks, the sentinel plants were brought to Corvallis for manual harvest of the seeds from the panicles; resultant seedlings were tested in a greenhouse setting for resistance to glyphosate spray. Similarly, panicles from the resident plants were collected in the field, seeds were manually harvested and the seedlings produced by those wild plants were also tested in a greenhouse setting. Two week old seedlings were sprayed with a field rate of glyphosate. Survivors of the field rate of glyphosate were re-sprayed with 2× the field rate. An immunological lateral flow test strip (TraitChek[TM] Test for *CP4 EPSPS*, Strategic Diagnostics, Newark, DE) was used to confirm the presence of the protein that conferred resistance to glyphosate. Plants that survived the herbicide tests and that also had a positive TraitChek were then subjected to molecular analyses (PCR and sequencing of their DNA) to confirm the presence of the *CP4 EPSPS* gene.

Previously we have reported pollen-mediated hybridization as far as 21 km from the perimeter of the control district in sentinel plants of *A. stolonifera*, 8 km in resident *A. stolonifera*, and 14 km in resident *A. gigantea* (Watrud et al. 2004). Based on the number of positive plants found per km^2, the spatial distribution of *CP4 EPSPS* positives was shown to be in the direction of prevailing winds, and also to be influenced by topography. The atmospheric wind conditions that enabled the long-distance pollen-mediated gene flow have recently been reported (Van de Water et al. 2007). Beginning in the summer of 2004, we sampled mesic non-agronomic habitats within 4.8 km of control district perimeter to determine if any GM plants that expressed the *CP4 EPSPS* gene had become established in non-agronomic environments. After testing over 20,400 leaf samples using a bulk sampling method in the field with TraitChek tests, we found nine *CP4 EPSPS* positives (Reichman et al. 2006). Confirmatory molecular lab tests that included PCR, sequencing, and use of nuclear ITS and chloroplast *matK* markers helped us to determine parentage and to distinguish potential inter-specific from intra-specific *Agrostis* sp. hybrids. Six of the *CP4 EPSPS* positive plants were intra-specific *A. stolonifera* hybrids and three were feral GM crop seedlings. The transgenic plants became established 0.2–3.8 km outside the control area; their spatial distribution was primarily in the direction of prevailing winds.

The objectives of the ecological field studies presented below were to: (1) survey publicly accessible *Agrostis* spp. habitats up to 4.8 km beyond the perimeter of control district, (2) compare *Agrostis* and non-*Agrostis* communities in similar habitats, and (3) compare GM and non-GM *Agrostis stolonifera* locations at non-agronomic sites: i.e., locations that were not cultivated agronomic fields.

Materials and Methods

Plot Design and Data Collection

At each location, a linear transect was established that consisted of four $50 \text{ cm}^2 \times 100 \text{ cm}^2$ quadrats at intervals of 5 m parallel to waterways i.e., rivers, creeks, canals, pond edges or roadside ditches. Plant community data collected from each of the quadrats included percent cover by species and percent bare ground. Soil samples (2.5 cm wide × 20 cm deep) were collected from each quadrat and pooled for soil chemistry and soil moisture analyses. Soil samples were collected between June 27 and September 8, 2006. Plant community and soil chemistry data from eight GM sites (two of the nine sites reported previously were less than 5 m apart and for purposes of these analyses were considered as one site) were compared to 26 non-GM sites that had *A. stolonifera* either as the sole *Agrostis* species or in combination with *A. gigantea*.

Soil Analyses

Chemical analyses of soil were carried out at the Oregon State University Central Analytical Laboratory. Iron, copper, manganese, and zinc content of soil samples were based on analyses made with a Perkin-Elmer OPTIMA 3000DV ICP spectrophotometer. Carbon and nitrogen content were based on analyses performed on a LECO CNS 2000, while nitrate, ammonium and phosphorus were analyzed on an Alpkem RFA 300. Percent organic matter was determined by loss-on-ignition. Gravimetric analyses for soil moisture content were conducted in-house.

Statistics

Forward stepwise logistic regression analysis and logistic regression with all combinations of one, two and three variables were performed to compare the soil chemistry, bare ground, and *A. stolonifera* cover data between the GM and non-GM sites, and a correlation matrix was also generated between all the soil variables. The t-test was used to make paired comparisons of plant and soil data at sites where GM plants did or did not become established. P values <0.05 were considered to be statistically significant. JMP statistical analysis software, version 6, 2005 (SAS Corporation, Cary, NC) was used for the statistical analysis.

Results

With covarying soil and habitat variables and only 8 GM sites, a uniquely best regression model was not possible. The best two-variable logistic regression model included Mn and *A. stolonifera* cover and explained 90% of the variance in the probability of a site having GM *A. stolonifera*. In the forward stepwise logistic regression analysis, three variables, Mn, Cu and Fe, explained 100% of the variance in the probability of the presence of GM *Agrostis* (P) (1).

$$\ln(P/(1-P)) = 200\text{-}50 * Mn + 16 * Cu\text{-}0.8 * Fe \qquad (1)$$

GM *Agrostis* sites had significantly lower Mn content than the non-GM sites (Fig. 1). Logistic regression analysis of all combinations of variables (excluding Cu because of its unusually high values at only two of the GM sites) generated 12 equations that explained 100% of the variance. Equation (2) was selected as the most ecologically meaningful and because it included the three variables most commonly found in the other 11 equations.

$$\ln(P/(1-P)) = 511\text{-}34 * Mn - 22 * Ast - 70 * NH_4\text{-}N \qquad (2)$$

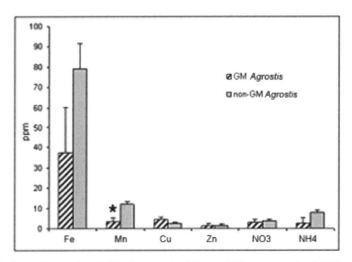

Fig. 1 Soil nutrients. *Asterisk* indicates significant difference at the 0.05 level

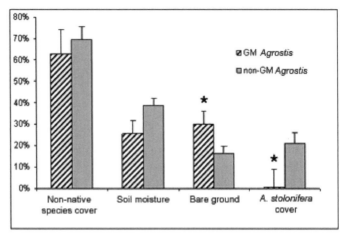

Fig. 2 Site characteristics. *Asterisk* indicates significant difference at the 0.05 level

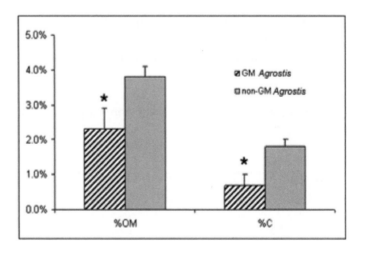

Fig. 3 Soil organic matter (OM) and carbon content (C). *Asterisk* indicates significant difference at the 0.05 level

The second model indicates that GM *Agrostis* sites had lower Mn, lower *A. stolonifera* cover and lower NH_4-N than the non-GM sites. Pairwise comparisons indicated that soil moisture and plant species composition (with regard to the types and abundance of non-native plants) were not significantly different between GM and non-GM sites. However, the percent of bare ground was significantly higher ($P < 0.05$) at sites where GM plants had become established as compared to sites that had wild resident *Agrostis* spp. (Fig. 2). Sites where GM plants became established had significantly lower ($P < 0.05$) percent *A. stolonifera* cover, organic matter and carbon than sites with wild resident *Agrostis* (Fig. 3).

Discussion

Many soil chemistry and plant community features were common to both GM and non-GM *A. stolonifera* in our central Oregon study area. There was no significant difference in the amount of soil moisture between GM and non-GM sites, and both were found close to water sources such as canals, rivers, creeks and ditches. Soil nutrient levels at both GM and non-GM *Agrostis* sites were generally lower than those typical of agronomic fields, except for iron which was high, possibly due to anoxic conditions created by high soil moisture along the waterways. Both GM and non-GM plants were found over a broad range of soil pH (5.9–9.1), however GM plants were found at sites with generally lower nutrient concentrations. The finding that GM *Agrostis* became established in sites that had more bare ground and lower *A. stolonifera* cover than where non-GM *Agrostis* occurred, suggests that lack of competition and disturbance may favor the establishment of GM *Agrostis* plants regardless of whether they are feral GM crop seeds or hybrids produced between a GM crop and either its compatible wild or cultivated relatives. Establishment of the GM plants in sites with a higher percentage of bare ground is consistent with observations that have been noted with conventional grasses i.e., seedling establishment is poorer in established turf as compared to plowed or otherwise disturbed ground (Cattani and Nowak 2001). *A. stolonifera*, classified as a competitive ruderal species in the CSR functional classification of plants (Grime 1977) is considered to be weedy in many countries (see MacBryde 2005). Particularly in the presence of herbicide selective pressure, herbicide resistant *A. stolonifera* could have a selective advantage as an early colonizer of disturbed soils. Our results suggest that low soil nutrient levels are of lesser importance to the establishment of *Agrostis* seedlings than is the presence of bare ground. The significantly lower percentages of organic matter and carbon at the GM sites may simply be attributed to the presence of less plant biomass at the sites with more bare ground. Finally, the higher Cu concentrations observed at two of the GM sites may simply reflect site-specific characteristics indicative of disturbance. That is, the higher Cu values were observed at a semi-urban location where old metal appliances and containers had been dumped into a pond.

Many of the same conditions favoring the establishment of GM *Agrostis* seedlings in non-agronomic environments would likely also be expected to favor the establishment of non-GM *Agrostis*. However, in order to determine if any causal relationships exist between the establishment potential of GM plants in soils with e.g., lower Mn, carbon, organic matter or NH_4 and more bare ground, systematic empirical studies need to be carried out with GM and non-GM, preferably isogenic crop lines. Comparative studies of GM crops and also of hybrids of GM crops with their wild and cultivated compatible relatives should preferably be carried out in different soil types and environments in the presence and absence of selective pressures. For example, testing should be carried out in the presence and absence of an herbicide or insect pest or pathogen to which the crop has been engineered to resist. Ideally, these types of studies should be carried out using a range of fertility, water, disease and insect herbivory levels. The relative ecological fitness of plants with a GM gene, i.e. their ability to germinate, establish, produce vegetative biomass and fertile progeny, and to spread and persist may also need to be considered. To determine the potential ecological consequences of establishment of GM plants and crop/wild hybrids on plant communities, multi-year studies are needed that allow access and freedom to operate with proprietary GM plants and genes. Collaborations between researchers in the private and public sectors are encouraged to help ensure the long-term environmental safety of GM crops.

Acknowledgements

This research has been wholly funded by the US Environmental Protection Agency. This document has been formally released by the US Environmental Protection Agency. Its content does not necessarily reflect the views of the Agency. Mention of any trade names or commercial products does not imply endorsement.

References

Cattani DJ, Nowak, JN (2001) Interseeding in creeping bentgrass: a viable option or wishful thinking? Golf Course Mgmt 69: 49–54
Grime P (1977) Evidence for the existence of three primary strategies in plants and its relevance to ecological and evolutionary theory. Am Nat 11: 1169–1194

MacBryde B (2005) White Paper: perspective on creeping bentgrass, *Agrostis stolonifera* L. http://www.aphis.usda.gov/about_aphis/printable_version/cbg-wpFinal.pdf US Department of Agriculture, Animal and Plant Health Inspection Service, Biotechnology Regulatory Services, Riverdale, MD

Oregon Administrative Rules (2002) Oregon Secretary of State, Oregon State Archives, item 603-052-1240. Oregon Department of Agriculture, Salem, OR. http://arcweb.sos.state.or.us/rules/OARS_600/OAR_603/603_052.html

Reichman JR, Watrud LS, Lee EH, Burdick C, Bollman M, Storm MJ, King GA, Mallory-Smith C (2006) Establishment of transgenic herbicide-resistant creeping bentgrass (*Agrostis stolonifera* L.) in non-agronomic habitats. Mol Ecol 15: 4243–4255

Van de Water P, Watrud LS, Lee EH, Burdick C, King GA (2007) Long-distance GM pollen movement of creeping bentgrass using modeled wind trajectory analysis. Ecol Appl 17: 1244–1256

Watrud LS, Lee EH, Fairbrother A, Burdick C, Reichman JR, Bollman M, Storm M, King G, Van de Water PK (2004) Evidence for landscape-level, pollen-mediated gene flow from genetically modified creeping bentgrass with *CP4 EPSPS* as a marker. Proc Natl Acad Sci USA 101: 14533–14538

Author Index

Printed in the United States of America